AF904321

Telematics Solutions in Maritime and Inland Waterway Transport

For Maisie, my little granddaughter, with hope that humanistic values such as friendship, love, kindness and faith in other people will remain important in your world of augmented and virtual reality, systems of telematics and artificial intelligence

Telematics Solutions in Maritime and Inland Waterway Transport

By

Ryszard K. Miler

Cambridge
Scholars
Publishing

Telematics Solutions in Maritime and Inland Waterway Transport

By Ryszard K. Miler

Peer-review
Professor Andrzej Makowski

Translation
Małgorzata Wilska

Layout, graphics and typesetting
Michał Janczewski

Front cover – courtesy of Rolls-Royce Ship Intelligence

Acknowledgement
Author wishes to thank to the Rector and Chancellor of
the WSB University in Gdańsk, Poland
for internal grant and financial support

This book first published 2019

Cambridge Scholars Publishing

Lady Stephenson Library, Newcastle upon Tyne, NE6 2PA, UK

British Library Cataloguing in Publication Data
A catalogue record for this book is available from the British Library

Copyright © 2019 by Ryszard K. Miler

All rights for this book reserved. No part of this book may be reproduced, stored in a retrieval system, or transmitted, in any form or by any means, electronic, mechanical, photocopying, recording or otherwise, without the prior permission of the copyright owner.

ISBN (10): 1-5275-3282-8
ISBN (13): 978-1-5275-3282-3

TABLE OF CONTENTS

INTRODUCTION

Since the very beginning of mankind, both seas and rivers have been the catalyst of trade, colonisation processes, and conquests of new territories. The immanent features in which maritime transport processes take place, provide advantageous conditions for shipping large volumes of cargo for long distances. Rivers, as the lines of communication, play a complementary role in this global activity, however inland waterway transport has become more significant due to its economic and environmental features. The world's biggest port locations in river estuaries highlight this potential for further development of maritime and inland waterway transport cooperation. Therefore, the potential for increasing the modal share of inland waterway transport is still growing.

Considering the conditions of global competition, maritime shipping, inland waterway transport as well as port operations are performed in accordance with strict technological and operational standards. All these activities are subject to severe time pressure, the imperative of economisation, and to the most important principle of all: providing these operations with the highest possible level of safety and security.

The increasing worth of cargo transported on various types of vessels (oceangoing ships and barges), their significance in economic processes represented by the growing volume of cargo, and finally growing congestion have already resulted in a considerable increase in security hazards (terrorism, piracy) and safety hazards (accidents at sea and rivers, collisions, increased anthropogenic impact causing environmental loss). Hence, efficient counteraction against these hazards becomes a crucial issue. It is no longer just a local or even an international challenge; its character has become continental and much more global.

This simple observation has become a basis of some definitely more complicated concepts pertaining to safety and security of maritime and inland waterway transport. In a holistic approach, maritime safety and security come as conceptual aspects which include two most important areas where hazards appear: nautical hazards of probabilistic nature, and conscious hazards of a determined character which are caused by anthropogenic factors. The level of safety/security comes as a derivative of the ability to identify hazards, and consequent perseverance presented by the national – and nowadays more often – international community in developing integrated counteraction systems as a solution. An indispensable instrument used for providing safety and security is a number of telematics systems of maritime and inland waterway transport which have been developed over the recent years.

Telematics refers to integrated and systemic solutions in the field of ICT (Information and Communication Technology). The most important functions performed by telematics systems are the functions related to the information operating (mega-data). It involves acquiring (sensors), processing (analytical and decisional systems) as well as distributing and transmitting operations (effectors). These are the processes implemented in a determined way (e.g. automatic control) as well as the processes which result from *ad-hoc* discretionary situations (decisions made by administrators, dispatchers, operators and independent users of specific infrastructure).

Considering the context of the presented study, a particularly interesting example which illustrates the application of the *telematics* term is modern maritime and inland waterway transport. Assuming the broadest approach, modern (intelligent) transport is supported with integrated systems of performance measurement, telecommunication, IT, information and indispensable automation (and finally connected to autonomous transport solutions). This part of intelligent transport which consists of dedicated systems and, particularly, its equipment (including software) along with services provided by such equipment, is referred to as *transport telematics*.

Telematics solutions are usually characterised by open architecture, they can be extended, complemented and modernised (we say they are scalable). Their basic aim is to allow the particular elements of a system (sub-systems) to cooperate and to interact with users (stakeholders). Such cooperation provides a considerable increase in the level of transport safety and security, operational reliability, efficient use of the infrastructure and, consequently, better economic results (optimisation), limitation of ecological risks and pressure exerted on natural environment (lowering anthropopressure).

Maritime and inland waterway transport must meet numerous requirements and challenges, and it must undergo some processes involving the implementation of intelligent systemic solutions. Application of telematics solutions is a step in the right direction. Each stage of transport processes and port operations has been provided with sensor systems used for acquiring information (e.g. cloud of points). In this way, vast amounts of information are acquired and processed, hundreds of ICT applications supporting data processing and data visualisation are developed (giving us the tracking and monitoring ability). It can be assumed that the saturation level of using telematics tools in maritime and inland waterway transport has been already achieved. Unfortunately, as there is no proper coordination, the architecture of the systems is of a dispersed and inconsistent nature; the systems are multiplied, and a problem of information entropy can be observed. As a consequence, the implemented architecture does not provide a satisfying level of integration with the low potential of harmonisation. Therefore, the first attempts at integrating such systems and at achieving economies of scale have started to appear (compatibility of data and systems and systemic integration in the future).

This study comes as an attempt at presenting a holistic approach to the maritime and inland waterway transport telematics. Identifying a role performed by contemporary maritime and inland waterway transport in the modern economy, hazards occurring in transport operations and an imperative to present an intelligent response to the identified challenges leads to the implementation of transport telematics solutions. Hence, "an inventory" of telematics systems in maritime and inland waterway transport concludes the presented study. The pragmatic approach to the achievement of such an aim divides this book into four chapters.

Indicating its place in the branch structure, the first chapter highlights the role of maritime and inland waterway transport in the contemporary regional and global economy. To emphasize the significance of maritime and inland waterway transport in the economy, the elements of their potential are documented in figures; also, further anticipated directions of their development are outlined as the derivatives of the identified regional/global geo-political and economic trends. The first chapter also introduces the concept of key parameters pertaining to maritime and inland waterway transport, based on their immanent operational features, such as universality, accessibility and degression of unit costs. These features are closely linked to the taxonomy and morphology of maritime and inland waterway transport (including the lighter aboard ship – LASH technology). To describe the modality of the discussed means of transportation, the containerisation challenges are pointed out and a concept of a container as a loading unit in maritime and inland waterway transport is presented. Finally, all types of containers used in maritime and inland waterway transport operations are described in detail. The identification of all these factors allows the Author to introduce safety/security challenges as the main determinants in the implementation of telematics systems in maritime and inland waterway transport.

Consequently, the second chapter refers to two fundamental groups of factors: safety and security in maritime and inland waterway transport. The risk factors in maritime and inland waterway transport are clearly identified, leading to a discovery of an explicit phenomenon of shifting the balance point from direct counteraction against hazards towards active prevention against them. Another important aspect discussed in this chapter refers to problems connected with the efficient management of safety/security processes in maritime and inland waterway transport, particularly to the questions of a unified structure of the management system, and precise indication of instruments which support this process (including emission control). The second chapter also presents the architecture of safety management systems in the maritime and inland waterway domain, and their transition into an institutionalised/quasi-integrated structure. Additionally, all the regularities and cause-and-effect schemes are pointed out as well.

Defining hazards makes it possible for the Author to exemplify them, to identify risks and to present the mechanisms and instruments of their assessment. It leads us to further considerations which refer to systemic mechanisms and instruments of preventing hazards in maritime and inland waterway transport. And finally, it gives us an opportunity to present the origin and significance of telematics systems in maritime and inland waterway transport (introduction/genesis of transport intelligence). All the systemic aspects of maritime and inland waterway transport safety/security defined at the local, national, regional and global levels have their internal, sensory, analytical and decisional components. These components enable us to map, to monitor (to follow the regularities and detect anomalies) and to provide the analysis of the processes in the function of risk assessment, and to assume appropriate preventive and counter-measure solutions. The indication of a typical architecture of telematics systems introduced in maritime and inland waterway transport safety/security comes as the conclusion of this chapter.

Having indicated tasks and morphological elements of telematics solutions, the third chapter presents a taxonomic division of these systems, based on the original identification and the criteria which have been assumed particularly for maritime transport. The main attention is paid to the VTMIS (Vessel Traffic Monitoring and Information Systems) with their basic sub-systems of the VTS, AIS, SRS, MAS types and the DGPS, as well as INMARSAT, which are the systems supporting their basic operations.

The consideration of the subsequent criteria referring to the division of the monitoring systems of maritime transport safety/security has made it possible for the Author to identify a group of partially integrated systems including, among others: IMDatE (EMSA – EU), BRITE (as an experimental tool of the MSA concept developed by NATO); complementary systems which include the following international systems: GMDSS (Global Maritime Distress and Safety System), MSSIS (Maritime Safety and Security Information System), NAMESIS (Naval Merchant Shipping Information System), VMS (Vessel Monitoring System), and national systems, e.g.: PHICS (Polish Harbours Information and Control System). The common character of the standards has made it possible for the Author to provide some examples of full commercialisation of some ICT solutions. Hence, some commercial monitoring systems of maritime transport safety have been identified (represented by such applications as: AIS Live, Marine Traffic, Vessel Tracker, exactAIS™ as well as based on a satellite signal: AIS-S, Gatehouse AIS Display System). Another group of systems which has been defined on the basis of the assumed division criteria includes terminal and ship owner systems which comprise TOS (Terminal Operation System), applications supporting processes which take place at the level of fleet management (for example: AMOS – Asset Management Operating System), CBM applications (Condition-based Monitoring) used for monitoring the parameters of propulsion operation, fuel consumption and GHG emission and GSOP (Geo-informational System of Port Protection) which meet the ISPS Code requirements.

The last identified category includes monitoring systems of container cargo units in maritime transport – solutions of the ECTS type (Electronic Container Tracking Service).

The variety and comprehensiveness of the application fields and processes referring to the commercialisation of standards pertaining to telematics systems in maritime transport safety/security (for example the AIS at the sensory level) have resulted in a massive dispersion of systems. Subsequently, the dispersion has resulted in some measurable negative effects, such as the lack of coordination, difficulties in communication and data use, multiplication of initiatives and incurrence of multiplied costs (the same systems are multiplied by particular operators, countries and regions). Observed particularly in the EU seas, this situation has released the potential for conceptual work on the development and implementation of a unified, integrated telematics system of maritime transport safety/security. Therefore, the development of the concept of CISE (Common Information Sharing Environment) as well as the IMDatE instrument (Integrated Maritime Data Environment) have been pointed out as a window of opportunity for further systemic integration.

Starting with the same methodology, the fourth chapter presents an attempt at the identification of telematics systems in intelligent inland waterway transport. The classification of telematics and vessel/barge monitoring systems in inland waterway transportation processes opens a discussion on their role and significance in this particular mode of transport. Taking into account the specific transportation environment of inland waterway transport, the morphological elements and taxonomy of applied telematics and vessel monitoring systems (including river terminal operating system Improved Port Ship Interface – IPSI) have been described. The main focus is on the RIS (River Information Service) as an EU-originated telematics system of inland waterway transport management. Apart from the identification of the main tasks of the RIS, the organisational and operational structure is widely described. Furthermore, the main functions (functionalities) of the RIS are identified for some better understanding of the advantages coming with this telematics system. It allows the Author to provide readers with some examples of benefits resulting from the application of the RIS system in inland waterway transport management. One of the main features of the RIS system is an application of the Inland ECDIS (Electronic Chart Display and Information System for Inland Navigation), which equips skippers with a much more accurate and reliable navigation tool. In order to avoid the same mistakes (a massive dispersion observed in maritime transport telematics) in inland waterway transport telematics systems, the problem of integration and compatibility has been indicated since the very beginning. Thus, a concept of the Navigation and Inland Waterway Action and Development in Europe (NAIADES II) and the PLATINA supporting platform clearly indicates how important the integration and cooperation of the VTMIS (maritime) and RIS (inland waterway) transport systems becomes.

While working on this study and carrying out his scientific research, the Author has benefited from consultations and bibliographic materials provided by various foreign civilian and military institutions, the most supportive of which have been: International Maritime Organisation (IMO), European Maritime Safety Agency (EMSA), NATO Headquarters of Maritime Command in Northwood (a unit of NCAGS – Naval Cooperation and Guidance for Shipping), NATO Shipping Centre (NSC), Eurostat, Rolls-Royce (Ship Intelligence Division), Scottish Canals, Kongsberg and software companies such as Tide Works, SpecTec, Avante, StarCom, Orca.

The considerations included in this study come as a continuation of the research on maritime transport safety which has been published as a monograph *Maritime Transport Safety*. It should be also noted that expert literature of the subject does not provide any synthetic monograph study on the issues presented in this book. Considering dynamic changes in the area of maritime shipping and inland waterway transport monitoring systems and systemic support of managerial decisions that took place during the last decade (dynamic development of the telematics systems), it has been so far difficult to follow these transformations in a comprehensive way. At present, the sophistication level of advanced systems allows us to state that it all comes as the near-ultimate functional architecture; therefore, an attempt at their comprehensive presentation is well-justified. The Author hopes that his publication will fill the gap in expert literature, and also that "an inventory" of telematics systems with a synthesis of their architecture presented in this study will prove to be useful in better understanding of intelligent maritime and inland waterway transport processes.

CHAPTER 1

MARITIME AND INLAND WATERWAY TRANSPORT

1.1. A notion of maritime and inland waterway transport

1.1.1. The domain of maritime transport

Since the beginning of our civilisation, transport has been an element which allows people to travel and to move their products and raw materials indispensable for the development of societies. The *transport* term comes from a Latin word *transportere* which means to move, to relocate. The term is ontologically related to another notion, namely, *communication* (*communicatio* in Latin) which means providing information, talking, a form of communication but also – transporting.[1]

The notion of transport is defined as ... *a chain of actions related to movement of people and material goods with the use of relevant means; it includes translocation from one place to another place, as well as any other actions required to reach the intended aim, i.e. loading actions (loading, unloading, cargo handling) and manipulation operations.*[2]

At present, transport is implemented practically in all the fields of human natural environment, and considering this fact, it presents some distinctive immanent features attributed to the particular environment and the specificity which affects the division made in the field of transport. Expert literature diversifies transport by applying various features which constitute the criteria for transport classification. The most recognised and commonly accepted classification is the one which takes the criterion of vertical and

---

[1] I. Kamińska-Szmaj, M. Jarosz [et al.] *Słownik Wyrazów Obcych*, Wydawnictwo Europa, Wrocław 2001, p. 390 and 820.

[2] http://encyklopedia.pwn.pl/haslo/transport;3988780.html [accessed on 12th October 2015].

horizontal division into consideration. The first criterion divides transport into transport modes; the second one divides transport into types.[3]

Considering the vertical classification, the division criterion is the specificity of the environment in which means of transport are operated, and it is determined by the type of a transportation way and a technique of moving along such a way, which ultimately affects the type of the applied means of transport.[4] The first level of the vertical classification divides transport into land, water and air transport (three elemental types of environment). Then, the division takes place in accordance with the specific character of the particular environment (considering a possible transportation way and means of transport), and there are six main modes of transport. These are namely: road transport (vehicles), rail transport, air transport, inland waterway transport, maritime transport and pipeline transport.[5] During the last decades, as a result of civilisational advancement and "conquest" of space, some development in cosmic transport could be observed; however, it has not been commonly recognised as a typical mode of transport yet. The full picture of the vertical classification of transport is presented in Figure 1-1.

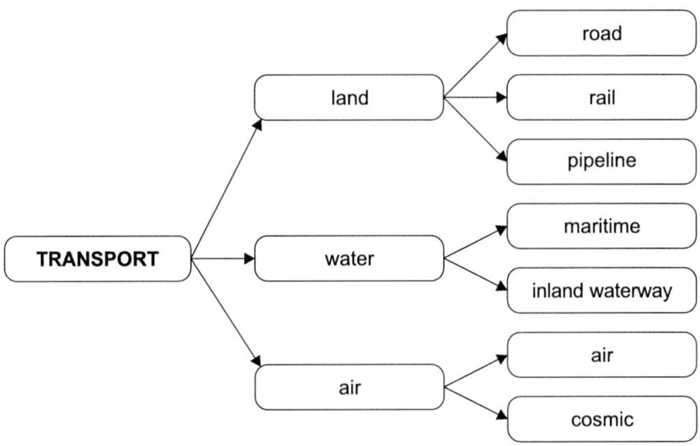

Figure 1-1. Vertical classification of transport

Source: adapted from A. Koźlak, *Ekonomika transportu. Teoria i praktyka gospodarcza*, Wydawnictwo Uniwersytetu Gdańskiego, Gdańsk 2008, p. 13.

[3] W. Grzywacz, J. Burnewicz, *Ekonomika Transportu*, WKŁ, Warszawa 1989, p. 46.

[4] Such a division is of a technical and operational character, however, in expert literature it is often referred to as the mode of transport division – the Author's note.

[5] A. Piskozub, *Gospodarowanie w transporcie*, WKŁ, Warszawa 1982, pp. 19–20.

The horizontal division includes a number of additional criteria which allow us to make further specifications for operational, statistical and economic requirements.[6] The division of transport with the consideration of some diversified horizontal criteria is presented in Table 1-1.

Table 1-1. Horizontal classification of transport

Division criterion	Categories
considering the object of transport	passengers and cargo
considering functional and organisational aspects	scheduled (line) and non-scheduled (charter)
considering continuity of the transportation process	direct and indirect
considering the geographical criterion and transportation distance	national and international
considering transportation distance	short-range, middle-range and long-range transport
considering availability for users	public, sectoral and own
considering ownership forms	state, public, cooperative, private

Source: adapted from A. Koźlak, *Ekonomika transportu. Teoria i praktyka gospodarcza…*, op. cit., pp. 14–17.

An inherent attribute of transport is also its anthropopressure, because each transport activity involves exploitation of environment in the form of emission, noise, occupation of land (for transport infrastructure), congestion and accidents. The structure of such negative phenomena which result from transportation processes is presented in Figure 1-2.

One of the oldest and most frequently used mode of transport, as stated by the vertical (branch) classification, is maritime transport. Benefits obtained from this mode of transport have been appreciated since the beginning of our civilisation, hence it has been described in a Latin adage: *navigare necesse est* – navigation is necessary.[7] Maritime transport has been present on almost all the waters of the world (referred to as the World Ocean), including oceans, mediterranean and marginal seas, straits, sea and river channels,

[6] R. Miler, *Bezpieczeństwo transportu morskiego*, WN PWN, Warszawa 2015, p. 24.

[7] *Navigare necesse est vivere non est necesse* (*Pompeius*, circa 60 BC),– *navigation is necessary but life is not* – translated from Greek by Plutarch; Pompeius said that during a storm to the helmsman on a ship which was carrying grain to Rome during a great famine. He probably meant: *We have to leave right now, even if it means risking our lives.* – the Author's note.

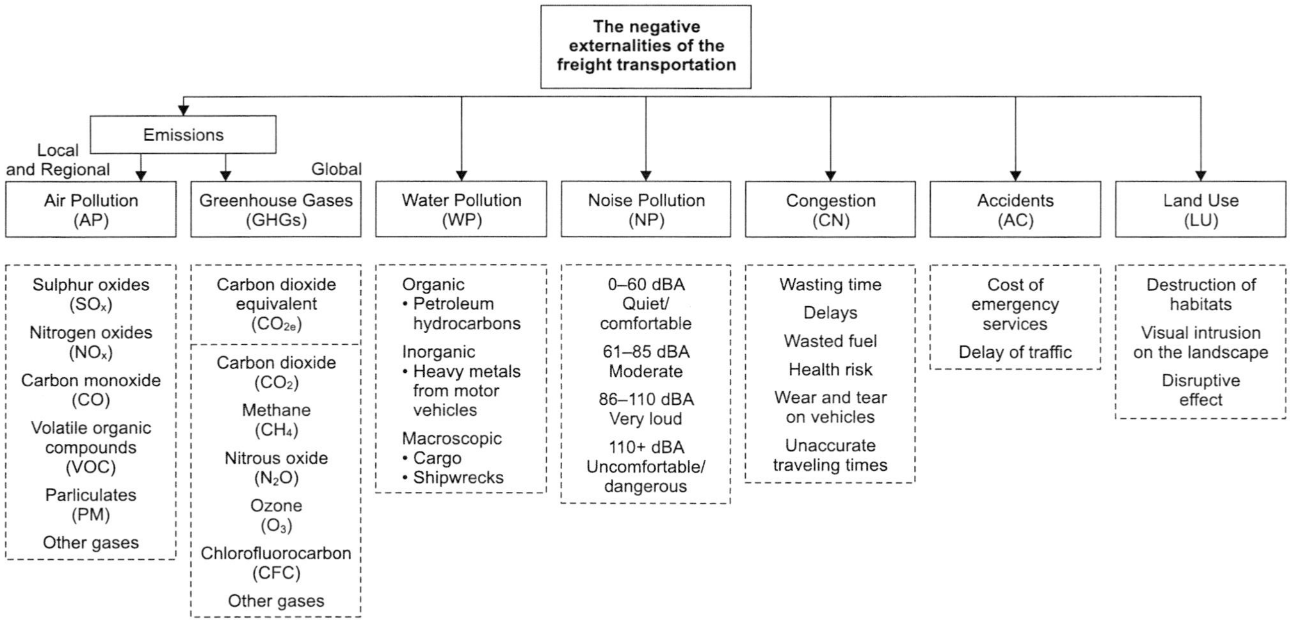

Figure 1-2. Negative externalities of transport

Source: adapted from https://tomvanwoensel.com/2015/02/26/externalities-of-freight-transportation/ [accessed on 27th July 2017].

estuaries, artificial sea waterways and enclosed seas.[8] It particularly refers to transportation of passengers and cargo by vessels across the sea in order to make profit.[9] It has been assumed that contemporary maritime transport is composed of two factors, namely: maritime navigation and seaports. In both these environments it is possible to observe complicated transport and logistic processes of a postulated level of technological advancement, which naturally tend towards unification and standardisation. They occur under strong competition and a growing level of risk related to broadly understood safety/security of these processes.

Summing up, for the requirements of this publication, it is possible to present a definition of maritime transport which describes it as a process consisting of a set of activities related to translocation of passengers and material goods by means of maritime transport (oceangoing vessels), which include maritime navigation and all the necessary port operations (loading, unloading, cargo handling and manipulation operations), implemented by specialised entities (maritime transport operators), equipped with telematics systems in the environment of the World Ocean (telematics systems of maritime transport).

Implementation of maritime transport (considering the aspect of maritime navigation) is based on a fundamental legal principle of freedom of the seas and a related concept of *freedom of navigation* (FoN).[10] The principle of freedom of navigation constitutes two key terms in this aspect: *high seas*, which refers to all international conventions, and *territorial seas* with internal waters, which are characterised by the complete sovereignty of a coastal state.[11] Figure 1-3 presents a method applied to the qualification of the *high seas* term.

---

[8] Cf.: R. Kacperczyk, *Transport i spedycja*, Wydawnictwo Difin, Warszawa 2010, pp. 74–75.

[9] Cf.: W. Rydzkowski, K. Wojewódzka-Król (eds), *Transport*, PWN, Warszawa 2007, pp. 201–202.

[10] The *freedom of the seas* term is based on a generally accepted law stating that no state is entitled to rule over any part of high seas exercising its sovereign authority, and freedom of navigation at high seas comes as the crucial principle of the contemporary law of the sea, the principle included in the United Nations Convention on the Law of the Sea (UNCLOS), referred to as the Constitution of Seas and Oceans. The convention consists of 320 articles and 9 appendices. It comprehensively regulates the use of seas and oceans through the codification of common maritime standards and principles. It states new concepts and legal regimes and provides regulations for further development of the particular fields of that law – the Author's note.

[11] In accordance with Article 2 of the United Nations Convention on the Law of the Sea (UNCLOS) – the Author's note.

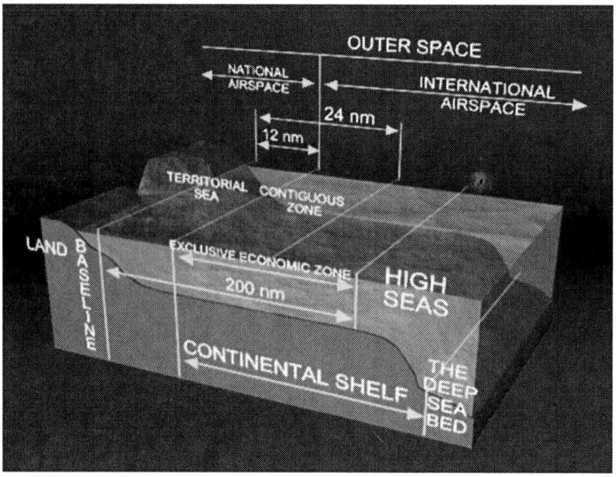

Figure 1-3. Division of sea areas in accordance with the range of legal regulations and jurisdiction of the coastal states

Source: adapted from *Global Maritime Security Cooperation In The Age Of Terrorism And Transnational Threats At Sea*, Maritime Security Primer, Copenhagen 2008, p. 6.

An important feature of maritime transport is its ecological aspect, especially considering waters of the European Union. We can clearly observe formal and legal support provided to the development of pro-ecological transport solutions in maritime navigation. Despite the fact that maritime transport is the mode with the lowest level of anthropopressure, mechanisms of higher control and charges for greenhouse gas emission to the atmosphere are still being developed.

Maritime transport occurs in the specific environment of the World Ocean which significantly affects the potential, efficiency and safety/security of this mode of transport. The basic features of maritime transport which are of immanent nature, that is: attributed to the mode of transport performed in maritime environment and positively affecting global economy, are as follows:[12]

- access to global transport and logistics centres located at large seaports and terminals (e.g. container terminals),
- strong cost degression per unit in translocation of goods, which is related to the function of the road and the volume of cargo,

[12] T. Szczepaniak (ed.), *Transport i spedycja w handlu zagranicznym*, PWE, Warszawa 2002, p. 30.

- high accessibility of transport resulting from the specific character of transport vulnerability pertaining to most cargo shipped by maritime transport and high supply of maritime transport services,
- the distance range that can be covered by vessels, which is unrestricted by any barriers of linear infrastructure.[13]

Apart from the undeniable and significant positive features mentioned above, and with the consideration of the low operating speed of vessels and time-consuming cargo handling operations, at present maritime navigation does not meet transport requirements for small batches of cargo and goods which must be delivered quickly.[14] Hence, it is efficiently replaced by other modes of transport, for example, road transport or, more and more often, by air transport.

Despite the above-mentioned limitations, and with the consideration of the progressive processes of globalisation and standardisation of transport (containerisation), maritime transport comes as the main mode in the developing field of containerised general cargo shipping (inter- and multimodal), especially in intercontinental trade.[15] Table 1-2 presents the representative features of maritime transport.

Table 1-2. Representative features of maritime transport

Features	
Positive	**Negative**
capability of transporting any kind of cargo	relatively low punctuality and low operating speed
the cheapest mode of transport for long distances	high susceptibility to weather conditions
easy access to global economic markets	necessity of using additional means of transport (the logistics structure of a hub and spokes which imposes cargo handling to smaller vessels, e.g. feeders, at hub ports)

[13] Oceangoing vessels can freely navigate at seas and oceans, choosing the shortest way to their destination ports. All the limitations to maritime traffic are defined by the principles of navigation in port approach fairways, natural and artificial channels and straits – the Author's note.

[14] Maritime transport is characterised by a catalogue of quality properties which are described in detail in the publication edited by I. Urbanyi-Popiołek (ed.), *Ekonomiczne i organizacyjne aspekty transportu morskiego*, Wydawnictwo Uczelniane Wyższej Szkoły Gospodarki w Bydgoszczy, Bydgoszcz 2012, pp. 69–70 – the Author's note.

[15] Cf.: D. Marciniak-Neider, J. Neider (eds), *Podręcznik spedytora*, Wydawnictwo Polskiej Izby Spedycji i Logistyki, Gdynia 2011, p. 35.

Table 1-2. cont.

Features	
Positive	**Negative**
capability of covering long distances at a large degression of the unit cost	risk undertaken during transport of cargo susceptible to humidity
capability of transporting heavy cargo in large quantities and oversized cargo	high costs incurred at the loss of cargo during unfortunate accidents (general average)
global character	high risk of piracy, terrorist attacks and asymmetric threats
strategic character	necessity of providing international regulators and legislation

Source: adapted from M. Michałowska (ed.), *Transport w gospodarce opartej na wiedzy*, Wydawnictwo Akademii Ekonomicznej, Katowice 2009, pp. 242–245; S. Markusik, *Infrastruktura logistyczna w transporcie*, Wydawnictwo Politechniki Śląskiej, Gliwice 2009, p. 152.

Summing up, maritime transport has been used by people for thousands of years, and it can be easily stated that it is a mode of transport which is deeply rooted in global economic and social processes. At present, maritime transport is still in the stage of dynamic development, and it has been attracting a lot of interest because of the above-mentioned processes of regionalisation (Europeaisation) and globalisation. However, further dynamic development of maritime transport will be impossible without focusing our attention on the elements of complementary transport. One of the most natural types of transport which are complementary to maritime transport is inland waterway transport.

1.1.2. The domain of inland waterway transport

Undoubtedly, rivers are the oldest transport routes in the world. Inland waterway transport has been developing on all the continents inhabited by people. Due to the development of river systems and their supplementation with artificial waterways, in some countries a dynamic development of inland waterway transport can be observed. As a result, this mode of transport becomes competitive to other modes of transport, particularly for land (rail and road) transport. Similarly to maritime transport, in inland waterway transport the cargo is shipped in the specific environment (rivers, lakes and channels, including artificial ones), with the use of specific vessels (self-propelled barges, non-self-propelled barge convoys with a tugboat) and technologies characteristic for this mode of transport (including port technologies) and telematics systems. Hence, the definition of inland waterway navigation is understood as any translocation of vessels along a particular navigable inland waterway.

However, in contrast to maritime transport, the immanent features of inland waterway transport allow us to identify this mode of transport as definitely much more local (regional) and of less economic significance (the volume of cargo shipped in this way is relatively lower than the volume of cargo shipped by maritime transport). On the other hand, however, when compared to the land modes of transport, inland waterway transport of cargo can be more advantageous because pushed barge convoys can transport more cargo, when converted to a distance unit (tonne-kilometer – tonkm). Table 1-3 presents the key parameters of inland waterway transport (IWT) and of land (rail and road) transport.

Table 1-3. Comparison of the parameters characteristic for inland waterway transport and land (rail and road) transport

Parameters	IWT	Rail	Road
Energy efficiency: 1 kW can move what weight of cargo [kg]	4,000	500	150
Fuel efficiency: 1 liter of fuel can move how much freight [tonkm]	105	85	24
Equivalent single unit carrying capacity	1 barge	15 rail wagons (standard)	60 trucks (standard)
Air pollution	Low	Medium	High
Land acquisition	Low	High	High
Capital required	Low (relatively low for bigger scale investments)	High	High

Source: adapted from http://blogs.worldbank.org/endpovertyinsouthasia/india-great-yet-unexplored-potential-inland-water-transportation [accessed on 27th July 2017].

Furthermore, the carrying capacity of the largest barges used in inland waterway navigation equals the carrying capacity of hundreds of trucks and railway carriages, which may bring about considerable savings in transportation costs and reduce road and rail traffic (decrease congestion). The comparison of the parameters referring to the size and carrying capacity of a pushed convoy consisting of four river barges (7k tonnes net) in relation to its equivalent in rail and road transport is presented in Figure 1-4.

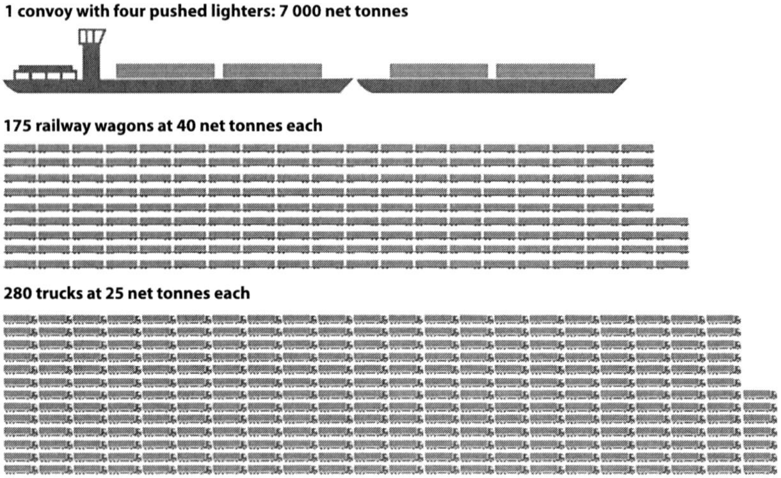

Figure 1-4. Comparison of the parameters characterising the carrying capacity of a pushed convoy of four river barges (7k tonnes net) to its equivalent in rail and road transport

Source: adapted from *2015 Special Report No 01 Inland Waterway Transport in Europe,* European Court of Auditors, Luxembourg: Publications Office of the European Union, 2015, p. 11.

Inland waterway transport is also energy-efficient because a vessel of inland waterway navigation may carry one tonne of cargo almost four times further than a truck at the same energy consumption level (the distance is respectively 370 km for an inland vessel, 300 km for a train and 100 km for a truck). In this way, vessels of inland waterway navigation achieve very good results in the field of the lowered GHG (greenhouse gases) emission, noise and a higher level of safety. Figure 1-5 presents a comparison of the anthropopressure parameters of inland waterway transport, road and rail transport.

Considering the fact that inland waterway transport is more time-consuming (slower and presenting a higher rate of distance extension) than road transport, it is broadly used for transportation of goods which do not require quick delivery. However, it should be noted that shipping containerised cargo with the use of inland waterway transport has been increased over the last several years (in the USA, India and Europe, especially in the Rhine basin). Inland waterways, however, have a limited geographical range. Moreover, whenever any problems occur along a transportation route, such as

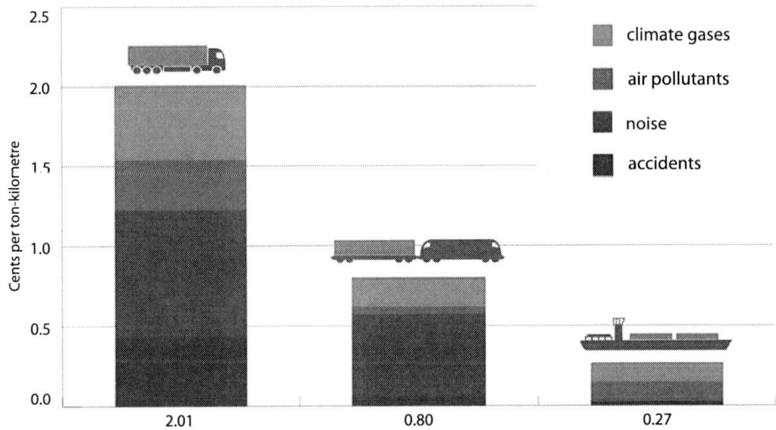

Figure 1-5. Comparison of the anthropopressure parameters characteristic for inland waterway, road and rail modes of transport

Source: adapted from *2015 Special Report No 01 Inland Waterway Transport in Europe*, European Court of Auditors, Luxembourg: Publications Office of the European Union, 2015, p. 11.

accidents, bad weather conditions, too low or too high levels of water, it is hardly possible to change that particular shipping route of a vessel for an alternative one. Moreover, the advantages resulting from lower costs can occur only in some particular circumstances – depending on the distance covered by the goods transported by waterways and the distance of a sender or a receiver to the cargo handling point, where the goods are unloaded from a barge or loaded on a barge. Hence, the unit cost is decreased with an increase in the distance covered by inland waterway transport, with simultaneous attempts made to minimise the distance covered by complementary transport, e.g. road transport in comprehensive door-to-door relations (Table 1-4).

Table 1-4. An example of transport costs when converted to a tonne-kilometer [ton-km] for the transportation distance of 200 km and 1000 km (presented in Eurocents)

Freight transport mode	**200 km**	**1,000 km**
Road	14.30	8.80
Rail	16.04	7.40
IWT	2.73	1.95

Source: adapted from Planco Consulting GmbH, *Economic and Ecological Comparison of Transport Modes: Road, Railways and Inland waterways*, November 2007.

The specific character of inland waterway transport defines the structure of demand for water transport and relations in which this transport mode becomes particularly competitive; hence, at the same time it defines the basic field of its application which traditionally refers to the following types of cargo shipping:

- large batches of bulk cargo between their sending and receiving locations based in the vicinity of waterways, including transport handling services of:
  - seaports (sea-river transport, including ro-ro),
  - mines and other places where raw materials are acquired (e.g. wood) and industrial plants located in the vicinity of waterways,
  - larger agglomerations (services provided to cities and city logistics centres, including supplies, waste collection, urban transport, etc.),
  - industrialised agriculture (transport of plants, fertilisers and biofuels);
- cargo the transportation of which excludes the use of any other transport modes, e.g. sand dredged from a river bed (during dredging operations);
- oversized and heavy cargo, shock-susceptible cargo;
- containerised cargo related to a considerable increase in the turnover of regional and international logistics centres (often located at inland ports, considering the fact that they are conveniently connected with the hinterland, and they have land reserves for further development).

Summing up, for the requirements of this monograph, it is possible to provide a definition of inland waterway transport presenting it as a process involving a set of operations related to translocation of material goods (less often passengers) with the use of inland waterway means of transport (river barges), including river navigation and all the necessary port operations (loading, unloading, manipulation operations) performed by specialist entities (inland waterway transport operators) equipped with telematics systems (telematics systems of inland waterway transport) in the river environment. Table 1-5 presents the representative features of inland waterway transport.

At present, despite the indication of numerous immanent features of inland waterway transport which clearly distinguish the inland waterway domain from the maritime domain, the complementarity features of both these modes of transport have become particularly significant. Hence, an important attribute of inland waterway transport is its complementarity in sea-river shipping. As a result, the share of inland waterway transport in services provided at seaports has already reached a considerable level, and for example, it has

Table 1-5. The representative features of inland waterway transport

Features	
Positive	**Negative**
a low level of emitted pollution, low energy consumption, low noise, high efficiency in the occupancy of additional area	high susceptibility to hydrological and hydro-meteorological conditions (high and low level of water, ice covers, currents)
the cheapest mode of transport in the regional scale which is correlated with maritime transport (maritime-river transportation)	relatively low punctuality and low operational speed
possibility to use rivers as natural communication routes	necessity of providing international regulators and legislation and separate legal regulations
covering long distances at a considerable degression of the unit cost	a high rate of distance extension
capability of transporting heavy and oversized cargo	low security/safety of cargo and vessels
a small number of collisions and related costs of eliminating accident damage recovery	mostly total loss coverage insurance, additional premiums for other types of insurance
high durability and longevity of vessels and infrastructure	very high costs of waterway maintenance

Source: adapted from M. Michałowska (ed.), *Transport w gospodarce opartej na wiedzy*, Wydawnictwo Akademii Ekonomicznej, Katowice 2009, pp. 242–245; S. Markusik, *Infrastruktura logistyczna w transporcie,* Wydawnictwo Politechniki Śląskiej, Gliwice 2009, p. 152.

been recorded at the level of almost 80% in Rotterdam which handles the largest oceangoing vessels.

The current tendencies in the development of transport in Europe have been directed towards the growing significance of the role performed by inland waterway transport in a new field of application, namely, in the field of handling container turnover of seaports/terminals, because:

- the share of containerised cargo transported by sea has been growing;
- larger container vessels are able to enter European seaports (19–22 k TEU (Twenty Equivalent Unit)), and as a result, cargo handling processes have faced serious challenges;
- road congestion has been growing (road transport as an element of the implementation of feeder transport processes has almost reached its capacity limits);

- rail transport at seaport hinterlands has almost reached the limits of its capacity.

Having defined the specificity of the environment, immanent features, strengths and weaknesses of maritime and inland waterway modes of transport, it is possible to move on to another field of their characteristics, namely to the identification of their taxonomic features and morphological elements.

1.2. Taxonomic and morphological elements of maritime and inland waterway transport

1.2.1. Taxonomy and morphology of maritime transport

Morphological elements, that is namely the main elements of the systemic structure pertaining to each mode of transport, include transport infrastructure, superstructure and means of transport.[16]

Broadly understood, transport infrastructure (identical for all the modes) can be divided into the following groups, considering its spatial location:[17]

- nodal infrastructure – including transportation nods and hubs, all the equipment and facilities (such as ramps, warehouses, stock yards, cargo handling equipment, ports, runways, terminals, stops, etc.);
- linear infrastructure – any type of transportation roads, industrial facilities (e.g. power lines) required for translocation (roads, railways, rivers and any other navigable waters with necessary facilities);
- informational and IT infrastructure – any type of informational systems and auxiliary IT systems required in order to provide efficient and effective transport management (e.g. ITS – intelligent transport systems, navigation monitoring systems, air and railway traffic monitoring and management systems, generally: transport telematics).

Similarly to other modes of transport, maritime transport includes linear and nodal elements of infrastructure and elements of informational and IT infrastructure (including satellite systems). The first group includes natural and artificial routes of maritime communication, along which vessel traf-

[16] A. Piskozub (ed.), *Ekonomika transportu*, WKŁ, Warszawa 1979, p. 200.

[17] A. Koźlak, *Ekonomika transportu. Teoria i praktyka gospodarcza…*, op. cit., p. 34.

fic is partially formed by people with the use of traffic systems; the second group includes seaports and their terminal infrastructure and hinterland.[18]

Considering the specific character of maritime transport, it is difficult to precisely define the linear infrastructure because of the environment in which this mode of transport is operated, namely waters of the World Ocean, in accordance with the doctrine of thalassocracy. In the areas of open sea it is possible to move freely, and there are no determined traffic systems. Routes for vessels are chosen on the basis of safety/security analysis which involves the navigable depth of a particular area, its convenience for vessel manoeuvering and expected weather conditions (hydro-meteorological).[19] Obviously, economic factors are also taken into consideration (time and distance).[20]

The elements of linear infrastructure include waterways (routes, lines), transportation junctions (waterway crossroads), anchorages and roadsteads.[21]

Along the communication routes in linear infrastructure, wherever it is required because of some security reasons (intensive vessel traffic, limitations in the free choice of the route because of some navigation obstacles, insufficient water depth, permanent bad weather conditions), there are regulation and separation systems of vessel traffic implemented.[22] Such systems take various forms – such as recommended very simple one-way navigation routes or complex areas of traffic separation, where several two-way navigation routes are intersected, understood as a separate water area suitable for vessel traffic.[23]

Diversification of maritime communication routes results from numerous factors which mainly include the level of (regional) land development and the demand for maritime transport services that they generate, climate and oceanographic conditions. Hence, maritime transport routes appear across oceans, marginal and mediterranean seas, along natural and artificial connections between seas and oceans, that is namely wherever transportation

---

[18] Cf.: W. Rydzkowski, K. Wojewódzka-Król (ed.), *Transport...*, op. cit., pp. 207–208.

[19] R. Miler, *Bezpieczeństwo transportu morskiego...*, op. cit., p. 40.

[20] Cf.: W. Rydzkowski, K. Wojewódzka-Król (ed.), *Transport...*, op. cit., p. 209.

[21] Some authors consider on-ground facilities dedicated to the systems of vessel traffic control to be the elements of a linear structure, as they are of the nodal character in the technical sense (VTS stations, AIS monitoring, etc.) – the Author's note.

[22] Cf.: S. Markusik (ed.), *Infrastruktura logistyczna w transporcie (vol. 3. Infrastruktura liniowa)*, Wyd. Politechniki Śląskiej, Gliwice 2013, pp. 43–45.

[23] R. Krystek (ed.), *Zintegrowany System Bezpieczeństwa Transportu.* Vol I. *Diagnoza bezpieczeństwa transportu w Polsce*, WKŁ – PG, Warszawa – Gdańsk 2009, p. 294.

needs appear, justified by economic reasons. The strategical natural connections, among others, include the Danish Straits, the Turkish Straits, the English/La Manche Channel, the Strait of Gibraltar, the Strait of Malacca, the Torres Strait, the Magellan Strait, the Bering Strait. There are also artificial, anthropogenic connections, such as the Suez Canal, the Panama Canal and the Kiel Canal.[24] Considering their location, maritime routes can be divided into local, international and strategical ones. International trade is implemented with the use of intercontinental oceanic routes. Such activities are mainly aimed at providing transportation between regions where raw materials are acquired (first of all, strategical ones, such as hydrocarbon raw materials) and regions where they are processed and consumed.[25] Figure 1-6 presents a map of the main maritime transport routes.

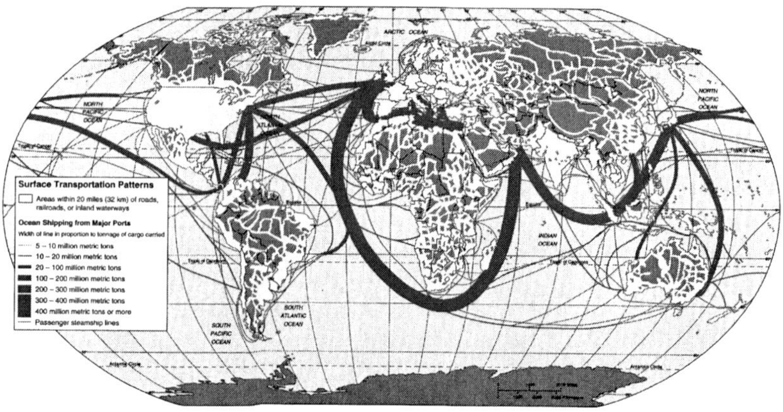

Figure 1-6. A graph of the main maritime routes

Source: adapted from http://co-opcoast.blogspot.com/2010_10_01_archive.html [accessed on 2nd June 2014].

Another key element of maritime transport infrastructure refers to seaports, understood as nodal objects of the infrastructure connected with the transport network and other elements of logistics and transport environment.[26] These are complex, logistical and multifunctional economic facilities and

[24] W. Rydzkowski, K. Wojewódzka – Król, *Transport…*, op. cit., p. 180.

[25] See more in: K. Szczyrbowski, *Zmiany w strukturze ochrony wybranych podmiotów gospodarki morskiej na przykładzie wprowadzenia kodu ISPS*, unpublished materials AMW, Gdynia 2008, pp. 11–12

[26] W. Rydzkowski, K. Wojewódzka – Król, *Transport…*, op. cit., pp. 182–183

transport and cargo handling nods which perform commercial, transportation, industrial and city-forming functions.[27] Performing transportation functions, they are used for handling cargo from vessels and means of transport used in other transport modes, such as road, rail and inland waterway transport involved in the transportation process (a transport chain).[28]

As an important link in sea-land transport chains, a seaport is a point of cargo concentration (consolidation) in export on its sender's side, and a point of deconcentration (deconsolidation) in import on its receiver's side.[29] Such processes are implemented with the use of homogenous international transport (maritime transport) and a diversified system of domestic transport (with the use of means of various transport modes) which fit into the logistical concept of a hub and spokes.[30]

Among all such ports, terminal ports are characterised by the highest concentration of services and logistical processes which take place there at the highest level of intensity, as a result of operations carried out by logistical groups (referred to as logistical mega-operators). Logistical groups and mega-carriers have intensified efforts to perform their operations in the area of container terminals. Terminals have become a driving force for the cooperation of entities operating at ports which is aimed at the optimisation of logistical processes implemented also at broadly understood hinterland. In this way, organisation of ports has been transformed into "a community of terminals".[31] Figure 1-7 presents the above-mentioned relations.

The field of external (vertical) adaptation provides a convenient system of transport relations with the hinterland and the foreland of a port. The relations with the foreland involve formation of the main navigation connections. The hinterland involves mainly road, rail, pipeline (conveyor system) connections and, more and more often, inland waterway connections as well.[32]

---

[27] K. Misztal (ed.), *Organizacja i funkcjonowanie portów morskich,* Wydawnictwo UG, Gdańsk, 2010, p. 11.

[28] S. Markusik (ed.), *Infrastruktura logistyczna w transporcie (vol. 3, Infrastruktura liniowa),* Wyd. Politechniki Śląskiej, Gliwice 2013, pp. 54–55.

[29] K. Misztal (ed.), *Organizacja i funkcjonowanie portów morskich...*, op. cit., p. 13

[30] R. Miler, *Bezpieczeństwo transportu morskiego...*, op. cit., p. 44

[31] A. Tubielewicz, R. Miler, *Determinanty rozwoju logistyki morskiej oparte na zmieniającej się roli portów morskich...*, op. cit., p. 140

[32] W. Rydzkowski, K. Wojewódzka - Król, *Transport,* 4th amended edition, PWN, Warszawa 2006, pp. 194–200.

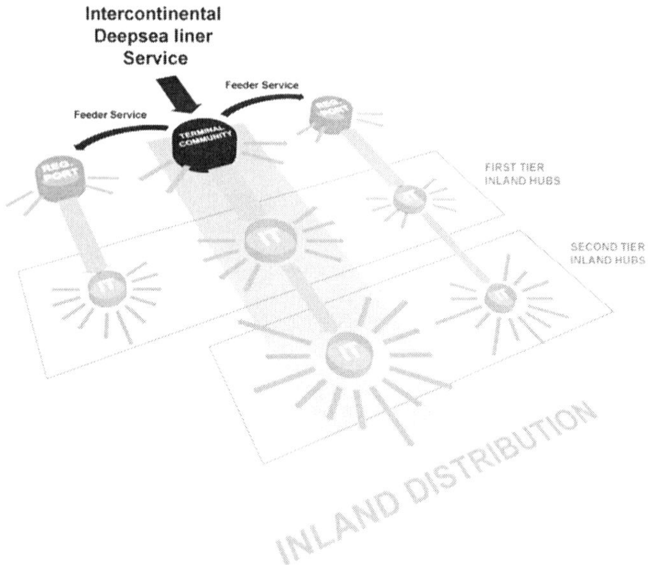

Figure 1-7. The stage of seaport terminalisation

Source: adapted from K. Vanroye, B. van Mol, *Zmieniająca się rola portów morskich UE w globalnej logistyce morskiej – możliwości, wyzwania i strategie*, an expert opinion provided by *Buck Consultants International* for the European Parliament and European Commission, Brussels 2009, p. 68.

Such a specific analysis of the key parameters of contemporary maritime transport clearly indicates that its infrastructure is closely related to its superstructure. The superstructure of maritime transport includes facilities, mobile technical and traction devices of maritime transport which are indispensable for transportation of cargo, cargo handling facilities and machinery and loading units.[33]

An extremely important element of the maritime transport system refers to the means of maritime transport, namely to oceangoing vessels.

Considering the basic types of vessels which belong to a particular category, the classification of vessels (as the means of maritime transport) allows us to indicate cargo vessels (container vessels, general cargo vessels, tankers, bulk carriers, reefer vessels), passenger vessels (liners, ferries,

[33] Ibidem, p. 32

cruisers), fishing vessels of different types dedicated to various tasks (cutters, factory mother vessels, trawlers, luggers), auxiliary vessels of a growing role and growing significance (tugboats, including AHTSs,[34] lighter barges, icebreakers, pushers, etc.).[35]

Another division results from the specificity of transportation tasks performed by particular vessels. It includes the group of conventional general cargo vessels, divided into general cargo ships used for transportation of conventional break bulk cargo, such as ore carriers or coal carriers, specialised bulk cargo vessels, such as tankers, gas tankers, chemical tankers, semi-container vessels, multi-purpose-general cargo vessels, such as COBO vessels (coal-ore-bulk-oil)[36] and ships for transportation of unitised general cargo, such as full container vessels lo-lo, ro-ro vessels and "classic" ferries, including car and train ferries, con-ro vessels and LASH – lighter aboard ship vessels. Figure 1-8 presents the structure of vessels used in maritime transport.

Considering the above-mentioned classification, it is possible to provide a more specific division which includes types of vessels used for ferry-container transportation:[37]

- roll on/roll-off vessels suitable for transportation of vehicles and rolling cargo;[38]
- con-ro vessels – a combination of a container ship and a ro-ro vessel on which the "con" part of the load area (most commonly located in the bow section with holds equipped with the cell guides, as it is on a typical container vessel) is dedicated exclusively to transportation of containers, and the "ro" part (most commonly the stern section with the holds dividing the load area into tweendecks) dedicated exclusively to transportation of rolling cargo;

---

[34] AHTS (Anchor Handling Tug / Supply vessel) – a vessel dedicated to work at offshore installations, handling anchors, providing supplies, which is often referred to as a type of a tugboat – the Author's note.

[35] The classification is available in the following publication: *Przepisy klasyfikacji i budowy statków morskich, Part I, Zasady klasyfikacji*, PRS, Gdańsk 2013, pp. 9–16; Cf.: R. Kacperczyk, *Transport i spedycja...*, pp. 75–77; Shipping Statistics Yearbook 2007. Institute of Shipping Economics and Logistics (ISL), Bremen 2008, p. 234.

[36] Cf.: J. Neider, *Transport międzynarodowy*, the 2nd revised edition, Polskie Towarzystwo Ekonomiczne, Warszawa 2012, pp. 36–37.

[37] See more in: *Przepisy klasyfikacji i budowy statków morskich...*, pp. 9–16; J. Neider, Transport..., pp. 37–39; W. Rydzkowski, K. Wojewódzka - Król, *Transport*, 4th revised edition, op. cit., pp. 123–124.

[38] See: R. Miler, *Bezpieczeństwo transportu morskiego...*, op. cit., pp. 48–50.

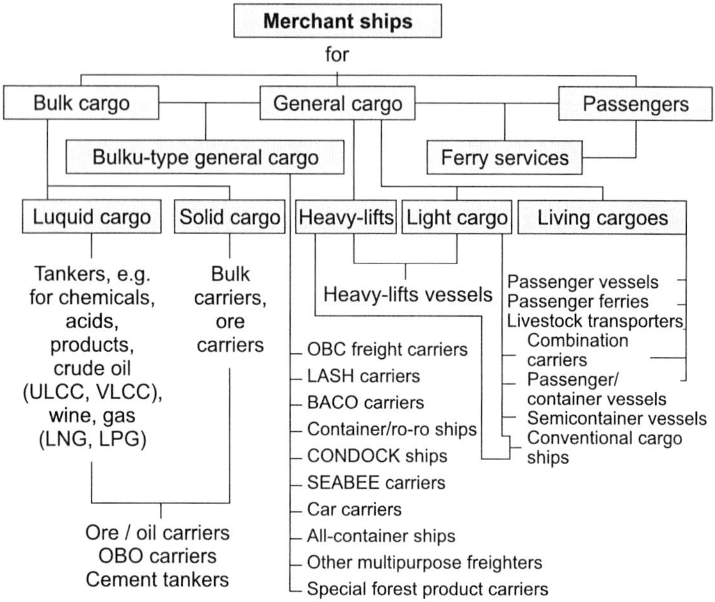

Figure 1-8. The structure of means used (division of vessels by type) in maritime transport

Source: adapted from http://www.containerhandbuch.de/chb_e/stra/index.html?/chb_e/stra/stra_01_03_01_01.html [accessed on 23rd May 2016].

- ro-lo vessels where a part of the whole load area is adjusted to transportation of cargo both in the ro-ro system (in the horizontal system, rolling cargo) and in the lo-lo system (lift on/lift off in the vertical system, palletised and containerised cargo);
- sto-ro vessels on which the cargo is loaded with the use of loading wheeled vehicles (there is an element of cargo handling in the ro-ro system), and then it is stowed and stacked there;[39]

[39] Typical sto-ro vessels are not equipped with any external or internal ro-ro ramps; they only have sheel doors with lifts/side-loaders; port machines which move along the wharf put a unit of cargo (a roll of paper, a pallet, etc.) on a "table" which is extended outside the ship together with a conveyor belt, at the level of which a platform of cargo lift stops and relocates the cargo to the level of the ship deck. The cargo is taken from the platform by a fork lift or a stacker working at a hold or a deck and transported to its ultimate location – the Author's note.

- ro-pax vessels – which are a type of passenger-car vessels with an increased load area (for the ro-ro cargo handling processes) and a decreased surface dedicated to passenger cabins;[40]
- hybrid vessels – a type of cargo vessels with elements characteristic for cargo handling in the ro-ro system with some additional "hybrid" configurations and lo-lo and sto-ro combinations.[41]

Figure 1-9 presents basic types of container vessels.

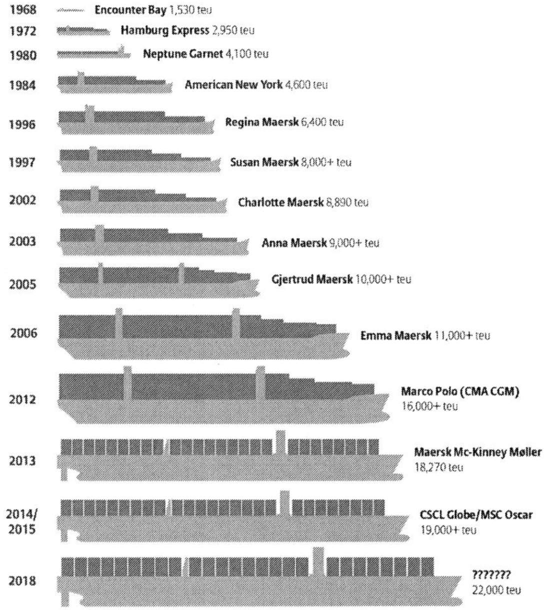

Figure 1-9. Basic types of container vessels (capacity in TEU, years 1968–2018)

Source: adapted from http://www.worldshipping.org/about-the-industry/global-trade/ [accessed on 15th April 2018].

[40] Typical passenger-car ferries are sometimes erroneously referred to as ro-pax vessels; ro-pax vessels have evolved from classic passenger-car ferries when transport of passengers (and their vehicles) started to lose its significance along numerous navigation routes; at the same time the importance of cargo transported exclusively in the ro-ro system started to grow – the Author's note.

[41] For example, vessels equipped with side-loaders are also equipped with a stern door; almost each ro-ro vessel (with the weather loading deck) can be completely handled in the ro-ro system, and additionally (it exclusively refers to the upper deck) in the lo-lo system – the Author's note.

The most frequently omitted structural elements in the morphology of maritime transport are informational and IT elements which take the form of tools (applications) used for supporting the monitoring processes of a vessel, cargo tracking, optimisation of cargo handling operations performed at a port and/or a terminal and management of transport processes and their security/safety (telematics of maritime transport). The other structural elements attributed to the morphological structure of maritime transport with the consideration of its informational and IT aspects are mainly the elements of the e-navigation environment, strictly related to the notion of navigational infrastructure. It has been assumed that it is a set of basic, permanent and properly located navigation objects and systems indispensable for providing a proper level of maritime safety/security, including security/safety of maritime transport which takes place in such an environment.[42]

The division of navigational infrastructure into types and their elements, with consideration of their functions is presented in Table 1-6.

Identified and described above, the immanent features of two basic morphological elements of maritime transport (maritime navigation and seaports) allow us to define some fundamental characteristics and capabilities related to the division of maritime navigation.

Depending on its operational area, maritime navigation can be divided into:[43]

- maritime navigation (including oceanic and short-sea),
- coastal navigation (including short-sea, cabotage),
- inland waterway navigation (in river-sea systems).

Maritime navigation is performed without any regional limitations at across all the seas and oceans of the world, with the use of oceangoing means of maritime transport. The characteristic features of maritime navigation are, among others, the lack of agreements monopolising the market (however, there is a tendency towards consolidation), the use of vessels characterised by various levels of deadweight (with a tendency towards maximisation), a relatively high frequency of sailings, lower operational costs resulting from the economies of scale, the functioning of ship-owners who have a higher number of vessels,

---

[42] Z. Kopacz, W. Morgaś, J. Urbański, *The Infrastructure of Maritime Navigation, to Content, Designing and Development Trends*, "Zeszyty Naukowe WSM Szczecin", No 65, Explo-Ship 2002, p. 14.

[43] See more in: Ficoń K., *Logistyka morska. Statki, porty, spedycja*, Wydawnictwo Belstudio, Warszawa 2010, pp. 9–11.

Table 1-6. Navigational infrastructure, its types and elements

Bathymetric infrastructure		Signal, warning and visual positioning infrastructure	Infrastructure of radio navigation positioning		Infrastructure of vessel traffic monitoring, information and navigation support	
Linear elements of infrastructure	Field elements of infrastructure		Ground-based infrastructure	Cosmic/satellite infrastructure	Infrastructure of navigation support (aid to navigation)	Infrastructure of vessel traffic monitoring and information
Fairways	Roadsteads	Permanent signal and warning signals	Local satellite augmentation systems, such as the DGPS system	The GPS system	The VTS systems	Automatic identification systems (AIS)
Deepwater fairways	Anchorages	Floating navigation signs	Short-range radio-navigation systems etc.	The GLONASS system	Systems of port and harbour navigation services (pilotage, vessel traffic at ports, etc.)	Monitoring and assistance systems
Systems of vessel traffic separation	Turning basins	Lighthouses	The GALILEO system			
	Shelters for damaged tankers, etc.	Leading lights	Regional satellite positioning systems, such as the EGNOS system			
					Systems of long-range identification and tracking (LRIT) – in development since 2006, etc.	

Source: adapted from Z. Kopacz, W. Morgaś, J. Urbański, *Infrastruktura nawigacyjna. Jej rodzaje, kryteria oceny oraz zasady projektowania…*, op. cit., p. 37.

large dependency on seaports of the hub type in the logistical network.[44] In accordance with a definition presented in the report of the European Commission of 29th June 1999, short-sea shipping (SSS) comes as […] *maritime shipping between ports located within the geographical borders of Europe and between such ports and ports located outside Europe which have their coastal lines at seas adjoining Europe.*[45] The range of short sea navigation is presented in Figure 1-10.

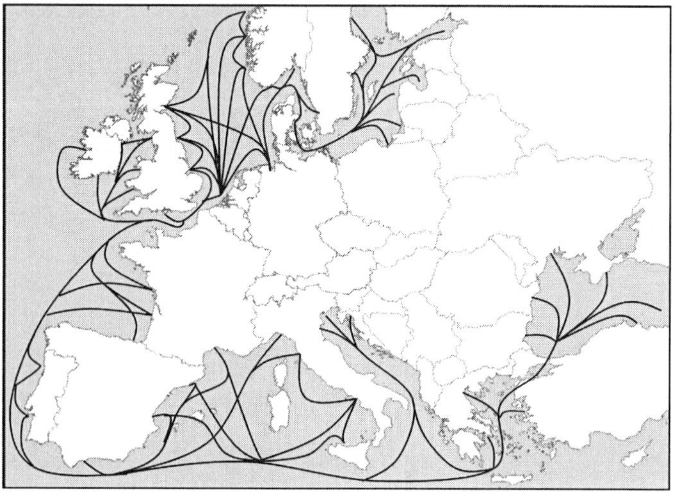

Figure 1-10. Routes of short sea navigation in the EU waters

Source: adapted from K. Vanroye, B. van Mol, *Zmieniająca się rola portów morskich UE…*, op. cit., p. 46.

Coastal navigation is limited to the area of coastal waters, and it is performed by vessels of low navigability which most often results from their constructional limitations (e.g. small size, small draught, low seaworthiness).[46]

[44] R. Miler, *Bezpieczeństwo transportu morskiego…*, op. cit., p. 37 and more in: E. Załoga (ed.), *Europejska przestrzeń transportu. Infrastruktura i środki transportu.* Zeszyty Naukowe Uniwersytetu Szczecińskiego no. 743, series Problemy Transportu i Logistyki no. 20, Szczecin 2012.

[45] E. Czermański, *Żegluga bliskiego zasięgu w rejonie Morza Bałtyckiego*, online edition, https://www.portalmorski.pl/referaty/2004/23.pdf, p. 2.

[46] K. Szczyrbowski, *Zmiany w strukturze ochrony wybranych podmiotów gospodarki morskiej na przykładzie wprowadzenia kodu ISPS…*, op. cit., p. 35.

Considering some other classification factors and their character, maritime navigation can be divided into cabotage, liner, tramp and shuttle navigation.[47]

Cabotage navigation involves shipping between ports located in the same country, however not necessarily on the coasts of the same sea. Therefore, if it takes place in the area of the same sea, it is referred to as small cabotage; if it takes place in the area of various seas, it is referred to as great cabotage.[48] Liner navigation takes place along regular routes, in accordance with a specific schedule; tramp navigation or tramping refers to navigation without any regular schedules (a characteristic feature refers to transportation of the whole shipload cargo, which results from the current demand on the transport market). Shuttle navigation takes place exclusively between two particular ports or harbours, and it is implemented with the use of seagoing ferries (depending on the water area where a vessel operates, it can be an oceangoing, a coastal or an inland waterway navigating vessel).[49]

Described in this sub-chapter, the morphological and taxonomical structure of maritime transport allows us to identify all the factors which affect the processes taking place in this transport mode in a holistic way. Therefore, inland waterway transport must undergo the same kind of evaluation.

1.2.2. Taxonomy and morphology of inland waterway transport (including the LASH technology)

Similarly to other modes of transport, inland waterway transport also has its linear and nodal elements of infrastructure, elements of informational and IT infrastructure (including telematics systems), superstructure and river vessels which are specific for this mode of transport. Linear elements include natural (rivers and lakes) and artificial (canals) routes of inland waterway communication (waterways) along which vessel traffic (including river barges) is partially formed by people with the use of various traffic systems. Nodal elements include inland ports and their terminal

[47] Cf.: J.T. Hołowiński, *Ekonomika transportu morskiego*, WM, Gdynia 1961, p. 7

[48] Coastal navigation is generally referred to as cabotage. The term originates from an English version of the surname of Giovanni Caboto: John Cabot (born in 1450 in Genoa, died in 1499; an Italian sailor who served under the English flag) – the Author's note.

[49] R. Miler, *Bezpieczeństwo transportu morskiego...*, op. cit., pp. 38–39

infrastructure and hinterland areas.[50] Therefore, it is possible to assume that navigable inland waterways refer to water routes which are not parts of any sea waters, and which – as a result of their natural features or changes implemented by people – are suitable for navigation performed by inland waterway navigating vessels.[51]

Considering the specific character of inland waterway transport, the possibilities of development of this mode of water transport are not determined by the length of routes and density of their network, but their quality and spatial configuration. The quality of waterways depends on two groups of factors:

- factors determining navigability of a waterway, namely, natural conditions of rivers which determine the scope of necessary investment into infrastructure which is indispensable for the development of inland waterway navigation;
- navigational conditions of a particular waterway – they come as a resultant of natural conditions and the development of a particular waterway (also considering available telematics systems).

Other elements of linear inland waterway infrastructure (except for waterways) include the following:[52]

- auxiliary canals (including canal basins and bank linings, sills, groynes, berms, towpaths and service roads), river bank reinforcement, canal-carrying aqueducts, siphons and conduits, canal tunnels, service basins, water level control facilities, flow gauges, records of water levels and warning devices;
- water dams and weirs (facilities constructed across the river bed to maintain the sufficient depth of water for navigation and to reduce the flow speed) and any other associated facilities (fish ladders, relief channels);
- navigation locks, lifts and inclined planes, (e.g. Falkirk, Scotland); the Falkirk Wheel is presented in Figure 1-11.

---

[50] Cf.: W. Rydzkowski, K. Wojewódzka-Król (eds), *Transport…*, op. cit., pp. 207–208.

[51] Regulation (EC) No. 1365/2006 of the European Parliament and Council of 6th September 2006 on statistics referring to inland waterway transport of cargo, repealing the Directive of the Council 80/1119/EEC – the Author's note.

[52] L. Tołkacz, *Infrastruktura transportu wodnego*, vol. 1 *Infrastruktura transportu śródlądowego*, West Pomeranian University of Technology in Szczecin Press, Szczecin 2010, pp. 123–124.

Figure 1-11. The Falkirk Wheel in Scotland

Source: adapted from left: http://sunseaboats.com/how-does-the-falkirk-wheel-work/, right: http://www.architravel.com/architravel/building/falkirk-wheel/ [accessed on 4th June 2018].

Informational, IT and navigational infrastructure includes the following elements:[53]

- devices used for marking canals, signalisation, security/safety and lighting;
- devices used for traffic control, including sensory elements of telematics;
- devices used for collecting fees.

Considering the abovementioned conditions of linear infrastructure, the notion of a waterway becomes especially significant. The elements of a waterway are:

- a fairway – a water lane dedicated to navigation;
- minimal waterway depth safe for navigation – the minimal depth along a particular navigable section of a waterway.

Figure 1-12 presents an intersection of an inland waterway and its characteristic parameters.

Interconnected inland waterways form navigable routes. The most important inland waterways in the world are:[54]

- the system of the Great Lakes of America and the St. Lawrence River which provides connection to the great agglomerations of the USA and Canada (Chicago, Detroit, Cleveland, Toronto, Montreal, New York);

---

53 L. Tołkacz, *Infrastruktura transportu wodnego*, vol 1 *Infrastruktura transportu śródlądowego*, Wydawnictwo Zachodniopomorskiego Uniwersytetu Technologicznego w Szczecinie, Szczecin 2010, p. 45.

54 W. Rydzkowski, K. Wojewódzka-Król (eds), *Transport...*, op. cit., p. 201–203.

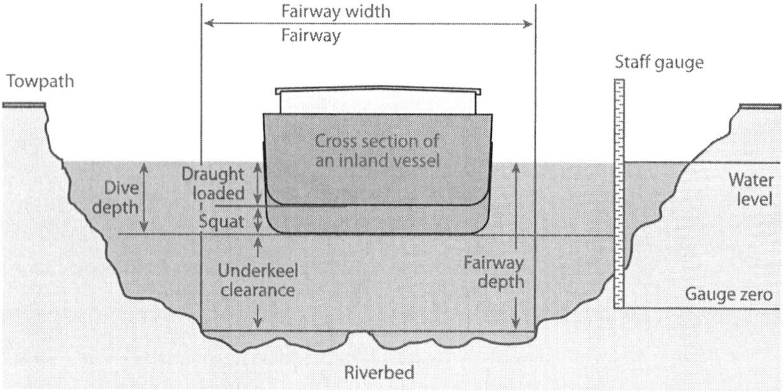

Figure 1-12. Fairway parameters (schematic presentation)

Source: adapted from *2015 Special Report No 01 Inland Waterway Transport in Europe*, European Court of Auditors Luxembourg: Publications Office of the European Union, 2015, p. 10.

- the system of rivers in Western Europe, the main axis of which is the Rhine. It is connected by a network of canals with the Seine, the Rhode, the Elbe and the Danube. The system connects the North Sea with the Mediterranean Sea;
- the system of the Danube – the river flows through Germany, Austria, the Czech Republic, Slovakia, Hungary, former Yugoslavia republics, Bulgaria and Romania (inland waterway navigation takes place mainly along the river section between Vienna and Budapest).

In order to provide waterways of proper navigable parameters (desired quality) to the extent much larger than in maritime navigation, there is some infrastructural investment indispensable in inland waterway transport. Basic infrastructural investment in waterways includes:

- dredging – it comes as a basic method of improving navigability of rivers, applied in narrow rivers; some difficulties for navigation can be solved by dredging operations or providing cuttings at places of insufficient depth;
- regulation – the aim of such investment is to improve navigable conditions by providing the minimal depth and width for safe navigation, a regular shape of navigable waterway in order to maintain the same conditions along the whole waterway as much as possible; hence, river regulation is straightening and deepening of the riverbed;

- elimination of islands, reinforcement of river banks, construction of embankments (including anti-flood embankments), groynes which narrow the riverbed (to prevent meandering), etc.;
- channelisation of rivers – the most radical method applied to improve navigable conditions of waterways; in other words it refers to constructions set across the riverbed (weirs or barrages) which increase the depth of some dammed waterway sections; after channelisation, the waterway acquires a step-like profile, and if there is a big difference between the water levels along some sections, it is necessary to provide such facilities as lifts and inclined planes.

Considering the specific character of the European transport system (the system of Pan-European transport corridors TEN-C), the Barcelona Convention has been implemented in the territory of all the European countries with the infrastructure necessary for inland waterway transport. The Convention states a principle of free navigation for all vessels along international and internationalised waterways.

Transport is implemented on the basis of an agreement referred to as a bill of lading. It regulates responsibilities of an inland waterway carrier and a holder of a bill of lading, as stated in the transport contract.

The European system of international waterways consists of the western European and the southern European parts. The western part includes the Rhine River which combines all the waterway networks of Switzerland, France, Germany and the Netherlands into one. The southern part includes the Danube River which connects Germany, Austria, Slovakia, Hungary, Croatia, Serbia, Romania, Bulgaria and Ukraine.[55] Figure 1-13 presents the main Pan-European inland waterway routes.

To provide efficient navigation, it is necessary to provide proper parameters of waterways. According to the European classification of waterways approved in 1992 by the United National Economic Commission for Europe and the European Conference of Ministers of Transport, inland waterways are divided into waterways of regional and international importance. In 1996 the European Agreement on the Main Inland Waterways of International Importance (also referred to as the AGN Convention) came into effect.

---

[55] *2015 Special Report No 01 Inland Waterway Transport in Europe,* European Court of Auditors Luxembourg: Publications Office of the European Union, 2015, pp. 10–11

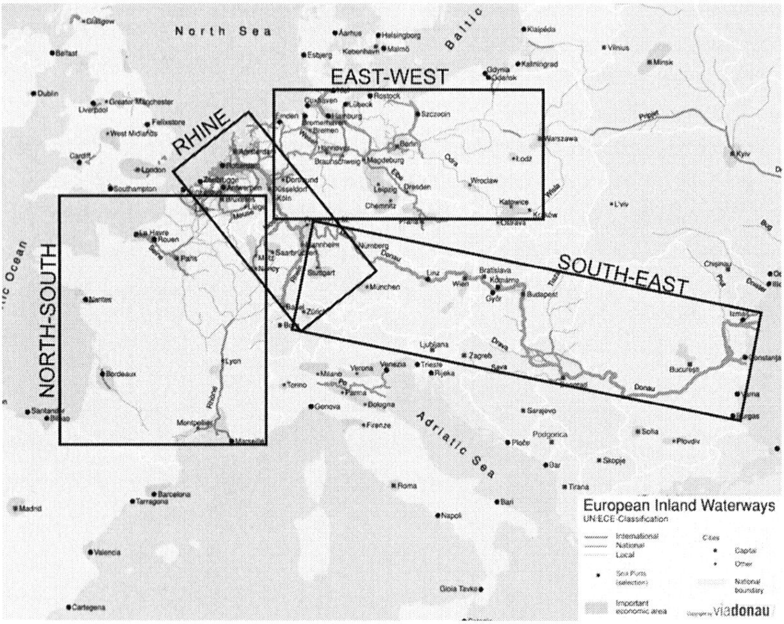

Figure 1-13. Main Pan-European routes for inland waterway navigation

Source: adapted from https://www.researchgate.net/figure/Overview-of-European-inland-waterways_fig1_273446772 [accessed on 15th April 2018].

The document provides a new classification of inland waterways. It considers the requirements of modern transport technologies, and it is aimed at the development of a network of waterways accessible for self-propelled vessels of the following parameters: length – 85 m, beam – 9.5 m, draught – 2.5–4.5 m, deadweight tonnage – 1250–2500 t. Marked with the "E" letter in the AGN Convention, waterways of international importance should connect seaports, provide connections in coastal transport relations and hinterland. The typology of inland waterways in Europe is presented in Table 1-7.

The nodal infrastructure of inland waterway transport includes, first of all, inland ports (and seaports which handle inland shipping) and inland harbours and company cargo handling wharfs.

An inland port is an area on a river bank, a canal bank or a lakeshore with the facilities which enable vessels, such as barges or tugboats, to arrive and to dock, to handle cargo and passengers and to perform usual operations

Table 1-7. Inland waterway classification in accordance with the AGN Convention

Type of inland waterways	Classes of navigable waterways	Motor vessels and barges. Type of vessel: General characteristics					Pushed convoys. Type of convoy: General characteristics					Minimum height under bridges
		Designation	Maximum length	Maximum beam	Draught	Tonnage		Length	Beam	Draught	Tonnage	
			L(m)	B(m)	D(m)	T(t)		L(m)	B(m)	D(m)	T(t)	H(m)
1	2	3	4	5	6	7	8	9	10	11	12	13
Of regional importance – To west of Elbe	I	Barge	38.5	5.05	1.80–2.20	250–400						4.0
	II	Kampine-Barge	50–55	6.6	2.50	400–650						4.0–5.0
	III	Gustav Koenigs	67–80	8.2	2.50	650–1 000						4.0–5.0
Of regional importance – To East of Elbe	I	Gross Finow	41	4.7	1.40	180						3.0
	II	BM–500	57	7.5–9.0	1.60	500–630						3.0
	III		67–70	8.2–9.0	1.60–2.00	470–700		118–132	8.2–9.0	1.60–2.00	1 000–1 200	4.0
Of international importance	IV	Johann Welker	80–85	9.5	2.50	1 000–1 500		85	9.5	2.50–2.80	1 250–1 450	5.25 or 7.00
	Va	Large Rhine vessels	95–110	11.4	2.50–2.80	1 500–3 000		95–110	11.4	2.50–4.50	1 600–3 000	5.25 or 7.00 or 9.10
	Vb							172–185	11.4	2.50–4.50	3 200–6 000	
	VIa							95–110	22.8	2.50–4.50	3 200–6 000	7.00 or 9.10
	VIb		140	15.0	3.90			185–195	22.8	2.50–4.50	6 400–12 000	7.00 or 9.10
	VIc							270–280 195–200	22.8 33.0–34.2	2.50–4.0 2.50–4.50	9 600–18 000 9 600–18 000	9.10
	VII							285	33.0–34.2	2.50–4.50	14 500–27 000	9.0

Source: adapted from *2015 Special Report No 01 Inland Waterway Transport in Europe*, European Court of Auditors Luxembourg: Publications Office of the European Union, 2015, p. 21 (Annex).

related to vessel maintenance (replenishment of supplies, waste collection, etc.). Hence, an inland port is an area of water (port aquatory) and land (port territory) determined by its borders, where cargo handling operations take place with the use of proper equipment (superstructure). Frequently, basins of inland ports are constructed in such a way that vessels navigating both up and down the river can enter the port area. It usually means that it is necessary to construct a system of locks and weirs in the vicinity of an inland port in order to regulate the water levels, depending on the river water flow, which may change very often because of the season of the year, precipitation, etc. Such facilities enable vessels to continue navigation despite differences in water levels. The main inland ports of Western Europe are presented in Figure 1-14.

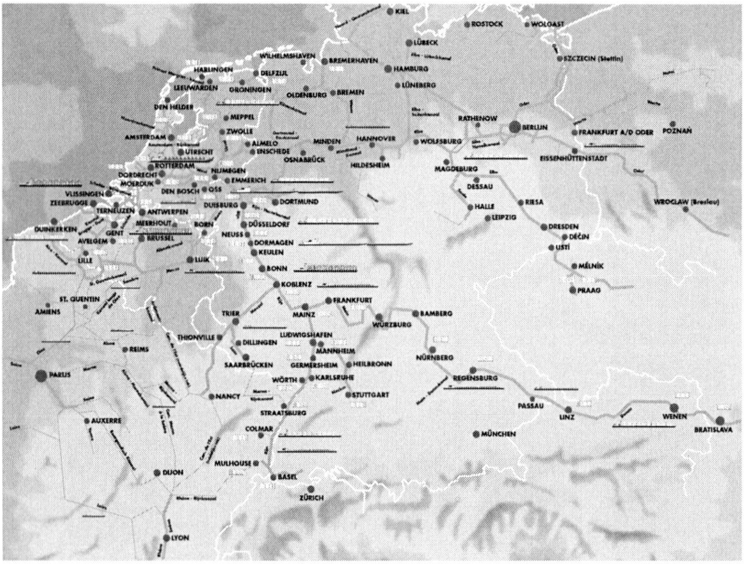

Figure 1-14. The main inland ports of Western Europe (including the system of waterways: their class and typical river barges which operate along the particular routes)

Source: adapted from https://www.bureauvoorlichtingbinnenvaart.nl/assets/images/Over/BVB_01l.jpg [accessed on 15th April 2018].

Considering this specific analysis of the key parameters of modern inland waterway transport, its infrastructure is closely related to its superstructure. The superstructure of inland waterway transport includes facilities, mobile technical and traction devices of inland waterway transport which are indispensable for transportation of cargo, cargo handling facilities and machinery and loading units.[56]

The loading and unloading of goods on and off a ship can be performed in various ways depending on the wharf facilities and the type of cargo. Dry bulk can be moved by a fixed hoist or a crane, a mobile crane, a tilting pier or a (dump) conveyor belt. A lorry can dump the cargo directly into the hold of a ship from a tilting pier. A fixed hoist, a crane or a mobile crane (with a claw), or a pneumatic suction conveying system can be used for the unloading of the cargo. This kind of equipment is used for the unloading of light bulk, such as grain. Some inland navigation vessels have their own loading and unloading systems; however, this is very uncommon because they take up a lot of room, and they are very heavy to have them on the deck of a vessel.

An extremely important element of the system of inland waterway transport is its watercraft. It mainly refers to the means of inland waterway transport (river ships, barges, tugboats, pusher boats). The term of an inland navigation vessel refers to a floating vessel designed for transportation of cargo or passengers, which mainly navigates along inland navigable waterways or in the area or direct vicinity of sheltered waters or in the area where port regulations are applied.

Over 90% of vessels used for inland waterway navigation are private motor barges which belong to one ship-owner; the largest ship-owner companies have over one hundred barges.

Inland carriers can be divided into contract carriers, who have entered long-term agreements with one consignor/shipper, for example an ironworks or a mine, and carriers who offer their services on a day-to-day basis (depending on the received orders). The basic types of vessels used for inland waterway navigation are: pontoons, flat top barges, lash barges, log barges, oil tanker barges, accommodation barges, project cargo barges, push/pull barges, crane barges, split hopper barges and various types and categories of tugs.

Barges are used for transportation of cargo and for general utility purposes. Typically, a flat-bed vessel, a barge is towed or pushed by another vessel (tug) along a waterway. Barge hulls are either box-shaped (vertical) or angled (raked). The major difference among most barge types is the type of the carried cargo. Figure 1-15 presents the types of river barges (considering their intended purpose).

---

[56] Ibidem., p. 32

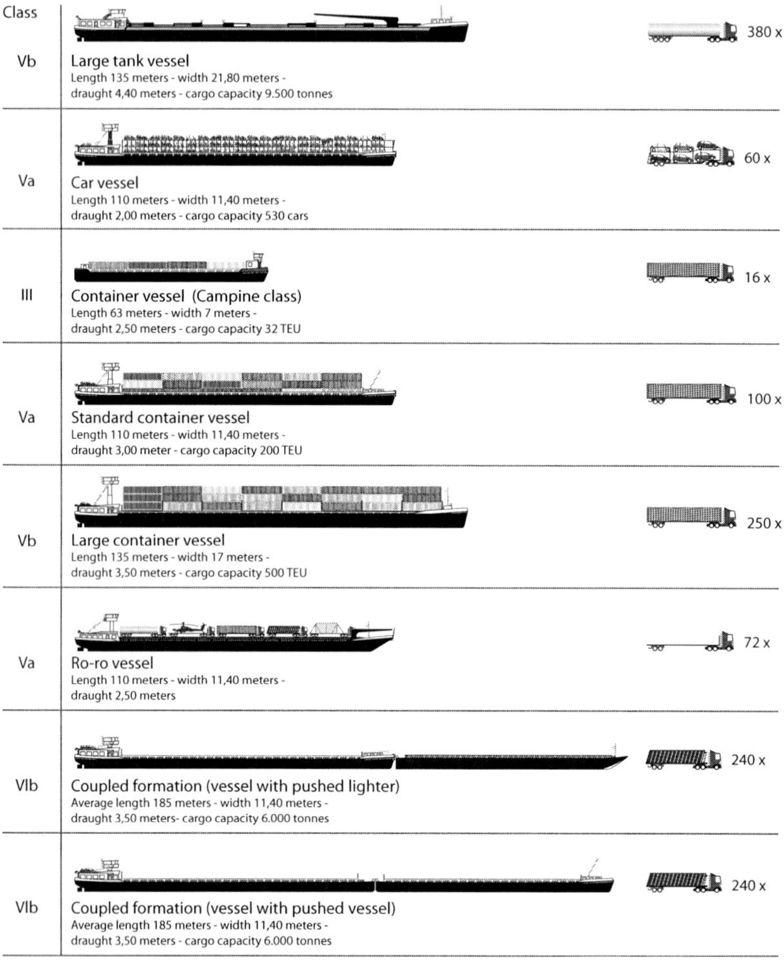

Figure 1-15. Types of river barges (divided by the criterion of their intended purpose)

Source: adapted from https://www.bureauvoorlichtingbinnenvaart.nl/inland-navigation-promotion/basic-knowledge/waterways [accessed on 15th May 2018].

Considering the above-mentioned arguments for the complementarity of maritime and inland waterway navigation, it is impossible not to mention the lighter aboard ship technology (LASH), which is also referred to as the barge carrier transport system. The operation scheme of the system involves inland waterway transport of cargo sent from the mainland with the use of a river barge to a sea-inland port and then loading the whole barge onto a special vessel, a barge carrier, which navigates across the sea to another sea-inland port, where the cargo is unloaded and transported along inland waterways to its receiver. The loading and unloading of barges on/off a barge carrier is performed with the use of three main systems (Figure 1-16):

- (A) the LASH system – the most common system which is operated in the following way: a barge is lifted out of water with the use of special cranes which move with the lifted barge along the vessel to put it in a particular place on the vessel;
- (B) the SEABEE system – a barge enters a special dock which is then lifted with the barge to a required level, namely, the deck of a barge carrier. On the deck the barge is moved with the use of a special frame;
- (C) the float on/float off system – a barge directly enters or exits the deck of a barge carrier without the assistance of any cranes or lifts. Barge carriers of that type belong to the class of semi-submersible or catamaran vessels.

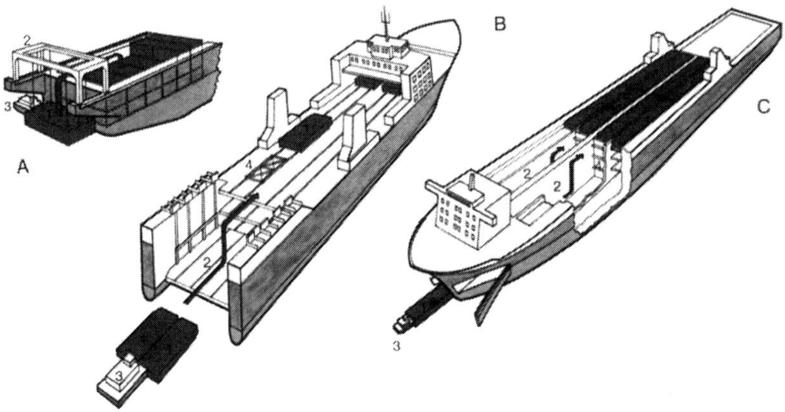

A – LASH, B – SEABEE, C – Float on–Float off

Figure 1-16. The barge carrier systems

Source: adapted from http://www.containerhandbuch.de/chb_e/stra/index.html?/chb_e/stra/stra_01_03_01_02.html [accessed on 15th May 2018].

On inland waters, barges that can be transported by a barge carrier navigate in integrated pushed convoys. At present, such a system is not commonly applied, however, it is possible to observe that the concept has returned in a modern function performed by lighter barges at sea-river ports. A convoy of LASH barges is presented in Figure 1-17.

Figure 1-17. A convoy and technology of LASH barges

Source: adapted from https://shipoftheday.blogspot.com/2007/04/rhine-forest.html [accessed on 15th May 2018].

Despite the fact that it is not commonly applied, the LASH technology, which has been discussed above, allows the Author to present the complementarity of maritime and inland waterway transport and to indicate the potential of both these modes of transport in global and regional economy.

1.3. The significance of water (maritime and inland waterway) transport for global economy

1.3.1. The potential and development lines of maritime transport

Observed over the recent years, dynamic development of maritime transport has been mainly related to an ongoing process of Europeanisation and globalisation of economic relations and to an increase in the volume of international trade turnover. The major part of that trade is implemented with the use of maritime transport. Consequently, such a situation requires a full spectrum of knowledge related to the status quo of that economic branch. Identification of the key elements which determine the potential of the maritime transport market, with special consideration of the aspects related to global transport, should include the following:[57]

- the specific character of maritime transport organisation,
- factors pertaining to supply and demand in maritime transport,
- determinants of costs and prices in maritime transport.

The relation between world trade and maritime transport is mutual and inseparable. Development of maritime transport, which results in the renewing of the commercial fleet, the implementation of more and more advanced technologies to the shipbuilding industry, directly affects the enlargement of markets where goods can be sold and the sources of raw material supplies.[58] Table 1-8 presents the potential of the maritime commercial fleet (with special consideration of ten countries – the flags of which are carried by the highest number of registered vessels).

Table 1-8. The maritime commercial fleet of the world, in accordance with the TOP 10 of registration countries (January 2013, DWT)

Flag of registration	Number of vessels	Share of world total, vessels	Deadweight tonnage (thousands DWT)	Share of world total (% DWT)	Cumulated share (% DWT)	National ownership (%)[a]	DWT growth 2013/12 (%)
Panama	8580	9.87	350506	21.52	21.52	0.14	5.03
Liberia	3144	3.62	198032	12.16	33.68	0.01	5.83
Marshall Islands	2064	2.37	140016	8.60	42.27	0.11	11.08
Hong Kong (China)	2221	2.55	129806	7.97	50.24	12.15	16.87

[57] Cf.: E. Załoga, T. Kwarciński, D. Milewski, A. Wysocka, *Sustainable development as the economical challenge* [in:] H. Witte (ed.), Nachlatigkeit, ökonomische Gleichgewichte und Wertschöpfung, Serie Extensión, Colección Estudios Sociales y Humanidades, Bahía Blanca, Argentina 2014, pp. 65–102 and R. Miler, *Bezpieczeństwo transportu morskiego*, op. cit., pp. 38–40

[58] I. Chrzanowski, *Współczesne problemy żeglugi morskiej*, PWE, Warszawa 2001, p. 51

Table 1-8. cont.

Flag of registration	Number of vessels	Share of world total, vessels	Deadweight tonnage (thousands DWT)	Share of world total (% DWT)	Cumulated share (% DWT)	National ownership (%)[a]	DWT growth 2013/12 (%)
Singapore	3339	3.84	89697	5.51	55.75	36.60	16.62
Greece	1551	1.78	75424	4.63	60.38	92.60	5.13
Bahamas	1446	1.66	73702	4.52	64.91	1.18	1.44
Malta	1794	2.06	68831	4.23	69.13	0.35	8.18
China	3727	4.29	68642	4.21	73.35	98.18	9.83
Cyprus	1030	1.18	31706	1.95	75.29	19.51	7.61

[a] The estimate of national ownership is based on the available information about commercial seagoing vessels of 1,000 GT and above.

Source: adapted from *Review of Maritime Transport 2013*, UNCTAD/RMT/2013, (data supplied by Clarkson Research Services), New York – Geneva 2013, p. 56.

A direct result of the fact that maritime transport has been considerably burdened in terms of the shipped cargo volume is the implementation of newly constructed vessels which are bigger, they can navigate faster, and their carrying capabilities are much higher than those of previously used vessels.[59] At the same time, it is possible to observe some tendencies towards the limitation on the number of crew members through the implementation of more and more automated workstations which minimise personnel costs (up to full autonomy), and towards the imperative of the maximal shortening of time when vessels stay at ports, which determines economic efficiency of transport processes.[60] According to the data provided by the UNCTAD, at the beginning of 2012, the global fleet was 104.3 thousand units (over 1000 DWT) of total weigh equaling net 1534 million tonnes (DWT – means without cargo, fuel and supplies).[61]

[59] The basic technical parameters of a vessel are its linear dimensions, namely length, beam, draught, depth moulded. Another group refers to its capacity dimensions, namely, gross tonnage (GT), net tonnage (NT), cargo capacity. There are also weigh dimensions of a vessel e.g. buoyancy and deadweight tonnage (DWT) – the Author's note.

[60] R. Miler, *Bezpieczeństwo transportu morskiego...*, op. cit., p. 50.

[61] *Review of Maritime Transport 2013*, UNCTAD/RMT/2013, p. 9.

The growing tonnage of oceangoing vessels goes along with the growth of their number. At the end of 2007, it was 100 243.[62] The structure of the global fleet is presented in Figure 1-18.

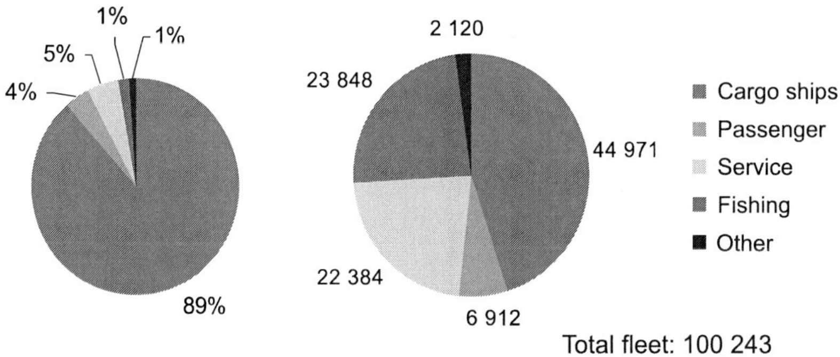

Figure 1-18. The structure and volume of the global fleet (2007)

Source: adapted from *Second IMO GHG Study 2009*, International Maritime Organization, London 2009, p. 13. [c] – [colour version shown in the centrefold]

Considering some obvious reasons, there is a very large difference in the carrying capacity of vessels. An average capacity of bulk carriers is 63.4 thousand DWT, whereas an average capacity of tankers is 45.3 thousand DWT,[63] of container vessels – 39.5 thousand DWT, of general cargo vessels – 5.2 thousand DWT and of other vessels – 1.7 thousand DWT. The global fleet is concentrated under the flags of over 30 countries. Greece, Japan, Norway, the United States of America and China handle 50% of the global fleet carrying capacity.[64]

[62] *Second IMO GHG Study 2009*, p. 13.

[63] *Knock Nevis* – the largest vessel in the history of shipbuilding industry, a super-tanker built at the Oppama Shipyard of the Sumitomo, a Japanese shipbuilding corporation, in 1979–1981. Its previous names were *Seawise Giant*, *Happy Giant* and *Jahre Viking*. It was 458 m long, 69 m wide, its deadweight tonnage (DWT) was 647 955 tonnes and when loaded, almost 650 000 m^3 (4,1 million oil barrels); its draught was 24.6 m. In 1986, as *Seawise Giant*, it was hit with an Exocet missile by an Iraqi fighter jet in the Strait of Hormuz, during the Iran-Iraq war, and it was declared a total loss. However, at the end of the war (1989) it was renovated and given a new name: *Happy Giant*. It was given another name once again in 1991, *Jahre Viking*, by its new owner. In March 2004 it was sold to a Norwegian company, Fred Olsen Production ASA, and it went to a shipyard in Dubai to be transformed into a floating production storage and offloading vessel (FPSO) at the coast of Qatar. Then it was given its last name before being scrapped – the Author's note.

[64] W. Rydzkowski, K. Wojewódzka – Król, *Transport…*, op. cit., pp. 178–184.

Dynamic changes which take place in maritime transport, its specificity and specialisation are manifested in the progressive containerisation of transport processes in modern land-sea transport chains and, more broadly speaking, in logistic supply chains. Maritime transport determines dynamic development of seaports which base their operations on container terminals. The implementation of the discussed loading unit, a container, is considered to be the third revolution in transport.[65]

Being placed in a container, the cargo is protected from theft, loss, damage and bad weather conditions. At the same time, the process of containerisation allows us to shorten the time when vessels stay at ports and to use the loading space of vessels and port storage yards more efficiently, because it is possible to store containers in stacks.[66]

A unit commonly applied for the monitoring and analysis of shipping and container handling operations is 1 TEU (Twenty-Foot Equivalent Unit), an equivalent of one 20 ft container (in practice 1 TEU refers to approximately 9 tonnes of handled cargo). Referring to the above-mentioned unit, it can be assumed that in 2009 container transport by sea reached the level of 123 million TEU.[67]

Considering the development observed in this field of transport, the size of vessels has been increased. In 2014 the first Triple-E vessel[68] appeared, with its potential of 18 000+ TEU, operated by a Danish ship-owner MAERSK. In the subsequent years, the volume was increased up to 19 000+ TEU (MSC Oscar). At present

---

[65] The first one was the invention of a wheel, and the second one was the implementation of the steam engine to the field of transport – the Author's note. See for more in: J. Neider, *Transport...*, op. cit., p. 115.

[66] K. Ficoń, *Logistyka morska...*, op. cit., p. 321.

[67] *Funkcjonowanie oraz perspektywy rozwoju rynku przewozów kontenerowych w Polsce do roku 2015 Balticon* – a report, on-line materials, Gdynia 2010, p. 5 [accessed on 15th May 2018].

[68] As the abbreviation of the series suggests, when developed to its full form: economy of scale, energy efficiency and environmentally improved, the EEE vessels shall definitely release less greenhouse gases to the atmosphere. The designers indicate that at the speed of 20 knots, the CO_2 emission shall be lower by over 1/3, and at the speed of 17.5 knots even by 50% lower than emission caused by the most environmentally friendly vessels which have been operated so far. In this way, the emission level shall be only 3 g/1 tkm, which means that maritime transport shall become even more environmentally friendly than any other modes of transport, for which the average CO_2 emission is respectively: 18 g/1 tkm for rail transport, 47 g/1 tkm for road transport and as much as 560 g/1 tkm for air transport – the Author's note.

(2018), the first vessels of the carrying capacity of 22 000 TEU are expected to appear. The basic parameters of a Triple-E vessel are presented in Figure 1-19.

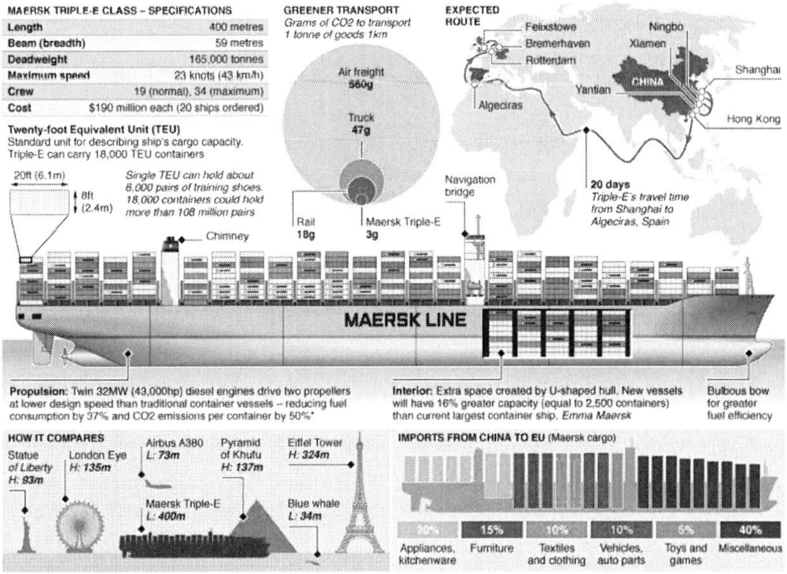

Figure 1-19. Basic technical and comparative parameters of an oceangoing container vessel (MAERSK)

Source: adapted from https://thelexicans.wordpress.com/2014/03/10/infographic-maersk-triple-e-class/ [accessed on 17th May 2018]. [c]

Container ship-owners (operators) enter global alliances. The largest ones are: CKHY Alliance, Grand Alliance and The New World Alliance. They do not relate ship-owners as closely as shipping consortia, but they refer to handling numerous shipping services, and therefore they provide a global character to such alliances. The alliances guarantee common operation of vessels which involves coordination of particular sailing schedules, choice of seaports and sailing frequency. On vessels which belong to ship-owners who have entered an alliance, jointly-operated freightage of available container slots is provided by standard; such a process is referred to as slot exchange.[69]

[69] E. Załoga, *W kierunku nowego paradygmatu rozwoju transportu w Unii Europejskiej*, "Logistyka" No. 2/2014, pp. 409–417 and R. Miler, *Bezpieczeństwo transportu morskiego…*, op. cit., pp. 58–59.

It means that a few ship-owners divide the loading space of their vessels into slots which are treated as common carrying capacity capacity. It is then divided among the particular partners who load their containers into the slots. Hence, shippers deal with various operators of the slots on the same vessel, who may even compete among themselves, offering different prices and service quality.[70]

Considering the presented analysis, there is another important question which refers to the model of the functioning of container maritime transport in the world. Operators do not connect particular seaports in a direct way, but they function on the basis of the *hub and spokes* system.[71] According to such a system, it is possible to define two basic types of shipping: oceanic (liner) shipping and feedering shipping. Oceanic shipping is implemented as:[72]

- regional (latitudinal) connections, understood as shipping services among the most developed regions of the world, e.g. Asia-Europe-Asia;
- shuttle connections of the pendulum type, relations between highly developed regions where two container routes are connected, e.g. Asia-Europe-the eastern coast of North America-Europe-Asia;
- round-the-world connections (RTW), understood as shipping services connecting highly industrialised regions, characterised by the fact that vessels operated on such routes navigate only in one-way direction across three oceans, e.g. to the east: the eastern coast of North America-Europe-Asia-the western coast of North America-the Panama Canal-the eastern coast of North America.[73]

Table 1-9 presents the main container shipping services in maritime transport.

Oceanic shipping is usually handled by large vessels (referred to as mother vessels) which may carry from 4 to 14 thousand TEU at one go. Considering their parameters (mainly their draught, length and beam), such vessels can be handled only at very few ports located on the particular continents (the hub ports). In the market of oceanic shipping, there are large ship-owner companies which operate sharing the market among themselves. At present,

---

[70] ftp://ftpf.uokik.gov.pl/analizy/rezultat5_v2.pdf [accessed on 17th May 2018].

[71] The solution has been adapted from air transport – the Author's note.

[72] I. Urbanyi-Popiołek (ed.), *Ekonomiczne i organizacyjne aspekty transportu morskiego*, Wydawnictwo Uczelniane Wyższej Szkoły Gospodarki w Bydgoszczy, Bydgoszcz 2012, p. 69–70.

[73] The implementation of post-Panamax container vessels – which, considering their size, cannot enter the Panama Canal – on a massive scale has resulted in further development of that kind of shipping and, at the same time, in an increase in the competitiveness of round-the-world services – the Author's note.

the largest one is APM-Maersk which operates 572 vessels of the total capacity of 2.1 million TEU (Table 1-10).[74]

Table 1-9. The number of the main container shipping services in global maritime transport (December 2012)

Route	Services
Far East – North America	73
North Europe – Far East	28
Far East – Mediterranean	31
North Europe – North America	23
Mediterranean – North America	21
Europe – Mid-East/South Asia	40
North America – Mid-East/South Asia	10
Far East – Mid-East/South Asia	72
Australasia	34
East Coast South America	26
West Coast South America	48
South Africa	24
West Africa	60
Total	490

Source: adapted from http://www.worldshipping.org/about-the-industry/global-trade/trade-routes [accessed on 14th April 2018], notice: services may be counted on more than one route [based on World Shipping Council information].

Table 1-10. The largest container operators in the world (in 2018)

Rank	Operator	TEU	Share [%]
1	APM-Maersk	4,155,991	18.9
2	Mediterranean Shp. Co.	3,226,454	14.7
3	CMA CGM Group	2,518,738	11.5
4	COSCO Shp. Co. Ltd.	1,944,496	8.9
5	Hapag-Loyd	1,587,793	7.2
6	ONE (Ocean Network Express)	1,514,506	6,9
7	Evergreen Line	1,088,509	5.0
8	OOCL	685,798	3.1
9	Yang Ming Maritime Transport Corp.	654,455	3.0
10	PIL (Pacific Int. Line)	416,777	1.9

Source: adapted from ASX-Alphaliner http://www.alphaliner.com/top100/index.php [accessed on 16th April 2018].

[74] *Funkcjonowanie oraz perspektywy rozwoju rynku przewozów kontenerowych w Polsce do roku 2015 Balticon...*, op. cit., p. 7.

Smaller ports located in the catchment area of hub ports are connected with them by a network of feeder services. Based on oceangoing vessels which come to the hub ports, the system is complemented with smaller container vessels (feeders) onto which some containers are loaded.[75] Owning to their parameters (the usual capacity ranging from 500 to 1500 TEU), feeder vessels can reach a considerably higher number of seaports and terminals located in a particular region. The structure of maritime connections of the above-mentioned type is presented in Figure 1-20, with the consideration of the global scale.

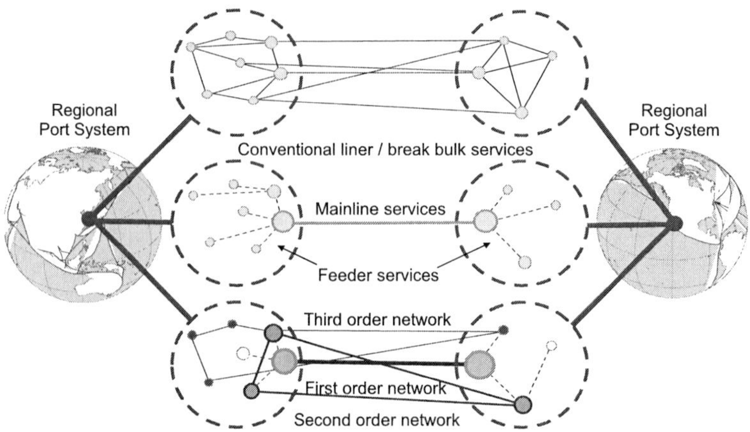

Figure 1-20. The network of global maritime connections (by the connection types and structure)

Source: adapted from J.-P. Rodrigue, Dept. of Global Studies & Geography, Hofstra University, non-published materials https://www.hofstra.edu, New York 2016, pp. 37 [accessed on 15th May 2016].

Since ancient times, the driving force of maritime transport has been international trade. At present, after a period of turbulence in the financial markets, which has resulted in a slight economic downturn, it is possible to observe a conspicuous increase in trade, particularly in intercontinental shipping, which perfectly fits into the concept of globalisation, implemented to a large extent by maritime transport.[76] It results from some dynamic

[75] At present, transshipment accounts for 26% of container turnover at sea ports in the world, based on: *Funkcjonowanie oraz perspektywy rozwoju rynku przewozów kontenerowych w Polsce do roku 2015 Balticon…*, op. cit., p. 7 – the Author's note.
[76] R. Miler, *Bezpieczeństwo transportu morskiego…*, op. cit., p. 31.

changes in the economic significance of the particular countries, groups of countries or continents in global economy, because of various rates of their economic growth, their macro-economic policies and their pursuit of reaching an international level of competitiveness.[77]

Considering the above-mentioned immanent features, maritime transport makes shipping of relatively cheap raw materials and goods a profitable operation for longer and longer distances in a relatively short time; frequently, the costs of transport do not exceed the level of 1% of the final price of the goods in retail sales.[78]

If the role, the function and the significance of maritime transport in global economy are understood in such a way, then an attempt at indicating some further potential lines for its development seems highly advisable.

Analysts foresee that in the future the global growth of international trade and maritime transport shall reach an approximate level of 3% annually, and the driving economies will be the following: the USA – 3.4%, Japan – 1.6%, and the Eurozone countries – 1.9%.[79] At the same time, there is some confidence that maritime transport will still continue to hold its leading position in the volume of transported cargo – 90% of the total volume (including almost 70% of the worldwide oil and natural gas production). This fact becomes of particular significance in the current geopolitical situation in the world, and it explicitly determines the necessity of providing the highest security/safety standards in maritime transport of strategical raw materials.[80]

These phenomena have been translated into the high rate of growth in the volume of maritime transport which has already outrun the dynamics of the world's GDP and industrial production of the OECD countries, however it has been lower than the dynamics of the global trade growth – in the years 2000–2013 it reached an average level of 8.9% annually.[81]

---

[77] Cf.: Grzelakowski A.S., *Model zarządzania polskimi portami morskimi jako element systemu regulacji działalności sektora portowego* [in:] *Kierunki rozwoju polskiego transportu morskiego – programy i rzeczywistość*, H. Klimek, J. Dąbrowski (eds), Fundacja Rozwoju Uniwersytetu Gdańskiego, Gdańsk 2010, p. 2.

[78] R. Miler, *Bezpieczeństwo transportu morskiego...*, op. cit., p. 32.

[79] *Review of Maritime Transport 2013*, UNCTAD/RMT/2013, p. 25.

[80] R. Miler, *Bezpieczeństwo transportu morskiego...*, op. cit., p. 33.

[81] A.S. Grzelakowski, *Model zarządzania polskimi portami morskimi jako element systemu regulacji działalności sektora portowego...*, op. cit., p. 10.

Presented in a dynamic approach, the mutual relations between those figures in the years 1975–2013 are displayed in Figure 1-21.

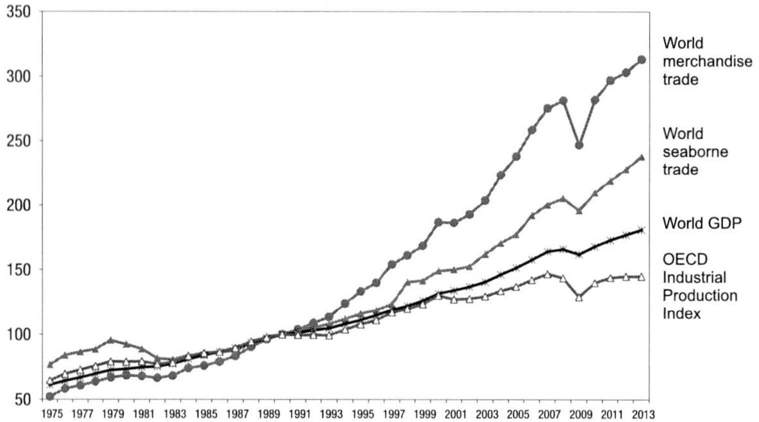

Figure 1-21. Dynamics of the growth of the global trade and maritime transport, the GDP and industrial production (based on the OECD index)

Source: adapted from *Review of Maritime Transport 2013*, UNCTAD/RMT/2013, p. 4.

The second element of the global system of maritime shipping refers to seaports. In 2009, they handled 471 million TEU altogether.[82] At present, the largest container ports of the world are the centres located in Asia, mainly in China. The largest container port of the world is Shanghai, where 25.9 million TEU were handled in 2009. However, this number was lower than the result obtained in the previous year by 7.5%. Similarly, in 2009 other largest ports went through a period of some significant decline in cargo handling operations. However, that trend was changed, and since 2010 there has been a considerable increase in trade (even by +19.8% YoY). In this way, the loss of 2009 was covered, and the expected total trade of ports was increased to the level of 535 million TEU. In 2012, 9.3 million tonnes of cargo were transported by sea.[83] The statistics explicitly indicate that a dominant group in the cargo structure in maritime transport consists of oil and oil derivatives, then subsequently there is dry bulk cargo, such as coal, iron ore and

[82] On average, 1 TEU is handled 3.8 times at seaports (it results from the *hub and spokes* model of the functioning of container shipping in the world) – the Author's note.

[83] A.S. Grzelakowski, *Model zarządzania polskimi portami morskimi jako element systemu regulacji działalności sektora portowego…*, op. cit., p. 10.

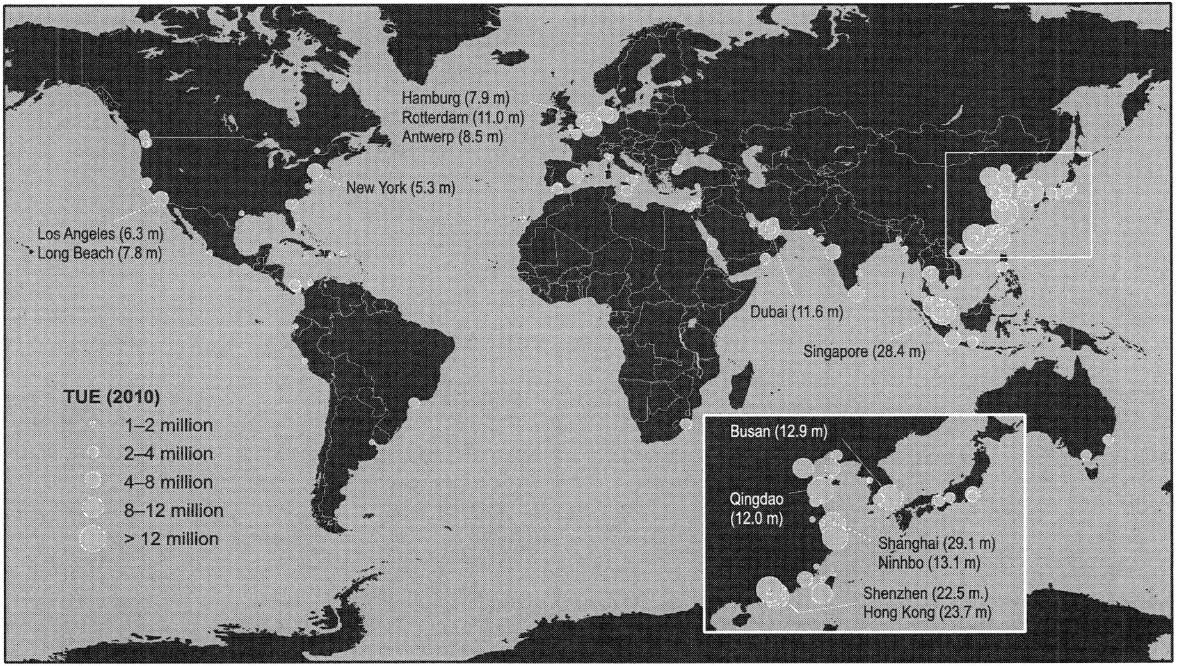

Figure 1-22. Location and turnover of the main container ports of the world (given in TEU million, in 2010)

Source: adapted from the report: *Safety and Shipping 1912–2012. From Titanic to Costa Concordia. An insurer's perspective from Allianz Global Corporate & Specialty*, Seafarers International Research Centre (SIRC), Cardiff University, Cardiff 2012, p. 58.

grain. In 2012 the volume of containerised general cargo (industrial goods and ready-made products in containers) transported by sea reached the level of 156 million TEU. Transport of containers between Europe, Southeast Asia and North America is characterised by the highest dynamics of growth. Hence, the global network of ports (with modern container terminals) becomes extremely significant for container shipping.[84] Figure 1-22 presents the distribution and turnover of the main container ports of the world.

Obviously, the dynamically growing turnover of containerised general cargo in the world does not refer only to one type of cargo transported by sea in the global trade. Figure 1-23 presents the volume of transported cargo divided by some basic cargo groups.

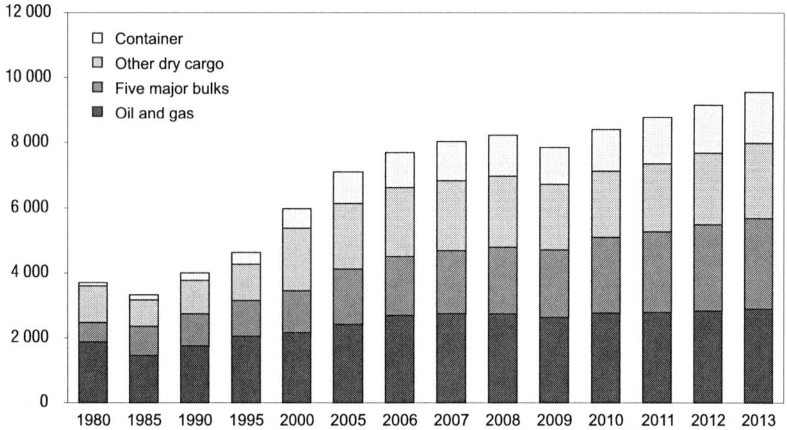

	1980	1985	1990	1995	2000	2005	2006	2007	2008	2009	2010	2011	2012	2013
Container	102	152	234	371	598	969	1076	1193	1249	1127	1275	1421	1480	1578
Other dry cargo	1123	819	1031	1125	1928	2009	2112	2141	2173	2004	2027	2084	2184	2300
Five major bulks	608	900	988	1105	1295	1709	1814	1953	2065	2085	2335	2486	2665	2786
Oil and gas	1871	1459	1755	2050	2163	2422	2698	2747	2742	2642	2772	2794	2836	2904

Figure 1-23. International maritime shipping presented in accordance with basic cargo groups (given in million loaded tonnes in the years 1980–2013)

Source: adapted from *Review of Maritime Transport 2013*, UNCTAD/RMT/2013, p. 7.

[84] R. Miler, *Bezpieczeństwo transportu morskiego...*, op. cit., p. 33.

Based on the characteristics of basic tendencies which can be observed in the area of the functioning of freight markets and in the regulations pertaining to that sector, the analysis of maritime transport in the global scale explicitly indicates that it is not a barrier to the development of global supply chains and global economy. Moreover, as it results from the analysis of the crisis of 2008–2010, it also indicates a high capability of absorption of the most negative results generated in the relations taking place on global commodity markets – both: the supply and the demand.[85] In the current stage of its development, maritime transport wins the status of a stable factor which supports development of global economy and logistical processes which stimulate it.[86]

1.3.2. The potential and development lines of inland waterway transport

Inland waterway transport (IWT) presents some definitely different features than maritime transport. Considering the specificity of inland waterway transport (its regional character, different organisational structure and morphology), the identification of its potential will be best presented on a regional (European) example. In Europe, IWT plays an important role as a transport system complementary to maritime shipping. There are 41.9 thousand kilometers of inland waterways going through the territory of the European Union (out of which over 37 thousand km are intensively exploited in economy). 21 EU member countries use rivers for navigation and 13 of them have developed an interconnected waterway network. 10 countries do not have access or do not make any economic use of inland waterways in their transport system. The highest infrastructural potential of inland waterways can be observed in the following European countries: Finland (8014 km), Germany (7675 km), the Netherlands (6237 km) and France (4996 km). In the EU countries the tendencies related to the development of inland waterways are highly differentiated. In the years 2000–2012 the most intensive increase in the length of inland waterways was observed in Germany (by 921 km), in Hungary (by 491 km), Croatia (by 297 km) and Finland (by 172 km). France and Poland are the leaders among the countries where the length of navigable inland waterways has been decreased (during

[85] Cf.: R. Miler, *Bezpieczeństwo transportu morskiego...*, op. cit., pp. 34–36.
[86] Cf.: A.S. Grzelakowski, *Wyzwania gospodarki globalnej*, Prace i Materiały Instytutu Handlu Zagranicznego Uniwersytetu Gdańskiego no. 31, Gdańsk 2012, p. 784.

the above-mentioned period of time a decrease was observed, respectively by 793 km and 154 km).[87]

The share of inland waterway transport in shipping which includes the modes of land transport was slightly increased: from 4.1% in 2000 to 4.4% in 2013. The most positive changes were observed in the following countries: Bulgaria (an increase from 4% to 14,8%), the Netherlands (an increase from 33,4% to 37,3%) and Romania (an increase from 8,2% to 20,4%). An increase in activities performed in the market of inland waterway transport in the EU could be better observed in the data pertaining to the shipping volume, which was increased from 134 billion tkm in 2000 to 153 billion tkm in 2013, that is by 14,1%. Such a level of growth is higher than the level achieved at that time by road transport (13,9%) and by rail transport (0,3%). The countries with the largest share of inland waterway transport in cargo shipping are the following: the Netherlands (37,3%), Romania (20,4%), Belgium (19,9%), Bulgaria (14,8%) and Germany (12,1%).[88] The percentage share of water transport in the general volume of shipping in the EU territory is presented in Figure 1-24.

Considering inland waterways transport in Europe at the macro-level, it should be observed that this mode of transport is highly fragmented. In Europe, there are approximately 5 700 river barge operators/owners registered, and more than 60% of them are based in the Netherlands (Figure 1-25).

The situation is different when the micro-level is considered. Barge operators/owners provide their services to a small number of clients, which is usually between 10 (small operators) to 100 (large operators) customers. 42 200 people have found their employment in the sector of inland waterway transport in the EU (the Annual Report 2016), 70% of whom in the Rhine states (Germany, France, Belgium, the Netherlands). The corporate sector has been dominated by companies which employ up to 10 people (97% in the Netherlands, 96% in France, 86% in Germany). A low percentage refers to large barge owners' enterprises which employ over 50 people (0,4% in the Netherlands, 1,8% in France, 2,1% in Germany). Family companies where there are one, two or several people employed, often referred to as particular

---

[87] L.M. Punter, W.J. Hofman, Report: *Digital Inland Waterway Area, Towards a Digital Inland Waterway Area and Digital Multimodal Nodes*, Luxembourg: Publications Office of the European Union, 2017, pp. 37–38.
[88] Ibidem, pp. 38–39.

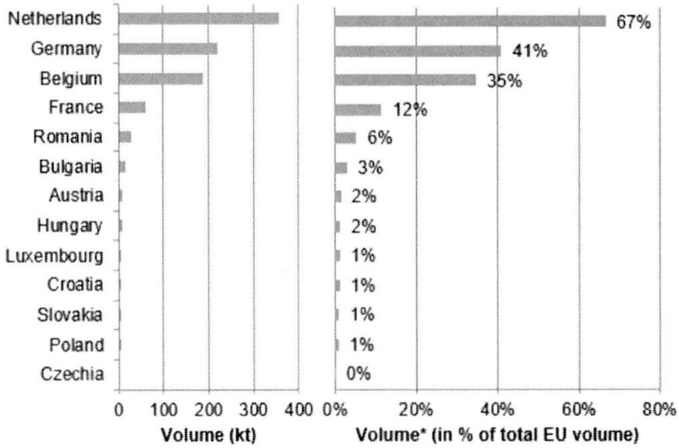

Figure 1-24. Volume of goods shipped by inland waterway transport in the EU [in kt, 2016]

Source: adapted from L.M. Punter, W.J. Hofman, Report: *Digital Inland Waterway Area, Towards a Digital Inland Waterway Area and Digital Multimodal Nodes*, Luxembourg: Publications Office of the European Union, 2017; http://ec.europa.eu/eurostat/statistics-explained/index.php/Inland_waterway_transport_statistics [accessed on 4th January 2018].

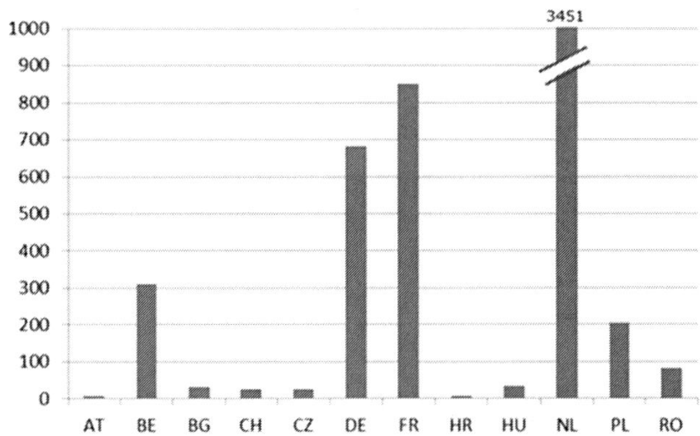

Figure 1-25. The number of barge operators per country [Europe]

Source: adapted from L.M. Punter, W.J. Hofman, Report: *Digital Inland Waterway Area, Towards a Digital Inland Waterway Area and Digital Multimodal Nodes*, op. cit., p. 34.

operators, come as a significant element in this sector of services. It is also important in the context of the capability of the sector to renew its fleet and to implement technological innovations.[89] Figure 1-26 presents the estimated number of companies in the IWT sector in Europe, considering the number of employees.

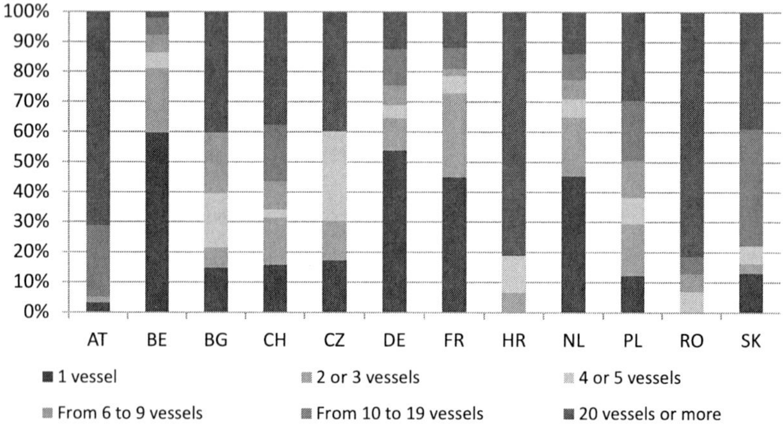

Figure 1-26. Numbers of vessels per company broken down to different classes per country [Europe, 2012]

Source: adapted from L.M. Punter, W.J. Hofman, Report: *Digital Inland Waterway Area, Towards a Digital Inland Waterway Area and Digital Multimodal Nodes*, op. cit., p. 48.

As far as inland ports are concerned, the reports indicate that there are approximately 150 inland ports used for inland waterway navigation in Europe. They are located in:[90]

- France: 20,
- the Netherlands: 25,
- Belgium: 15,
- Germany, Luxemburg, Switzerland, the Czech Republic: 60 (the Rhine/the Mosel/the Oder/the Elbe),
- Hungary, Austria, Slovakia, Croatia (the Upper Danube): 10,
- Romania, Bulgaria, Serbia. Moldova, Ukraine (the Lower Danube): 20.

[89] L.M. Punter, W.J. Hofman, Report: *Digital Inland Waterway Area, Towards a Digital Inland Waterway Area and Digital Multimodal Nodes*, Luxembourg: Publications Office of the European Union, 2017, p. 40.

[90] Ibidem, p. 19.

Inland ports differ among themselves in terms of their purposes and cargo handling/port operations performed there:[91]

- most inland ports are relatively small and have only one terminal operator and just a few berths. In addition, there are some industrial sites along inland waterways with dedicated berths which are not included in this overview;
- there is a small number of very large ports with multiple terminals. The examples include combined maritime and inland ports (Antwerp, Le Havre, Rotterdam, etc.) and large inland ports, such as the Port of Duisburg and the Haropa Port in Ile-de-France. Given their larger size, their operation is also usually more complex, e.g. it requires more advanced scheduling of calls and management of berths;
- most ports are specifically focused on either a specific type of bulk cargo (depending on the industry in the vicinity) or multimodal container operations.

Despite a dominant share of bulk cargo transport on the roads of Western Europe, unitised (containerised) cargo transport gains more and more significance. Container transport is stimulated by maritime transport also through an increase in the level of general cargo containerisation. It is, in turn, translated into shipping containerised units by IWT. In the EU, the main areas where container transport takes place along inland waterways are Germany and the Netherlands. When converted to TEU units, in 2009 there were 1.845 million containers transported. In 2010 the number of transported containers was increased to the level of 2.185 million TEU, and since that year some moderate growth of shipping was observed until 2015 when another decline was reported.[92] Table 1-11 presents the fluctuation in the market of container shipping in the IWT sector in the EU.

The main direction of container transport is the Rhine corridor (approximately 85% of the whole transport by the inland waterways in Germany). In the largest port of Europe, Rotterdam, the share of inland waterway navigation in feedering and receiving containers has been maintained slightly above the level of 30%. The general number of transported TEU is 10.783 million.

---

[91] L.M. Punter, W.J. Hofman, Report: *Digital Inland Waterway Area, Towards a Digital Inland Waterway Area and Digital Multimodal Nodes*, Luxembourg: Publications Office of the European Union, 2017, p. 19.

[92] J. Kulczyk, E. Skupień, *Znaczenie transportu śródlądowego w systemie transportowym UE*, Logistyka 4/2012 p. 474–482.

Table 1-11. EU transport of containerised goods by type, shipped by inland waterway transport – 1000 TEU km

		2012	2013	2014	2015				Growth rates 2014–2015 (%)			
		Total	Total	Total	National	International	Transit	Total	National	International	Transit	Total
EU-28		**984 749**	**1 014 568**	**1 075 384**	**239 126**	**535 165**	**239 466**	**1 013 757**	**-8.1**	**-3.1**	**-9.0**	**-5.7**
Belgium	BE	94 807	72 936	80 466	17 449	25 987	531	**43 968**	-57.5	-31.0	-69.4	**-45.4**
Bulgaria	BG	622	133	68	–	19	49	**68**	–	–	-28.3	**-0.1**
Czech Republic	CZ	–	–	–	–	–	–	–	–	–	–	–
Germany	DE	415 540	446 273	469 929	22 106	332 475	96 366	**450 946**	+5.5	-0.2	-16.8	**-4.0**
France	FR	67 999	64 240	69 363	56 652	7 100	18	**63 771**	-8.6	-4.0	+36.7	**-8.1**
Croatia	HR	–	–	–	–	–	10	**10**	–	–	–	–
Luxembourg	LU	–	–	41	–	9	165	**174**	–	–	–	**+329.8**
Hungary	HU	–	–	–	–	15	–	**15**	–	–	–	–
Netherlands	NL	404 508	430 748	455 370	142 919	169 473	142 327	**454 718**	+4.9	-2.4	-2.2	**-0.1**
Austria	AT	–	–	–	–	–	–	–	–	–	–	–
Poland	PL	–	–	–	–	0	–	**0**	–	–	–	–
Romania	RO	1 274	238	148	–	87	–	**87**	–	-41.0	–	**-41.0**
Slovakia	SK	–	–	–	–	–	–	–	–	–	–	–

Source: adapted from http://ec.europa.eu/eurostat/statistics-explained/index.php?title=File:Transport_performance_of_goods_in_loaded_containers_by_type_of_transport_-_1000_TEUkm.png [accessed on 7th February 2018].

Figure 1-27 presents the network of inland waterways and the number of container layers which can be transported by water.

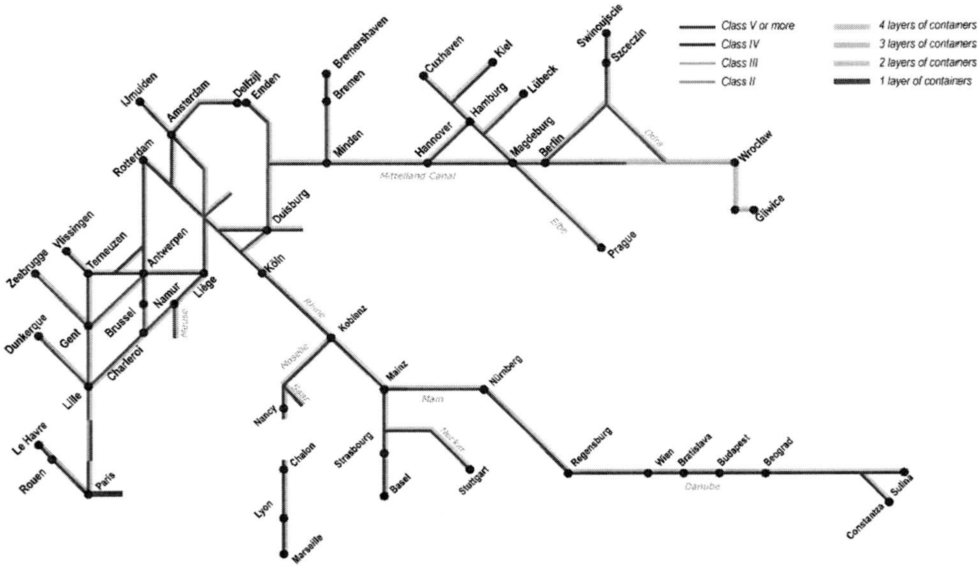

Figure 1-27. The network on inland waterways and the number of container layers which can be transported by water

Source: adapted from J. Kulczyk, E. Skupień, *Znaczenie transportu śródlądowego w systemie transportowym UE*, Logistyka 4/2012, pp. 474–482. [c]

Until 2005, container transport along the Rhine was characterised by constant growth. The years 2005–2007 brought some stabilisation in the number of transported containers. In 2009 there were about 16% fewer containers transported than in 2007. In comparison to 2008, a 10% decrease was reported. Such declines were considerably smaller when compared to the total figures in transport. In 2010 there was an increase in the number of containers transported by the Rhine. In comparison to 2010, in the first half of 2011 an increase by 1.1% was reported, and until the present moment some upward tendencies have been observed. Transport along the Rhine is based on regular container lines.[93] At the Port of Duisburg, eight operators provide transport from Duisburg to Rotterdam. Four operators provide transport 7 days a week. Other operators provide 21 sailings to Rotterdam. There are 5 container lines provided from Hamburg to other German and Czech ports along the Elbe, and there are 15 sailings a week. However, recently a stagnation in the number of containers handled by inland waterway transport has been observed. Apart from the crisis, a significant factor is insufficient adaptation of inland ports to the handling of containers onto vessels. At numerous existing terminals a lot of modernisation work is being performed in order to increase their container handling capacities and storage, as for example at the terminals in Nurnberg, Hannover, Braunschweig, Paris, Lyon, Stuttgart and Mainz. Figure 1-28 presents the fluctuation in the volume of cargo transported by inland waterways in the years 2016–2017.

Summing up, it should be stated that the development of modal technologies and a change of direction towards pro-ecological solutions have stimulated some acceleration in the development of IWT in Europe.

The conditions necessary for the development of inland waterway transport in the EU result from developmental paradigms, namely, from the *shift* paradigm (a shift of shipping from road transport to water and rail transport) and *decarbonisation* paradigm. They come as a manifestation of the policy of sustainable development which focuses its operations on efficient consumption of resources, decreased consumption of non-renewable fuels and limited CO_2 emission in all economic activities in the EU territory. Considering the political aims of the EU, such as shifts between the modes of

[93] L.M. Punter, W.J. Hofman, Report: *Digital Inland Waterway Area, Towards a Digital Inland Waterway Area and Digital Multimodal Nodes*, Luxembourg: Publications Office of the European Union, 2017, pp. 19–21

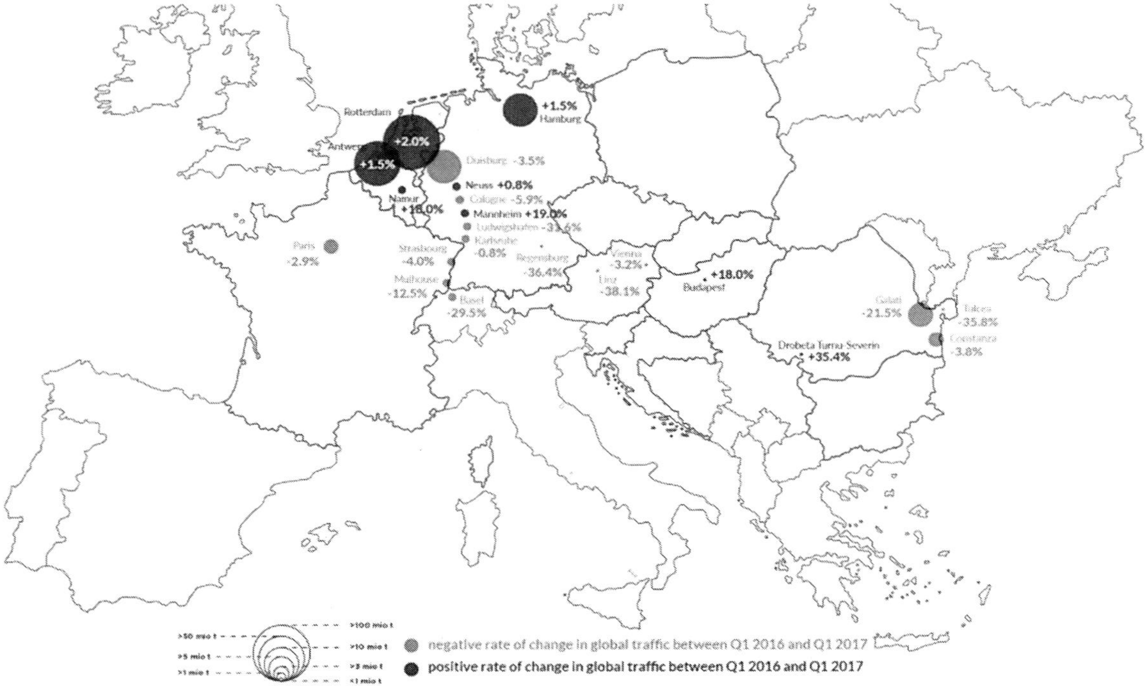

Figure 1-28. Total transshipment volume in Q1 2017, total transshipment volume in Q1 2016 and rate of change

Source: adapted from National Statistics Offices, Eurostat, CCNR.

transport, transport decarbonisation, all the activities which are aimed at an increase in the share of inland waterway transport provided to meet the demand for transport in the economy of the EU, including direct and combined shipping,[94] are very well-justified. It is directly translated into further development of containerisation processes in maritime and inland waterway transport.

1.4. Containerisation in maritime and inland waterway transport

1.4.1. A container as a loading unit in maritime and inland waterway transport

A basic loading unit for general cargo is a container (the term comes from a Latin word *contine*, which means "to keep together"). The application of such a loading unit is considered to be the third revolution in transport.

A container is a standardised, closed and universal box adapted for transportation of cargo, defined as [...] *a durable transporting appliance of multiple use and of special construction which allows it to transport cargo by several means of transport without cargo handling operations. Containers are equipped with special corner castings which provide a possibility of easy interlocking and releasing containers during all handling operations.*[95]

The cargo placed in a container is protected against theft, loss, damage and bad weather conditions. The implementation of containers on a massive scale started at the turn of the 1960s and the 1970s.[96] The use of containers in various modes of transport is presented in Table 1-12.

[94] E. Załoga, *Przesłanki i narzędzia promocji transportu wodnego śródlądowego w Unii Europejskiej*, PTiL 1/2017 (37), pp. 323–333.

[95] K. Ficoń, *Logistyka morska*, op. cit., p. 320.

[96] A momentous year for the development of international containerisation was 1956, when a transport company, the Sea-Land Services, started door-to-door shipping within the system of land-sea logistical supply chains. Optimising the process of cotton handling from trucks to vessels, the owner of the company, Malcolm McLean, implemented an innovation: the direct loading of semitrailers to vessels (after dismantling the wheels and reinforcing the semitrailer sides). Soon after, he implemented special boxes, 33 ft long, which allowed him to streamline the process and to lower the costs. In April 1956, a properly adjusted freighter, M/V *Ideal*, for the first time ever, transported 58 such boxes from Newark to Houston. This date is considered to be the starting point

Table 1-12. Diversification of containers, depending on a particular mode of transport

Container type	Volume (m^3)	Payload (kg)	Density (gm/cc)	Stack possibility	Max height of arrays (m)	Stack cracks (%)
Aircraft Units Loads	3.4–17	1,162–6,800	0.20–0.63	No	2.5	n/a
Intermodal Air/Surface	6.0–16	10,000–23,450	0.36–0.68	3 high	7.2	8
Intermodal (ISO)	33–105	24,000–30,000	0.33–0.66	9 high	24.5	8
Truck and semi-trailers	60–117	30,000	0.20–0.26	No	2.8	n/a
Railroad Boxcars	257–312	92,300	0.25–0.30	No	5.2	n/a
LASH barges	555	385,000	0.69	4 high	16.4	14

Source: adapted from https://cargofromchina.com/sea-air-rail/ [accessed on 19th May 2018].

The implementation of intermodal containers has drastically changed general cargo shipping in maritime transport, and it has significantly affected the process of containerisation in inland waterway transport. The process of containerisation has resulted in shorter stays at ports (oceangoing vessels and river barges), better use of loading space on vessels and port storage yards through multilayer stacking. The implemented shipping innovations have minimised the risk of cargo damage in the processes of transportation and handling operations, and they have also decreased the cost of cargo handling operations.[97]

The parameters of intermodal containers are standardised under the standardisation procedures implemented by the ISO (International Standard Organisation). A basic unit has been acknowledged as 1 foot, that is 304.8 mm. The maximal mass of containers has been limited with the consideration of their capacity as well as the lifting capacity of cargo handling facilities.[98]

of modern containerisation. In 1966, the first container ship arrived from the USA to Europe. It was M/V *Fairgrand,* which came to Bremen with 226 containers on board – the Author's note.

[97] K. Ficoń, op. cit., p. 321.

[98] S. Markusik, op. cit., pp. 37–41.

A basic intermodal container applied in transport and logistical operations is a closed, steel 1C 20' box, also referred to as a TEU. Its gross weight is 20 320 kg, its tare weight is 2 300 kg and its cargo carrying capacity is 18 020 kg. The loading space of such a container is 13.5 m^2 and its capacity is 30 m^3. A 20' (ISO) container is presented in Figure 1-29.

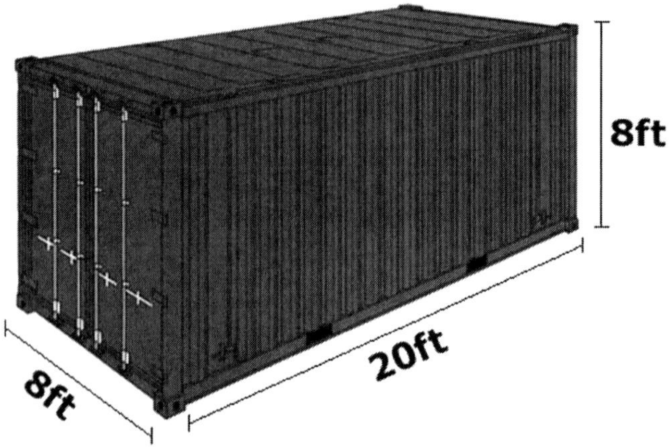

Figure 1-29. A 20' container (TEU)

Source: adapted from http://www.dimensionsinfo.com/20ft-container-size/ [accessed on 4th June 2014].

Universal containers are equipped with doors in their front walls. The doors are 2285 mm wide and 2135 mm high. When in their open position, the doors must be also properly secured, and the turn of the door wing must reach the angle of 260° at the minimum. Another significant feature of a container is its high durability, rigidity of construction and tightness which protects cargo against bad weather conditions. A container should transfer any dynamic forces which are caused by gravity acceleration up to 2g in the horizontal plane.[99]

Considering their constructional characteristics and their dimensions, containers require proper facilities and machines to provide their quick loading and unloading and safe transport. In order to handle containers properly, new-construction types of vessels have appeared in maritime transport, on

[99] S. Markusik, *Infrastruktura logistyczna w transporcie*, op. cit., p. 67.

which cargo is loaded in a vertical way: lift on-lift off (lo-lo vessels). Such vessels transport containers in their holds and on their decks. Considering the development of this mode of transport, the size of container vessels and container river barges has been successively growing.

The weight of the container and the size of the vessel (an oceangoing vessel or a river barge) make the cargo handling process dependent almost entirely on port facilities. At present, rubber tyred gantry (RTG) cranes and ship-to-shore (STS) cranes are most often used for handling containers. For further handling operations reach-stackers, that is: lift trucks with jibs or masts, and internal movement vehicles (IMV) with semi-trailers are applied. There are also chain slings, rope slings and spreaders used for handling containers. Unlike slings, the capacity of which is rather low, the technical construction of spreaders allows them to perform cargo handling operations quickly and efficiently; therefore, each gantry crane and hoist at any container handling terminal is equipped with spreaders. Due to the automatic change in the length of the carrying boom and translocation of the rotating spindles, modern devices are adapted to handle containers of various sizes. During handling operations, a container can be rotated by 90°, and stacking or placing it on a deck of a vessel is amortised by a hydraulic system.

Another important factor in shipping and handling containers is their safety which must be provided with proper stowage processes that guarantee not only the required stability of containers by placing heavy goods in their lower part and lighter goods in their upper part, but also with safe fastening on the vessel. The cargo placed inside the container must not pose any risk to the vessel, regardless of weather conditions. Containers are stacked on the vessel, depending on their weight: heavier containers are placed in the holds under the deck of the vessel in some more stable places where the roll amplitude is the lowest. Containers with valuable cargo should be particularly checked for any potential risk of damage in transport, in accordance with the specification stated in the documents. Stowage complying with the strict IMDG Code[100] standards must be provided to containers with hazardous cargo. The access to the aggregates and refrigerating appliances of containers with agricultural products, food and any goods requiring proper

---

[100] IMDG – *International Maritime Dangerous Goods Code* is discussed in the further part of the book – the Author's note.

temperature must be provided during the whole voyage. Another process which improves safety of containers and their cargo involves global systems of container monitoring, based on modern technologies of bar codes, RFID (radio frequency identification) tags and satellite communication systems (which are discussed in the further part of the book).

Container vessels must also meet the requirements pertaining to the particular safety rules. The fundamental condition refers to the maintenance of the statutory ship stability, the proper location of its meta-centre which affects proper distribution and formation of container stacks.[101] In order to streamline the processes of loading capacity management, there are some modern IT solutions available (stowage plans).[102]

Transposition of containers and any other necessary manipulation operations generate some particular categories of costs which are in direct proportion to the time of using and the type of containers.[103] Terminal costs involve all the factors related to services provided to containers and their cargo, namely loading, unloading, handling operations, consolidation, deconsolidation and costs related to the use of the cargo handling facilities at the terminal. Other costs involve fees for storage of containers in the area of the terminal, technical and commercial services provided to containers.[104]

According to the statistical data, at the beginning of 2011, the global fleet of oceangoing container vessels reached the number of 5961 ships of the loading capacity of 14 937 thousand TEU and the deadweight tonnage of 192 221 thousand DWT. Cellular container ships accounted for 81% of the fleet, that is 4846 vessels of the total loading capacity of 14 393 thousand TEU. At that time, the largest container shipping company, Maersk, had 586 vessels of the capacity of 2 176 thousand TEU.[105] The above-mentioned data present universal and common application of containers in transport, particularly in maritime transport and its complementary modes, including inland waterway transport. This situation, in turn, results in the fact that new types of containers have started to appear.

---

[101] K. Ficoń, op. cit., pp. 333–335.

[102] Ibidem, pp. 157–160.

[103] D. Dulęba (ed.), *Spedycja*, op. cit., p. 112.

[104] A. Salomon, *Spedycja w handlu morskim. Procedury i dokumenty*, Wydawnictwo Uniwersytetu Gdańskiego, Gdańsk 2003, p. 16.

[105] J. Neider, *Transport międzynarodowy*, op. cit., pp. 39–41.

1.4.2. Types of containers used in maritime and inland waterway transport operations

Universal and common application of intermodal containers in modern transport affects the criteria of the division and classification of containers which are almost always based on the ISO standards.[106] The type series of standard containers according to the ISO norms are presented in Table 1-13.

Table 1-13. The type series of standard intermodal containers, in accordance with the ISO norms

Type	Measurement	Length	Width	Height	Capacity [m^3]	Weight [tonnes]
1A	feet [m]	40 12,19	8 2,43	8 2,43	– 61,0	– 3,0
1B	feet [m]	30 9,12	8 2,43	8 2,43	– 45,5	– 2,0
1C	feet [m]	20 6,05	8 2,43	8 2,43	– 30,0	– 2,0
1D	feet [m]	10 2,99	8 2,43	8 2,43	– 14,3	– 7,0

Source: adapted from A. Daszkiewicz, *Kontenery i konteneryzacja w transporcie morskim*, Wirtualny Leksykon Transportu i Logistyki, on-line publication: logistyka.edu.pl, p. 9 [accessed on 15th April 2018].

As it has been already mentioned, a basic unit is a 1C, that is namely, a 20 ft container (TEU). Considering technical and operational parameters, the application of a 1A type of containers (40 ft containers – Forty-Foot Equivalent Unit, FEU) has become more frequent. For transportation of light, spatial and high cargo which cannot be loaded into a standard container, there are 20/40 ft containers of extended height, which is 9 ft 6 in. (2895 mm), also referred to as *high cubes*. As a result, about 10 m^3 of additional loading capacity has been achieved (in 40 ft containers).[107]

Apart from universal (ISO) containers, there are also specialised containers:[108]

- containers for transportation of liquid bulk cargo; such cargo is loaded into a cistern, the diameter of which fits into the container dimensions. Such multilayer plastic tanks, referred to as FLEXITANKS, are applied

[106] Ibidem, pp. 323–325.

[107] http://www.spedcont.com.pl/kont_abc2.htm [accessed on 26th July 2014].

[108] Cf.: A. Daszkiewicz, op. cit., pp. 3–6.

for transportation of various neutral, edible or chemical liquids which are not classified as hazardous substances, according to IMO Class/ADR. Such tanks are installed exclusively in universal maritime 20 ft containers;
- containers for transportation of livestock; such containers have side windows and feeders, they are made of slats or rods, and they are sheeted to protect animals from weather conditions and to provide them with proper ventilation;
- insulated containers which can be refrigerated, heated or refrigerated and heated with air conditioning; they are used for transportation of liquid nitrogen;
- flat rack containers used for transportation of bulk and pulverised materials, depending on the transported cargo, platform containers can be sheeted in order to prevent dust and soaking.

Summing up, considering their purpose and constructional solutions, it is possible to distinguish the following types of containers:[109]
- general purpose containers,
- open top / hard top containers,
- open sided containers,
- flat rack containers,
- platform containers,
- tank containers,
- bulk containers,
- dry bulk containers,
- ventilated containers,
- refrigerated containers,
- insulated containers,
- heated containers,
- thermal containers,
- mechanically refrigerated containers.

Considering the conditions and the specific character of inland waterway transport, there are some serious limitations related to the use of refrigerated containers and other types of thermal containers on river barges. Figure 1-30 presents the selected types of containers.

[109] http://www.spedcont.com.pl/kont_abc2.htm [accessed on 26th July 2014].

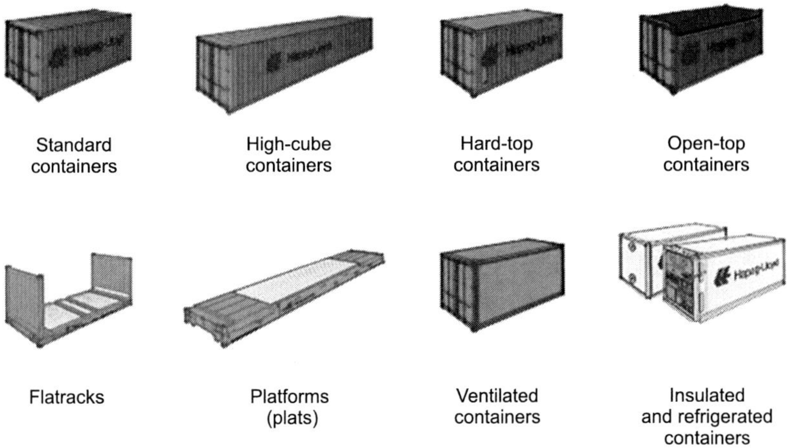

Figure 1-30. The selected types of intermodal containers

Source: adapted from http://www.spedcont.com.pl/kont_abc2.htm [accessed on 9th June 2014].

Despite the fact that there are many types of containers, the construction and dimensions of their corner castings have been also standardised to provide proper lifting, fastening and interlocking when containers are stacked.[110]

Due to the construction and durability of container frames, the weight of stacked containers is transferred onto their corner posts – not onto the container roof. It is possible to stack loaded containers into nine layers; empty containers can form a twelve-layer stack. An integral part of the frame are four bottom and four top corner castings which are used for fastening containers in stacks with a simple device, namely a twist lock and a lashing system.[111] The method of container stowage is presented in Figure 1-31.

Containers transported by sea are most often stowed along the vessel, because their side walls are stronger and the crosswise acceleration which affects the vessel during the voyage is higher. To avoid translocation of containers, they are stacked. There are stack blocks formed on a vessel by placing container stacks one next to another. Contrary to non-specialised ships, specialised container vessels do not require any additional fastening

110 S. Markusik, op. cit., pp. 42–43.
111 http://www.opensped.pl/kontenerownia/o-kontenerze [accessed on 26th July 2014].

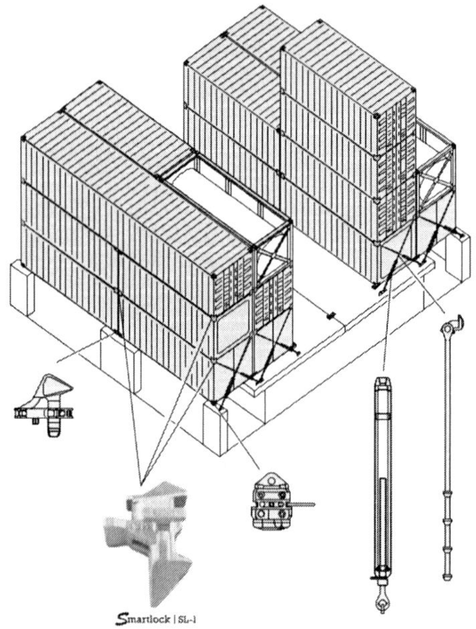

Figure 1-31. Devices used for container stowage on oceangoing vessels and container barges

Source: adapted from http://shipsbusiness.com/lashing.html [accessed on 28th May 2018].

systems, because they are equipped with cells and vertical cell guides which are adapted to the dimensions of containers.[112]

The positions of all the containers on a vessel are presented in a stowage plan. Stowed containers form blocks which are considered as holds and referred to as bays. Therefore, a stowage plan presenting the position of the containers on a vessel is also referred to as a bay-plan.[113]

---

[112] http://konteneryp olska.blogspot.com/2012/01/sztauowanie-kontenerow-na-statku.html [accessed on 26th July 2014].

[113] A bay-plan presents the cross-sections of the bays where the position of each container is stated and registered with 6 digits. The first and the second digit is the number of the bay, the third and the fourth digit is the number of the row, and two last digits refer to the number of the layer. In bays 20 ft and 40 ft containers can be transported at the same time. 20 ft containers are marked with odd numbers (starting from the bow): 1,3,5, etc., and 40 ft containers are marked with even numbers: 2,4,6, etc. A bay-plan should be available in all the documents required during maritime transport of containers – the Author's note.

Containers have obligatory and additional labelling. Obligatory labelling includes the following:

- identification marking,
- information about the country, the size and the type of a container,
- operational marking.

The identification marking of a container includes three code groups and a check digit, for example: SUDU 307007-9:

a) SUD – the owner's code (first 3 capital letters),
b) U – the code of the equipment category,
c) 307007 – serial number (6 digits), 9 – the check digit.

Figure 1-32 presents typical labelling of a container.

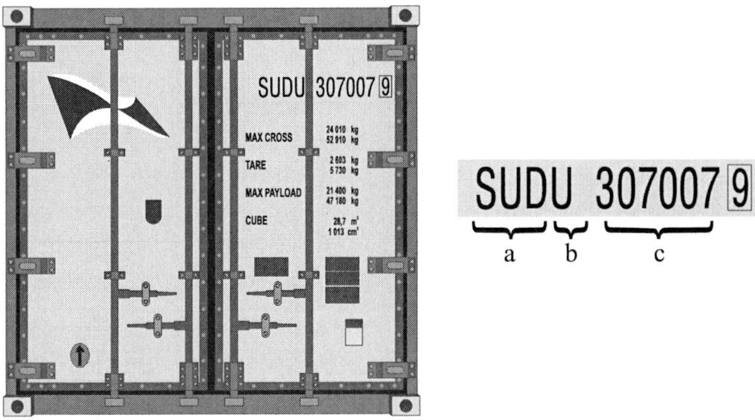

Figure 1-32. Identification marking of a container placed on the doors and presented in the code groups

Source: adapted from B. Wiśnicki, *Vademecum Konteneryzacji – Formowanie kontenerowej jednostki ładunkowej*, Link I, Szczecin, 2006, p. 90.

The owner's code must be unique, and it is registered at the International Container Bureau (Bureau International des Container – BIC), or at the affiliated registration organisations in particular countries. The check digit provides the possibility of code group verification. None of the IT systems which handle a particular container shall accept an identification marking when there is a conflict between the codes and the check digit.

The safety of the container construction is approved by a special plate presenting the data which comply with the requirements of the CSC 1972 (Convention on Safe Containers). The ACEP (Approved Continuous Examination Programme) means that a container is under constant supervision of a classification institution.

Identification of all types of containers used in maritime and inland waterway transport operations concludes the first chapter, where the role of maritime and inland waterway transport in contemporary, regional and global economy has been highlighted. To emphasize the significance of maritime and inland waterway transport in the economy, the elements of their potential are documented in figures. Moreover, further directions anticipated for their development are outlined as the derivatives of the identified regional/global geo-political and economic trends. The first chapter introduces also a concept of the key parameters of maritime and inland waterway modes of transport, based on their immanent operational features, such as universality, accessibility and degression of unit costs. These features are closely linked to the taxonomy and morphology of maritime and inland waterway transport (including the lighter aboard ship – LASH technology). To describe the modality of the discussed means of transportation, the containerisation challenges are pointed out and a concept of a container as a loading unit in maritime and inland waterway transport is presented.

Identification of all these factors allows the Author to present safety/security challenges as the main determinants referring to the implementation of telematics systems in maritime and inland waterway transport.

CHAPTER 2

SAFETY AS THE MAIN DETERMINANT FOR THE IMPLEMENTATION OF TELEMATICS SYSTEMS IN MARITIME AND INLAND WATERWAY TRANSPORT

2.1. A notion of safety in maritime and inland waterway transport

2.1.1. The essence of transport safety

Security (Latin *sine cura – securitas*) is most frequently interpreted as *the state of being untroubled, unconcerned, peaceful*, which is identified with a situation where there is no need to worry about any negative changes.[1] Dictionaries define security as *the state of peace, confidence and freedom from threats.*[2] Considering its social aspect as well as an individual approach (as freedom from existential anxiety), the state of freedom from threats becomes the most frequent reference point for the *security* term in its general sense.[3] Aggregating the notion of security to the level of abstract meanings, it is possible to interpret it as the opposite (alternative) to chaos. In social activities, an instrument applied to eliminate chaos is the imperative of development and implementation of routine activities (standard security procedures) which systematise relations among an individual, a society and its environment.[4] Achieved in such a way, peace and order

---

[1] R. Rosicki, *O pojęciu i istocie bezpieczeństwa*, "PP" no. 3/2010, p. 24

[2] Cf.: *Webster's Third New International Dictionary*, Könemann 1993, pp. 2053–2054.

[3] See more in: R. Zięba, *Pojęcie i istota bezpieczeństwa państwa w stosunkach międzynarodowych*, "Sprawy Międzynarodowe" 1989, no. 10, p. 49.

[4] R Miler, *Bezpieczeństwo transportu morskiego…*, op. cit., p. 71.

simultaneously develop more trust towards the state and its institutions, and they also affect the sense of social security.[5]

Hence, having followed that stream of reasoning a bit further, a concept presented by T. Hobbes assumes that security in a society is achieved through the process of a qualitative change in relations – it assumes leaving the conditions of nature (in an individual approach characterised by the lack of existential security) and entering the conditions of a state (a society) where the power of the state, taking the form of the law and legal regulations (legislature) of a positive character, becomes a guarantee to the functioning of a social agreement, that is – *de facto and de iure* – the conditions in which security is "guaranteed".[6] Therefore, we face a certain paradox in the social system: on one hand, there are some limitations to freedom of an individual; freedom which in the conditions of nature is usually related to existential anxiety and permanent danger. On the other hand, there are some conditions of social order which, by assumption, guarantee the sense of security and possibilities of development; however, this is what we have to "pay" for with the limitation to our individual freedom.[7]

Security and development come as two basic dimensions pertaining to the existence of individuals, fields of activities (e.g. maritime transport or inland waterway navigation) and entire societies, including societies organised into states, institutions or international organisations. These two dimensions are described with positive feedback: it is difficult to discuss development without security, and development facilitates and fosters security (sometimes it even comes as a condition necessary to provide security). Hence, such a relation determines mutual dependencies between security strategies and developmental strategies implemented by the state and its institutions with regard to all the interests, fields of activities (e.g. transport) or individuals.[8]

---

[5] Ibidem, p. 71.

[6] T. Hobbes, *Lewiatan*, PWN, Warszawa 1954, pp. 147–148.

[7] It is convergent with W. von Humboldt's views on security; he emphasizes the necessity of defining the range of activities performed by the state. It would also allow us to define the range of interventions undertaken by the state; security is the only objective and motive of the activity performed by the state, but that objective must be achieved with the use of proper legal regulations; security must be related to protection of other (belonging to the state) entities' freedom – the Author's note.

[8] S. Koziej, *Bezpieczeństwo: istota, podstawowe kategorie i historyczna ewolucja*, "Bezpieczeństwo Narodowe" II/2011/18, p. 19–20.

In a dynamic approach (understood as a process of achieving such a condition), security of a particular entity (a state, a system, a mode of transport, e.g. maritime transport or inland waterway transport)[9] refers to its activities aimed at providing a possibility of survival (existence) and freedom necessary to implement its own interests in a turbulent environment (hence, dangerous), especially through the pragmatics related to the application of mechanisms that pertain to the identification, control and reduction of risk (prevention based on monitoring and telematics systems), and counteraction undertaken against the effects of current direct threats.[10]

Considering the above-mentioned view on security (both in international and national aspects), it is possible to observe that [...] *the crucial notion categories in this field are the following:*

- *interests of a national entity (or a mission of an international organisation) and the consequent strategic and operational objectives in the field of security;*
- *security environment (conditions) – chances, challenges, risks and threats to the implementation of interests and the achievement of aims in the field of security; threats may involve crises, conflicts, major failures, catastrophes, natural and ecological disasters, asymmetric threats;*
- *strategic (long-term) and political (operational, current) concepts (principles and methods) of action aimed at the achievement of the assumed objectives under the particular conditions;*
- *security systems – some resources of an entity assigned for the implementation of the assumed concepts and tasks, which are properly organised and prepared for such tasks (in a systemic way, e.g. telematics systems and in an institutional way e.g. security service).*[11]

In the field of security, it is now possible to observe an explicit phenomenon of transferring the mass centre from the direct counteraction against threats towards active prevention. Therefore, prevention becomes much more important than defensive abilities, understood in their literal sense. Finding

---

[9] Such an approach has been applied in a suggested catalogue of basic notion categories, assumed for the requirements of the Strategic Review of National Security; online publication: http://www.bbn. gov.pl/portal/pl/577/2630/updated on_10_January_2011_r.html [accessed on 25th June 2016] – the Author's note.

[10] S. Koziej, *Bezpieczeństwo: istota, podstawowe kategorie i historyczna ewolucja*, "Bezpieczeństwo Narodowe" II/2011/18, p. 20.

[11] Ibidem, p. 25.

its reflection in numerous doctrines, such a tendency comes as a result of changes which have taken place in the awareness of societies, the character of contemporary requirements and dynamic development of technical and technological possibilities (e.g. monitoring systems). Obviously, it should be noticed that prevention has always been an inseparable element of security strategies and doctrines; however, in the previous periods it did not come as their crucial element. Such a qualitative change in a doctrinal approach towards prevention has resulted from a pragmatic (economic) attitude towards contemporary conflicts and threats.[12]

Economic conditions of such a doctrinal change come with the fact that – on one hand – the potential costs of standing up against threats have been dramatically increased, including defensive operations. It is determined, first of all, by the efficiency of modern weapon systems (e.g. weapon of mass destruction) and, unfortunately, by the efficiency of paramilitary operations and terrorist attacks. Hence, even a successful defensive campaign may entail severe loss and destruction which will throw general rationality of such actions into question.[13]

On the other hand, however, there are also some much greater possibilities of preventing threats. Modern systems of monitoring and accurate forecasting (provided within telematics systems) and efficient means which have various direct impact on other entities (e.g. economic sanctions available through integration and globalisation processes, the strength of military alliances) increase the efficiency of preventive and stabilising actions. Therefore, at present, it is far more effective to undertake some preventive effort in order to block the occurrence of threats rather than fight them when they have already appeared and then suffer even more severe costs incurred during the elimination of the consequences resulting from such operations.[14]

On the basis of such disproportions, another contemporary phenomenon can be observed in the field of security, namely asymmetry and asymmetric threats, commonly understood as the use of asymmetric (opposite to sym-

---

[12] R. Miler, *Bezpieczeństwo transportu morskiego…*, op. cit., p. 77.

[13] See more in: R. Miler, *Bezpieczeństwo transportu morskiego…*, op. cit., p. 76–77.

[14] Considering such a background, according to S. Koziej, there is a specific conflict in understanding preventive and preemptive operations. See more in: J.L. Gaddis, *Grand strategy in the second term*, "Foreign Affairs" I–II 2005, pp. 145–147 [in:] S. Koziej, *Bezpieczeństwo: istota, podstawowe kategorie i historyczna ewolucja…*, op. cit., p. 31.

metric, that is: "classic") fighting techniques and posing threats. However, understanding asymmetric threats exclusively as some particular methods or techniques (tactics) of action does not seem to be fully justified, because threats do not come as methods themselves, but they come with entities which apply them. In this way, asymmetric threats involve entities – organisations (recently identified as a "weaker" party of a particular conflict, without any state affiliation, however of criminal nature) which apply unconventional (from the point of view represented by their opponents – threatened entities, most often countries) means and techniques – not only just untypical, non-traditional methods of action.[15] Understood in this way, asymmetric threats most frequently involve:[16]

- international terrorism and piracy;
- international organised crime (including illegal immigration);
- the use of unconventional weapon by non-state entities (including biological or chemical weapons, "dirty bombs");
- hostile use of computer technologies (cyber-attacks).

A category which emphasizes "asymmetry" is also an economic category of cost-effect nature. Whenever an asymmetric threat appears, it involves disproportionately low costs incurred by the entity which poses a threat in relation to enormous costs which result from an asymmetric attack and which are incurred by the threatened entity.

Nowadays, the asymmetry of threats is huge, let us have some simulations provided by an American company, Booz Allen Hamilton[17] as an example. The simulations indicate that a terrorist attack with the use of explosives smuggled via a particular seaport can result in multi-billion costs which include preventive means implemented as a consequence of such an attack, such as reinforced control (systems, procedures, time) at the port, and a multi-billion loss in the entire economy (cooperating companies which operate at the port) treated as a derivative (aftershock) of the primary limitations at the port.

---

[15] M. Madej, *Zagrożenia asymetryczne bezpieczeństwa państw obszaru atlantyckiego*, Polski Instytut Spraw Międzynarodowych, Warszawa 2007, p. 23.

[16] Z. Ciekanowski, *Działania asymetryczne jako źródło zagrożeń bezpieczeństwa*, Bezpieczeństwo i Technika Pożarnicza no. 3, Warszawa 2009, available at http://czytelnia.cnbop.pl/czytelnia/13/162.

[17] T. Szubrycht, K. Rokiciński, *Gospodarka morska w świetle wybranych zagrożeń współczesnego świata*, Wydawnictwo Akademii Marynarki, Gdynia 2006, p. 107.

Similarly, a massive negative impact on the whole economy of a particular country and, consequently, on the whole region (e.g. the Netherlands and Europe) could be exerted by sinking a huge oceangoing vessel (e.g. bulk cargo or cement carrier) between the port entrance heads of a main hub port (e.g. Rotterdam) as a result of a terrorist attack. Clogging the port would be just a prelude to some serious economic loss for the whole European economy, whereas for the terrorists there would be another desired side-effect, that is fear and panic induced by global media (radio, television, multimedia, new media – the Internet). In our modern world, asymmetric threats come as the most serious ones or, at least, the most real threats to the stability of countries, their economies or processes, which occur disproportionally more often than "usual armed conflicts".

Being related to the notion of security, the phenomena described above fit perfectly into transportation processes as well (transport security). Understood as the overcoming of a temporal and spatial gap (covering a distance in a certain period of time), transport requires the use of adequate means and the performance of the subsequent sequences of various operations (organisational, executive, commercial handling, related to the maintenance of an acceptable level of risk and assumed economic outcomes).[18] Hence, the process of transportation comes as a closed cycle of operations indispensable to carry out a complex process of translocating people and cargo. During such a multi-element process, translocated objects go through the subsequent stages of the technological process of transportation.[19] An integral part of such a process is shipping (physical translocation of the cargo with a means of transport). Figure 2-1 presents a scheme of the transportation process.

Each of such operations must be performed under safe conditions which guarantee a high (or at least acceptable) level of a probability that the transport service shall be completed without any disturbances, according to the scheduled processes and time. Considering the specificity of water environment, the processes of maritime and inland waterway transport are very often combined. Figure 2-2 presents the sequence of operations understood in such a way.

---

[18] W. Grzywacz, J. Burnewicz, *Ekonomika transportu*, Wydawnictwa Komunikacji i Łączności, Warszawa 1989, p. 212.

[19] E. Mendyk, *Ekonomika i organizacja transportu*, Wyższa Szkoła Logistyki, Poznań 2002, p. 100.

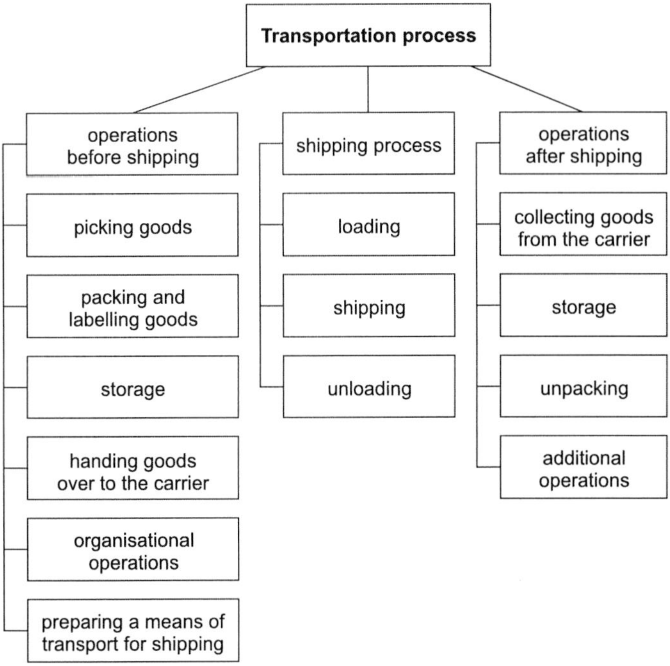

Figure 2-1. A scheme of the transportation process

Source: adapted from A. Koźlak, *Ekonomika transportu. Teoria i praktyka gospodarcza*, Wydawnictwo Uniwersytetu Gdańskiego, Gdańsk 2008, p. 132 and S. Marszałek, *Organizacja i zarządzanie w transporcie*, Wydawnictwo Śląskiej Szkoły Zarządzania, Katowice 2001, p. 73.

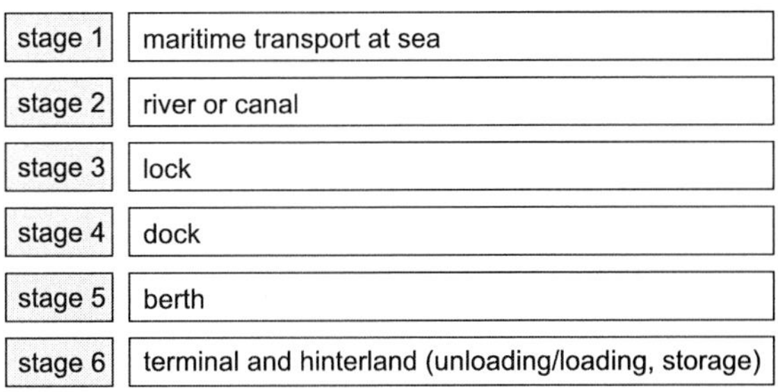

Figure 2-2. Decomposition of maritime transportation activities

Source: adapted from A. Miola et al, *External costs of Transportation, Case study: maritime transport*, Luxembourg: Office for Official Publications of the European Communities © European Communities, 2009, p. 14.

Despite such similarities, transportation processes in various modes of transport are characterised with a slightly different level of threats and entailed risks; hence, the process of their proper identification becomes significantly important. Considering transportation services provided in water environment (seas and rivers) with a considerable participation of port operations, maritime and inland waterway transport displays its immanent features which determine the scope and level of risks which occur in these modes of transport.

2.1.2. The risk factors in maritime and inland waterway transport

All the above-mentioned security determinants have also their practical justification and reflection in the maritime and inland waterway domains. For the requirements of this publication, water (maritime and inland waterway) transport security is defined as […] *a situation in a maritime or inland waterway area, where the international and national laws are respected, freedom of navigation is guaranteed, citizens, infrastructure, means of transport, environment and marine and water resources are protected.*[20]

Since the very beginning of maritime and inland waterway transport, the factors which generate various types of risk to these modes of transport have been observed there as well. Therefore, it is possible to assume that threats to water transport are of permanent nature, entirely non-limited and ever changing along with the development of our civilisation.[21] At the same time, the occurring threats are the main factors which determine the level of security in the maritime and inland waterway domains, considering a permanent necessity of their identification, analysis and assessment of the entailed risk.[22]

There are two fundamental areas which determine the notion of security in the domain of maritime and inland waterway transport: safety of navigation (nautical) and security of maritime (and inland waterway) economy

---

[20] *Strategia Unii Europejskiej w zakresie bezpieczeństwa morskiego*, a document by the EU Council 11205/14, Brussels 2014, p. 2.

[21] M. Winiarska, *Wybrane problemy dotyczące transportu morskiego w kontekście ochrony jego bezpieczeństwa*, Colloquium Wydziału Nauk Humanistycznych i Społecznych, IV/2012, p. 157.

[22] Cf.: R. Miler, *Bezpieczeństwo transportu morskiego…*, op. cit., pp. 76–77.

of a particular state (or a macro-region, such as the EU) which closely correspond to a slightly different division into the field of safety and security.[23]

Nautical safety refers to the maintenance of nautical conditions above the minimal acceptable level.[24] At the same time, is has been assumed that – as a result of navigation, maritime and inland waterway economy – such conditions directly affect human activities at sea and inland waters. Considering the anthropopressure character of maritime navigation (and definitely less anthropopressure nature of inland waterway navigation), nautical conditions also involve protection of maritime and fluvial environment against pollution which is generated as a consequence of navigation and broadly understood maritime and inland waterway economy.[25]

The basis of a definition referring to the notion of safety in the maritime (and inland waterway) economy of the state is constructed in a slightly different way. It has been assumed that this is freedom from threats at sea, threats from sea and threats at inland waters (rivers, lakes, canals) generated by the state, non-state, trans- or sub-national entities (including organisations of a criminal and asymmetric character), and safety of all the institutions and bodies of the state which constitute or protect its economic (maritime and inland water), political, military and social interests at its own territory as well as outside its borders.

The most significant conditions for the occurrence of threats to navigation safety, understood as a combination of nautical factors, are the following:[26]

- a dynamic (and often non-predictable as to its scale) change in conditions of maritime and inland water environment, which leads to the exceeding of the defined level ("*limes*") of conditions determined as safe for navigation at sea and inland waterways (e.g. bad weather and hydrological conditions);

---

[23] See more in: J. Urbański, Cz. Dyrcz, W. Morgaś, *Bezpieczeństwo morskie oraz bezpieczeństwo morskie państwa – podobieństwa i różnice*, ZN AMW; 172K/1/2008, pp. 31–40.

[24] The conditions resulting from human activities at sea which do not pose any threats to vessels, human life and to the environment; such conditions determine "*the limes*" – the level below which the lack of safety appears and proper calculation of risk is required – the Author's note.

[25] Cf.: R. Miler, *Bezpieczeństwo transportu morskiego…*, pp. 76–77

[26] Z. Kopacz, W. Morgaś, *Krajowy System Bezpieczeństwa Morskiego w zintegrowanej polityce UE…*, p. 65

- outdated or incomplete knowledge of conditions of maritime and fluvial environment in which navigation takes place (e.g. erroneous navigation resulting from the lack of navigational aids and charts);
- improper technical and operational features of transportation means (vessels, barges, etc.), characterised by their seaworthiness, and any shortage in their navigational, telematics and life-saving equipment (e.g. vessels in poor technical condition with insufficient equipment);
- operations performed by other users at sea, inland waterways or by various port and terminal operators who do not follow the standards of behaviour stated by and compliant to the current legal procedures and regulations (e.g. related to growing congestion at sea and inland waterways, including congestion and operational problems at sea and inland ports);
- threats resulting from the cooperation between a technical system and its operator in difficult working conditions (e.g. the lack of competences required for navigation and inland waters presented by some captains/skippers and their crews; inadequate super-structural equipment at ports/terminals);

The fundamental conditions favourable for threats posed to safety of maritime (and inland water) economy of the state are:[27]

- legal and illegal operations performed by institutions and services of the neighbouring countries in the territory of a particular state, which are aimed at its maritime institutions and economic interests in the field of maritime and inland water economy (e.g. hostile propaganda, sabotage, provocation);
- unfriendly and hostile operations performed by institutions and bodies of the neighbouring countries in their territories, which are aimed at the political, economic, military and social interests of a particular state (e.g. a wide variety of chicanery aimed at vessels navigating under the flag of a particular country and their crews);
- illegal criminal groups and structures (paramilitary organisations) which intrude upon the political, economic and social interests of a particular

[27] Identification of these factors and an attempt at their systematisation depends on numerous criteria, including a definition of asymmetic operations. See more in: K. Szczyrbowski, *Zmiany w strukturze ochrony wybranych podmiotów gospodarki morskiej na przykładzie wprowadzenia kodu ISPS…*, op. cit., p. 23; Z. Kopacz, W. Morgaś, *Krajowy System Bezpieczeństwa Morskiego w zintegrowanej polityce UE…*, op. cit., p. 67 – the Author's note.

state in its maritime and inland water territory, most frequently by asymmetric operations (e.g. terrorism and piracy);
- the defence and law enforcement institutions of the neighbouring countries threatening the maritime integrity of a particular country within a developing conflict, including maritime operations other than war (MOOTW), (e.g. power demonstration, intensified activities of intelligence services);
- the hostile use of inland waterways (especially border rivers) (groundless raising or lowering water levels by damming);
- hybrid operations[28] performed by the defence and law enforcement institutions of the neighbouring countries (including cyberattacks);
- direct, open maritime war (or inland water) operations.

Functional aspects which refer to the notion of safety in the domain of both modes of water transport include the following:[29]
- freedom from danger,
- freedom from unacceptable risk,
- freedom from personal harm,
- freedom from financial loss.

The systemic regulators referring to the notion of safety in the maritime and inland water domain are composed of three elements (sub-systems) which create a system: man – object (technology) – environment (MTE):[30]
- man (M) – most often: a navigator, a skipper;
- technical object (T) – most often: a vessel, a barge;
- environment (E) – understood as an area, some elements of linear infrastructure, hydro-engineering structures, hydro-meteorological conditions, navigational marking, other vessels/barges which operate in the same system, elements of traffic control, factors of geopolitical risks and systemic threats.

---

[28] Hybrid war involves a mixture of classic military and asymmetric methods – first of all, irregular armed operations (guerilla operations, sabotage, diversion, terrorist attacks) but also some elements of information economic and cybernetic combat (propaganda, disinformation) – the Author's note.

[29] Z. Burciu, *Niezawodność akcji ratowniczej w transporcie morskim*, Oficyna Wydawnicza Politechniki Warszawskiej, Warszawa 2012, p. 37.

[30] Cf.: R. Krystek (ed.), *Zintegrowany System Bezpieczeństwa Transportu. Vol. I. Diagnoza bezpieczeństwa transportu w Polsce*, WKŁ – PG, Warszawa – Gdańsk 2009, pp. 310–313.

Considering its specificity (importance and significance for global and regional economy, a one-off amount of volume of transported goods and anthropopressure potential), maritime transport transpires to be a branch which plays a much more significant role in the international and global systems of safety/security than inland waterway transport.

Considering a holistic approach towards safety/security of water (maritime and inland waterway) transport, it is a state of a system in which the acceptable level of risk for the basic operations of water transport (such as maritime and inland waterway navigation and port operations) is maintained with the use of specific tools (including monitoring and telematics); it results in increased efficiency of processes and decreased anthropopressure of these modes of transport.

Considering such a division, it is possible to indicate one more classification which is crucial while discussing maritime safety and protection of navigation, namely safety and security.[31] Safety refers to preservation of life, property and water environment, freedom from pollution generated by vessels of any kind. Security refers to protection of economic operations performed in the domain of water (maritime and inland waterway) transport against terrorism, piracy and any other similar threats and to counteraction against any harmful and forbidden activities performed in this field. It particularly refers to counteraction against illegal exploitation of water resources, illegal immigration, smuggling of drugs, people, weapon and other cargo that can be used by terrorists or pirates. More and more often, it also refers to operations from the field of cybersecurity.[32]

Considering a much broader spectrum of the notion and its direct translation into the scope and the functioning of telematics systems (a necessity of acquiring data and analysis of particular situations), both notions: safety and security in maritime and inland waterway navigation require some more clarification.

---

[31] In expert literature on safety/security of maritime transport in Eastern European countries, the problem of the relation between maritime security and maritime safety is relatively rarely discussed – the Author's note.

[32] J. Urbański, W. Morgaś, Z. Kopacz, *Bezpieczeństwo morskie i ochrona żeglugi oraz zarządzanie nimi; próba identyfikacji i specyfikacji przedmiotu…*, op. cit., p. 20.

2.2. The idea of division into safety and security in maritime and inland waterway transport

2.2.1. The field of *safety* in maritime and inland waterway transport (incidents and their consequences)

The field of safety in water transport is inseparably related to maritime and inland waterway navigation and other nautical factors, among which navigation appears to be a crucial element. Navigating in maritime environment (the World Ocean), where generally no local rules are applicable, results in the necessity of considering that homogenous factor in the process of international maritime transport.[33] The situation in inland waters and waterways is slightly different, as there are separate, local regulations applied for navigation (e.g. regulations stated by the Central Commission for the Navigation of the Rhine – Rhine Boatmaster Certificates).

Safety of navigation comes as an indispensable condition required for proper maritime and inland waterway navigation. Apart from port service, it becomes a fundamental element of maritime and inland waterway transport.[34] Maritime and inland navigation should meet two basic criteria resulting from its application scope:[35]

- it should be safe (also in terms of its standards),
- it should be efficient (also in terms of economy).

These two factors result in the fact that vessels/barges – regardless of the water area where they operate, their flags and the nationality of their crews – should respond to threats in the same way, they should cooperate with onshore systems and aim at the optimisation of their routes, fuel consumption and achievement of economic, energetic and ecological efficiency. It results from the definition of navigation which presents navigation as a process of safe (collision-free) and efficient (in terms of time and costs) sailing from one place to another.[36] Understood in such a way, the conditions of efficient

---

[33] Apart from navigational threats, safety of navigation involves danger related to fire, vessel stability, vessel floodability, cargo, etc. Safety of navigation and safety of maritime environment are included in the notion of maritime safety – the Author's note.

[34] In such a case, safety is understood as safety of a crew, passengers, a vessel and its cargo and broadly understood maritime environment – the Author's note.

[35] R. Miler, *Bezpieczeństwo transportu morskiego...*, op. cit., pp. 88–89.

[36] Z. Kopacz, W. Morgaś, *Krajowy System Bezpieczeństwa Morskiego w zintegrowanej polityce UE*, Zeszyty Naukowe AMW, no. 2(185), Gdynia 2011, p. 65; at present (since 2002)

navigation simultaneously become guidelines for the development of telematics systems (telematics systems should provide possibilities of monitoring, control and analysis of the above-mentioned factors of safe navigation).

Unfortunately, despite a tremendous effort made in order to provide safety in water (maritime and inland waterway) transport, incidents (such as collisions, stranding, etc.) of various levels of threat still occur. Safety accidents are most frequently related to nautical determinants which are included into the categories of human error, e.g. erroneous navigation, problems with navigational marking, such as translocation of navigational signs, unfavourable hydro-meteorological and hydrological conditions (violent storms, fog, icing, low water levels on a navigable river, objects carried by the water which might represent a serious threat) and faulty operation of vessel navigation instruments and port installations. As a consequence, accidents related to safety involve not only maritime collisions (including vessel sinking and loss – see: Figure 2-3) or stranding, but also ecological disasters which naturally result from such accidents (oil spills, emission of airborne hazardous substances, etc.). A sequence of such negative consequences may cause serious disturbances in ecosystems and real economic loss. Unfortunately, safety incidents involve also much more grievous occurrences, such as death of crew members, and they require immediate action (search and rescue operations, remedial actions), the scale of which is often exceeded even in the most pessimistic scenarios that were previously assumed.[37]

In the latest history of maritime transport and economic exploitation of the World Ocean, there have been numerous dangerous and tragic safety accidents. The most dramatic ones are presented in Figure 2-4 (including ecological disasters).

Considering its specificity, inland waterway navigation has not experienced such "spectacular" safety accidents. Nevertheless, there have been accidents along inland waterways caused by hydrological factors (low water levels) and human errors (insufficient training provided to crew members,

---

the applicable definition of maritime navigation, stated by the International Maritime Organisation (IMO) in its Resolution A.860(20) App. 1, presents it as *a process of planning, registering and controlling the traffic of vessels sailing from one place to another* – the Author's note.

[37] http://www.ekologia.pl/srodowisko/ochrona-srodowiska/najgrozniejsze-katastrofy-morskie-w-xx-i-xxi-wieku, 12330.html [accessed on 8th August 2014].

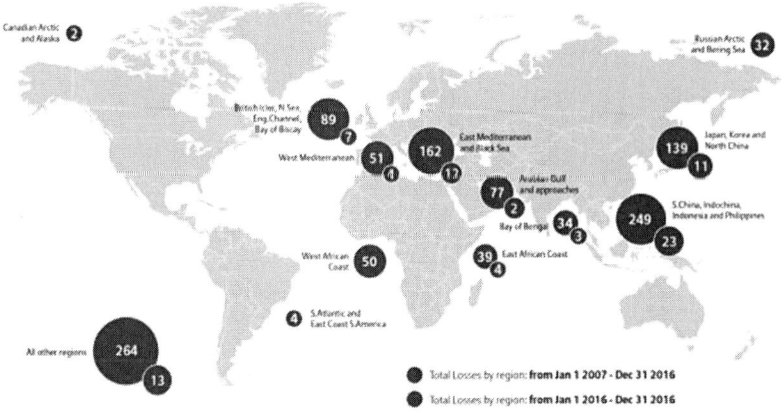

Figure 2-3. Total losses in maritime transport by top 10 regions [2007–2016/2016 separately]

Source: adapted from Lloyd's List Intelligence Statistics. Data analysis and graphics Alianz Global Corporate & Specialty,[38] https://www.ajot.com/news/shipping-accident-statistics-identify-loss-hotspot-in-asian-seas [accessed on 30th June 2018].

malfunction of vessels or equipment, including navigational equipment). Unfortunately, data and statistical information concerning safety incidents in the inland waterway transportation in the EU is fragmentary and partially unavailable. Based on some expert group findings (Expert Group E01036 Recognition and modernisation of professional qualifications in inland navigation)

[38] Shipping accident statistics identify loss hotspot in Asian seas. In its 2017 Safety and Shipping Review, AGCS analyzed 25,898 shipping incidents including 1,186 'total losses' between January 2007 and December 2016. Over this period, global shipping safety has generally significantly improved – total losses have dropped by 50% globally from 171 ships in 2007 to 85 in 2016. However, the number of total losses in these East/South East Asian waters have declined far less over this decade, averaging 39 per year which equals one third of all worldwide total losses (33%). Collision (involving vessels) is the fourth top cause of shipping losses over the past decade globally (72 losses) accounting for 6% of total losses, after foundering, wrecking and fire/explosion. In the waters where this incident has taken place, collision is the cause of over a quarter of all shipping incidents (including total losses) in general over the past decade (1,092 out of 3,915 in this overall region). Fire/explosion is the cause of 7% of all shipping incidents (including losses) in the East/South East Asia region over the past decade (269 out of 3,915). The tanker industry has made great strides in safety in recent years, enjoying an extended period of benign loss activity. It has been excellent at pursuing self-regulation and maintaining high safety standards. The tanker sector has seen just 15 total losses over the past decade around the world, according to AGCS analysis – the Author's note.

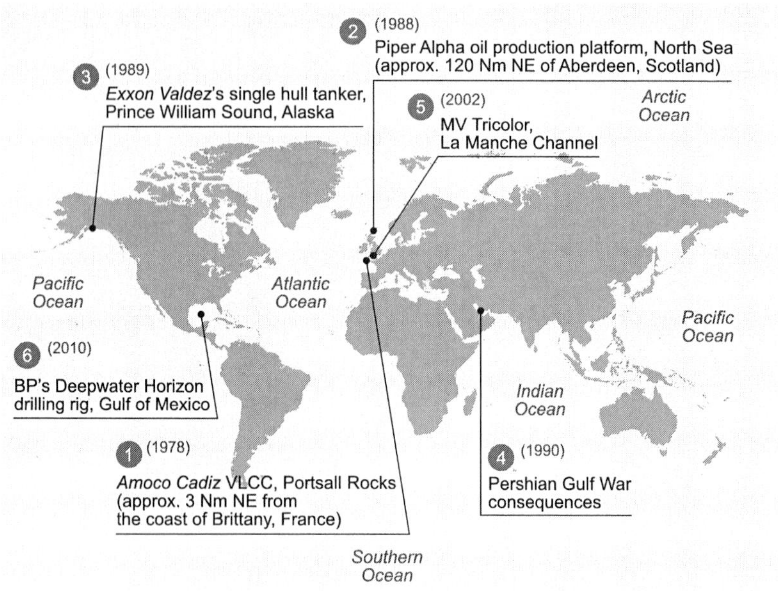

Figure 2-4. The most dramatic catastrophes (including ecological disasters) caused by safety accidents in the World Ocean over the last 50 years (excluding no. 4 which is a security incident)

Source: adapted from http://www.ekologia.pl/srodowisko/ochrona-srodowiska/najgroz niejsze-katastrofy-morskie-w-xx-i-xxi-wieku,12330.html [accessed on 28th May 2018].

it is possible to assess the number of collisions and incidents in European IWT in years 2004–2012 (Table 2-1).

Table 2-1. *Safety* incidents in European IWT (2004–2012)

Year	NL	DE	FR	CH	AT	SK	HU	UK
2012	161	n/a	26	n/a	40	2	7	n/a
2011	1072/159	767	26	7	38	2	7	36
2010	987/164	866	37	n/a	48	15	13	27
2009	903/121	838	39	n/a	33	8	6	30
2008	982/127	832	23	n/a	36	3	12	42
2007	795/150	890	34	n/a	37	3	n/a	49
2006	710/123	875	36	n/a	n/a	7	n/a	n/a

Source: adapted from *Evaluation of RIS Implementation for the period 2006–2011*, Main Report. Zoetermeer, July 2014.

Over the last two decades, in the Rhine Basin (one of the most important European inland waterway communication river system) the number of safety incidents has oscillated between 250–350 incidents per year, which is presented in Figure 2-5.

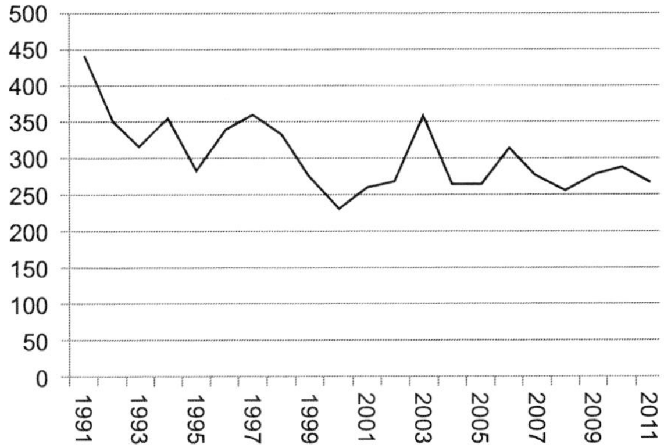

Figure 2-5. Safety incidents in the Rhine Basin (1991–2011)

Source: adapted from *Evaluation of RIS Implementation for the period 2006–2011,* Main Report. Zoetermeer, July 2014.

As transport congestion has been recently growing in water transport, congestion at sea and inland waterways has been intensified as well (most often observed as delays in port services and transshipment processes at ports and terminals); however, the advancement in the field of ICT fosters the implementation of modern monitoring and risk control instruments (which come as the examples of telematics solutions that are, in a way, derivatives of the implementation of solutions from the field of broadly understood safety of water transport).[39]

Over the last decades, a number of solutions and tools for providing safety have been implemented to the pragmatics of maritime and inland waterway transport processes. The most important ones in the field of safety include the following:

- standardisation of the navigational marking in the waters of the World Ocean and inland waterways (the IALA system),
- legislative packages Erica (I, II, III) implemented by the European Commission,

[39] R. Miler, *Bezpieczeństwo transportu morskiego...*, op. cit., p. 238

- the concept of e-navigation and full integration of information systems on a vessel (INS/IBS),
- regulations related to professional competences of crews (STCW and regional and national regulations in inland waterway transport),
- regulations related to the limited emission of greenhouse gases generated by maritime transport.

In relation to the growing volume of maritime transport and advancing globalisation of all economic processes, including maritime navigation as a part of logistical processes, at the end of the 1970s a necessity of unification and popularisation of the principles of safe navigation marking appeared. Waters of the World Ocean became more and more congested, and manoeuvering in restricted, cramped places became a reason for a significant risk in navigation with regard to the growing congestion. Some universal solutions of strategic nature which would be internationally acceptable became indispensable. Such a step in the field of pragmatics referring to operations which provide more safety of navigation and, at the same time, more safety of maritime transport, is the standardisation of navigational markings at all waters of the World Ocean.[40]

The International Association of Lighthouse Authorities (IALA) have stated the norms referring to navigational marking.[41] The whole world has been divided into two areas (regions) where the A system and the B system of navigational marking is obligatory.[42] The range of both these sub-systems is presented in Figure 2-6.

The accidents in maritime transport which became the reason for the implementation of such specific and strict regulations defining new levels of safety were the collisions of the MV/MT *Erika*[43] and *Prestige*.

---

[40] Ibidem, p. 170.

[41] In 1980 the IALA authorities and the International Hydrographic Organisation (IHO) along with 50 countries and 9 international non-government organisations accepted the principles for the standardisation of the system of maritime navigational marking, covering the whole area of the World Ocean, commonly known as the IALA Maritime Buoyage System – the Author's note.

[42] M. Jurdziński, *Podstawy nawigacji morskiej*, Wydawnictwo AM Gdynia, Gdynia 2003, p. 158.

[43] *Erika* was a single-bottom tanker registered in Malta, which sank at the French coast on 12th December 1999 and caused severe pollution of sea water and the seashore. In a sense, *Erika* was not an obvious candidate to sink: the vessel was granted all the obligatory certificates and passed various inspection tests. Experts concluded that single-bottom vessels

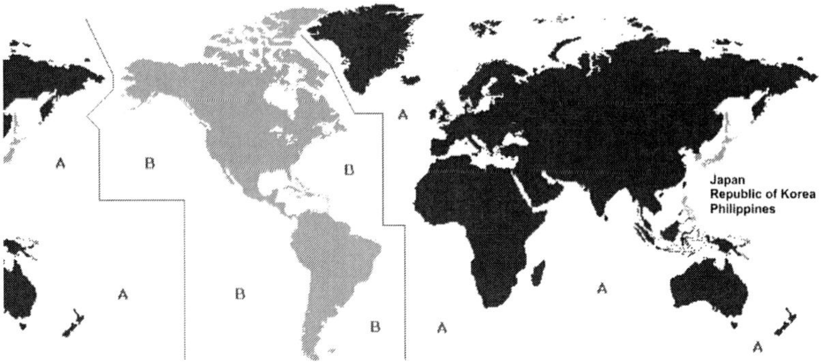

Figure 2-6. The A and B regions of the IALA system.

Source: adapted from http://www.iala.bp8.pl [accessed on 8th March 2015].

The response of the EU to the accident which involved *Erika* in 1999 was an implementation of two legislative packages by the European Commission, namely: *Erika I* and *Erika II.* They were intended to increase the standards of maritime safety through monitoring behaviour of vessels and equipping them with voyage data recorders (VDRs) which would provide information in case of investigation after a possible disaster. These regulations came into force in July 2003.[44]

Next two regulations which implemented instruments for improvement in safety at sea (2002) constituted the establishment of the European Maritime Safety Agency with its headquarters in Lisbon. The Agency was assigned with a task of providing scientific and technical resources in order to support the member countries in the implementation of the EC legislation and permanent evaluation of impact exerted by the above-mentioned legislation on the actual level of maritime safety. Its tasks also involved developing a central information system applied in the EU countries, which would extensively implement new concepts of navigation monitoring that could provide improvement in effectiveness and efficiency of the monitoring of maritime routes in the EU water areas. Coming to the European

did not guarantee a satisfactory level of safety, inspection methods were inadequate, and it was necessary to take a closer look at flags of convenience – the Author's note.

[44] Z. Bąbczyńska-Jelonek, *Na straży morskiego ładu – pakiet dyrektyw KE "Erika III"*, op. cit., http://sophico.onet.pl/2006/10/28/na-strazy-morskiego-ladu-pakiet-dyrektyw-ke-erika-iii/ [accessed on 27th July 2016].

waters, vessels had to be equipped with the identification systems that could automatically communicate with the coastal administration. The data about hazardous cargo became accessible more easily, and the member countries could intervene quicker in the case of an accident or pollution risk (in practice, it initiated conceptual and implementing work on the VTMIS and LRIT-class telematics systems and instruments such as Safe Sea Net).[45]

Another package of safety regulations was issued by the European Commission in November 2005. It is often referred to as *Erika III*, and it includes some directives on:[46]

- obligations of the countries which maintain their own flags (a draft directive on the compliance with the requirements of the flag state – FSC),
- state control of ports (a draft amendment to the Directive 95/21 on the port state control – PSC),
- surveillance of maritime navigation (a draft amendment to the Directive 2002/59 on the EC system of vessel traffic monitoring and information),
- surveillance of classification societies (a draft amendment to the Directive 94/57 on common requirements for organisations responsible for vessel inspections and surveys),
- uniform investigations of maritime catastrophes (a draft amendment to the Directive on investigation of maritime accidents),
- increased liability of ship-owners for accidents and water pollution (a draft amendment to the Directive on civil liability of ship-owners),
- an ordinance on liability and compensation for passengers for the loss of luggage.

A derivative of the directives included in the *Erika* packages is safe navigation. It comes as another considerable area in the field which refers to the strategy of safety improvement. The elements related to practical implementation of the assumptions referring to that system certainly include operations which minimise a negative effect of the human factor (i.e. most often

[45] See more on: http://www.europarl.europa.eu/highlights/pl/801.html, oraz Z. Bąbczyńska-Jelonek, op. cit. – the Author's note.

[46] Z. Bąbczyńska-Jelonek, *Na straży morskiego ładu – pakiet dyrektyw KE "Erika III"*, op. cit., http://sophico. onet.pl/2006/10/28/na-strazy-morskiego-ladu-pakiet-dyrektyw-ke-erika-iii/ [accessed on 27th July 2016].

errors made for different reasons by crew members responsible for navigation), such as systemic undertakings and solutions:[47]

- the database system GIS (Geographic Information System),
- satellite positioning systems (GPS, DGPS, EUROFIX, EGNOS, GLONASS, GALILEO),
- integrated navigation systems (ARPA, ECDIS, INS/IBNS),
- communication and search and rescue systems (COSPAS-SARSAT, INMARSAT, GMDSS, SAR),
- navigation monitoring and vessel traffic control systems (VTS, AIS, LRIT).[48]

One of the important elements related to accident monitoring in the field of safety is a module of navigational infrastructure which provides an access to the GIS database system; it has become an integrator of safety systems in broadly understood maritime environment.[49]

Traditional maritime charts and standard navigational equipment are still used for navigation, although it has been already decided[50] that in a few years new technical solutions will be extensively and obligatorily implemented, along with the forms of visualised maritime spatial information which will make automation of navigational processes and safety risk control possible. The systemic solutions in this field include:[51]

---

[47] Cf.: A. Nowicka, *Współczesna nawigacja morska oraz nawigacja przyszłości*, Wydawnictwo Akademii Morskiej w Szczecinie (Digital Library), Szczecin 2010, p. 6.

[48] The abbreviations stand for the following: AIS – Automatic Identification System, ARPA – Automatic Radar Plotting System, DGPS – Differential Global Positioning System, ECDIS – Electronic Chart Display and Information System, EGNOS – European Geostationary Navigation Overlay Service, GLONASS – Global Navigation Satelite System, GMDSS – Global Maritime Distress and Safety System, GPS – Global Positioning System, LRIT – Long Range Identification and Tacking, VTS – Vessel Traffic Service – the Author's note.

[49] See more in: T. Kubik, GIS. *Rozwiązania sieciowe*, PWN, Warszawa 2009 – the Author's note.

[50] The advisability of obligatory ECDIS on each vessel was considered by the Sub-Committee on Safety of Navigation (NAV) of the Maritime Safety Committee at the IMO; 54 proposals developed by the NAV were submitted for approval to the MSC 85 (on 5th December 2008) and as a result, the ECDIS implementation schedule for seagoing vessels came into force (with obligatory installation on each vessel, not later than by 1st July 2018) – the Author's note.

[51] Cf.: A. Weintrit, H. Nitner, *Czy jesteśmy już w epoce map elektronicznych?*, "Przegląd Hydrograficzny" no. 5, BHMW, Gdynia 2009, pp. 2–3.

- the concept of e-navigation which includes, among others:
  - implementation of the Electronic Chart Display and Information System – ECDIS,
  - implementation of the 3D chart,
  - full visualisation of navigational and hydrographic conditions (including the analysis of system reliability);
- the concept of the full integration of information systems on a vessel, within the framework of:
  - Integrated Navigation System – INS,
  - Integrated Bridge System – IBS.

As a result of work on the development of concepts related to the e-navigation system, its preliminary architecture has been approved (Figure 2-7).

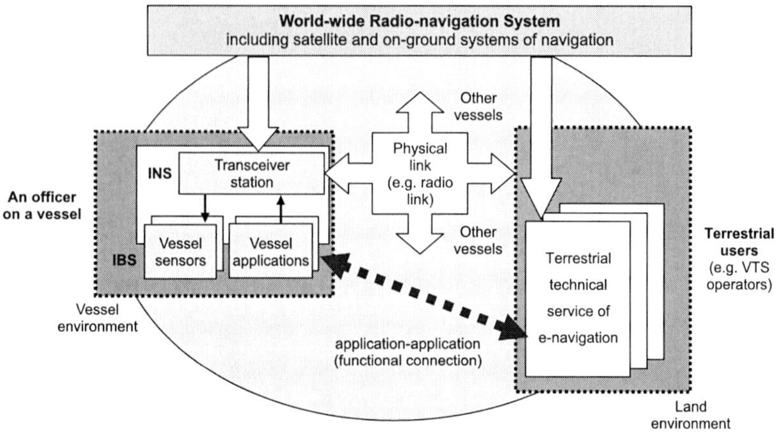

Figure 2-7. The architectural concept of the e-navigation system

Source: adapted from K. Korcz, *Radiokomunikacyjne aspekty planu implementacji strategii e-nawigacji*, "Elektronika" no. 11/2010, p. 12.

The basis for the "instrumental" concept of e-navigation is an electronic maritime navigational chart (ENC).[52] Obviously, the ENC itself (regardless of its type) does not make e-navigation possible. In order to use electronic charts, specialised IT and electrotechnical equipment is required. Devices

[52] K. Korcz, *Radiokomunikacyjne aspekty planu implementacji strategii e-nawigacji…*, op. cit., p. 12.

which are indispensable for the ENC reading, which meet the requirements stated by the International Maritime Organisation (IMO), the International Hydrographic Organisation (IHO) and the International Electrotechnical Commission (IEC), are referred to as the ECDIS, that is: the Electronic Chart Display and Information System.[53] A specific type of the ECDIS is the WECDIS (Warship ECDIS). It is a military version of the Electronic Chart Display and Information System of the standard developed by the NATO.[54]

The schedule for the obligatory implementation of ECDIS devices in seagoing vessels (in accordance with the recommendation of the IMO NAV-54) is presented in Figure 2-8.

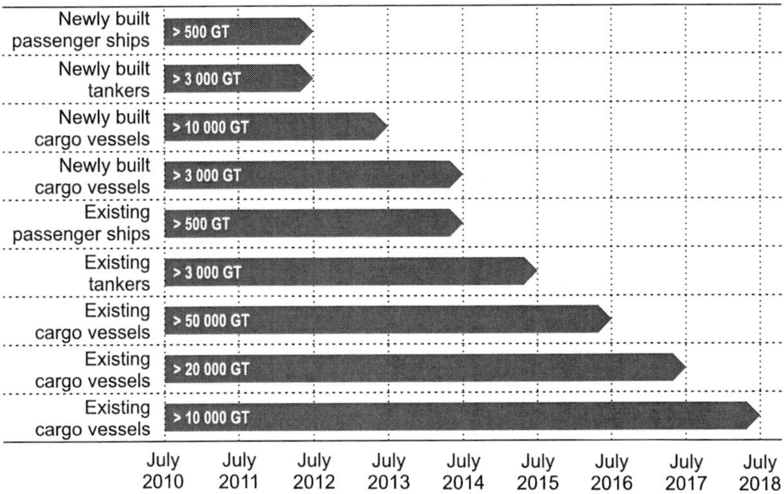

Figure 2-8. The schedule for the obligatory implementation of ECDIS devices in seagoing vessels

Source: adapted from http://www.furuno.com/en/business_product/merchant/product/ecdis_fmd/carriage/index.html [accessed on 5th May 2016].

[53] Cf.: A. Weintrit, H. Nitner, *Czy jesteśmy już w epoce map elektronicznych...*, op. cit., pp. 5–6.

[54] It has been assumed that the WECDIS system for warships must meet all the requirements for the ECDIS and additionally, also military requirements in terms of technical equipment as well as functional software and the products handled (except for the ENC). The NATO requirements for the WECDIS are specified in a standardising document STANAG 4564 – the Author's note.

For the popularisation of the ECDIS, it is important to have all the waters of the World Ocean covered by the ENC.[55] Figure 2-9 presents the area covered by the ENC at the end of 2008, as stated by the catalogue run by the regional PRIMAR Centre (it presents the whole accessible coverage, regardless of the ENC producer and distributor).

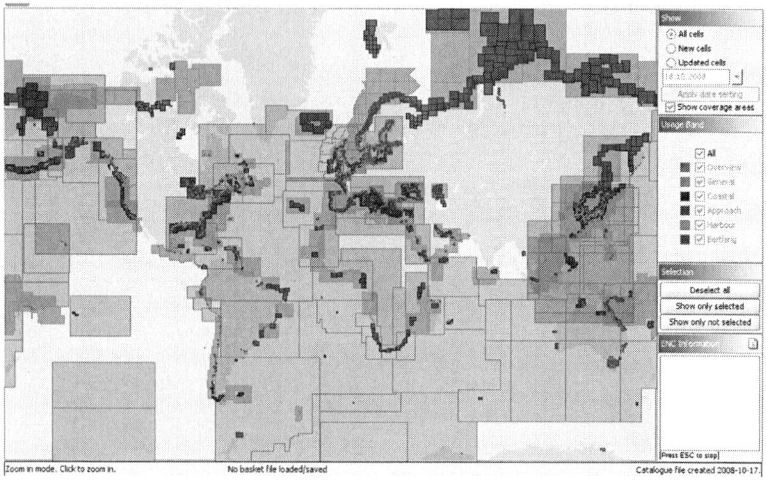

Figure 2-9. The areas covered by the ENC, according to the PRIMAR catalogue – all frequencies [2008]

Source: adapted from the PRIMAR catalogue, 2008. [c]

The ENC charts are also available for inland navigation, with the coverage of the European inland waterways. Figure 2-10 presents the register of inland ENCs for Europe.

The ECDIS system comes as a support for route planning and navigation, but it may be also used for the storage of archive information (an equivalent of a logbook). Additionally, it also makes integration and visualisation of navigational information obtained from various systems possible, e.g. a radar image from the ARPA overlaid on an ECDIS display (Figure 2-11).

[55] According to the concept of the World-wide Electronic Navigational Chart Database WEND approved by the IHO, the distribution of the ENC should be managed by the Hydrographic Offices at the Regional Electronic Chart Coordinating Centres RENC. Two such regional centres of electronic charts operate in Europe, PRIMAR in Stavanger and IC_ENC in Taunton (UK) – the Author's note.

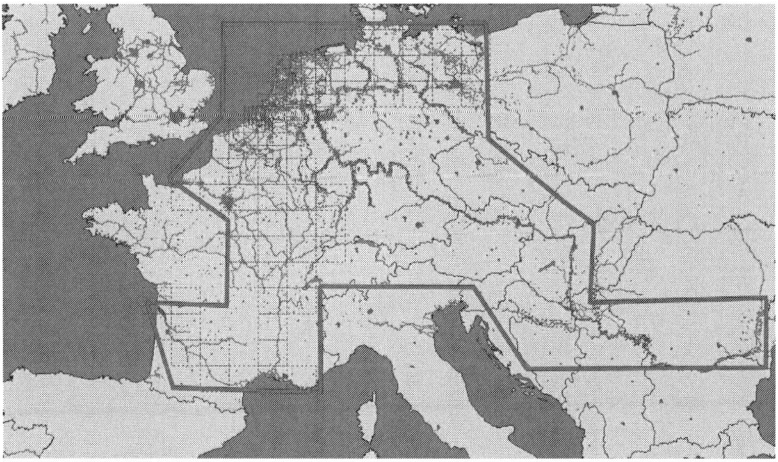

Figure 2-10. Electronic Navigational Charts Register and the digital parts of the Inland ENC standard – Inland ENC Register

Source: adapted from http://www.pcnavigo.com/charts/enc-and-ecdis/?lang=en [accessed on 9th February 2016].

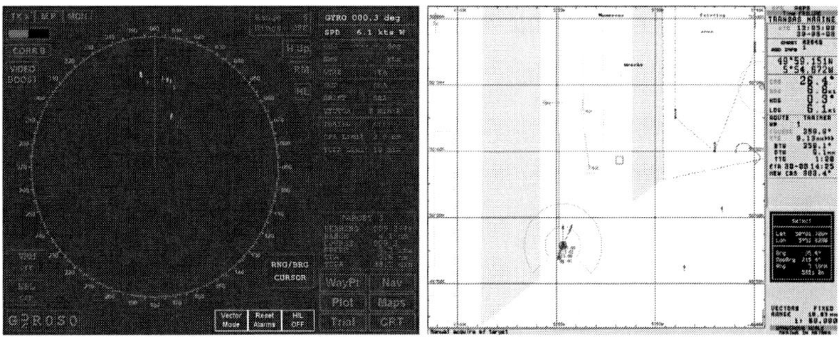

Left side – an ARPA image; right side: an integrated ECDIS display of a navigational situation

Figure 2-11. An example of a radar image overlaid on an ECDIS display

Source: adapted from https://www.am.szczecin.pl/userfiles/File/instrukcje_ecdis.pdf [accessed on 8th February 2016]. [c]

Apart from integration and analysis, proper visualisation comes as one of the basic tasks performed by the systems of electronic charts ECDIS. A new element which appears in the field of visualisation (data display) is a 3D mode of the ENC. 3D charts have become a standard in modern communication of contents contained in digital maps. Figure 2-12 presents a comparison of 2D and 3D images taken from the ENC of the Bay of Gdańsk, the Baltic Sea, in the dKart Navigator system.

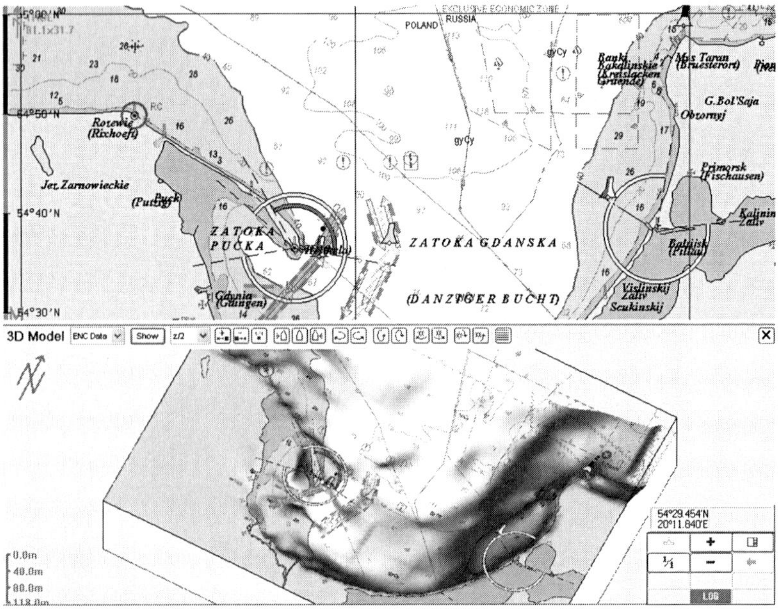

Figure 2-12. 2D/3D display taken from an ECDIS chart in the dKart Navigator system

Source: adapted from http://www.eti.pg.gda.pl/mapy_cyfrowe_2009_2010/Instrukcja_laboratoryjna_ECDIS_KB_v4.pdf [accessed on 8th February 2016]. [c]

The ECDIS makes further integration of systems supporting safe navigation in the form of INS/IBS possible. An integrated navigation bridge is a combination of navigational systems which allows the crew to access all the information coming from various sensors at one place: a command and control centre. The aim of such a combination is to increase navigation safety and efficiency of vessel management in the context of risk minimisation in safe navigation.[56] An integrated bridge is most often configured with:

[56] S. Brzóska, *Możliwości rozwoju mostka zintegrowanego na przykładzie statku "HORYZONT II"*, Zeszyty Naukowe Akademii Morskiej, Gdynia 2010, p. 27.

- an internal computer network (used for the control of the information flow from vessel receivers and sensors, e.g. radars, sonars, echo sounders, etc. to the particular components);
- an ECDIS chart database;
- an interface applied to mark the position of a vessel on an electronic map, with reference to its dynamic parameters (the course and speed);
- an interface applied for voyage planning;
- a system of analysis and traffic control which can adjust the speed and the course to follow the selected route, to avoid a collision course taken by other vessels, to maintain some intended economic and efficiency regimes, such as maintenance of economical speed.

Another sub-component of the pro-safety doctrinal solutions in maritime transport is the implementation of the regulations stated in the International Convention on Standards of Training, Certification and Watchkeeping for Seafarers (STCW).[57]

Professional qualifications and competences which refer to the particular levels of responsibility and methods of their verification are defined in the competence standards stated in the Convention, and they are confirmed by relevant certificates required by the Convention. It means that a person without the required professional competences confirmed by the Compliant Licenses and Certificates of Competency (CoC) and not trained in the field of life saving techniques (Basic Safety Training – BST) must not be employed on the board of a seagoing vessel which performs tasks related to maritime transport.

Considering inland waterway navigation, there are slightly less stringent regulations applicable. Therefore, it is often possible to hear an unfair opinion that helmsmen/skippers and their crews can have less experience and skills in inland waterway navigation. Generally, there are applicable regulations which refer to qualifications (diplomas and certificates of competences are required), and considering the specificity, traffic intensity and operational

[57] It was constituted in London on 7th July 1984; since then it has been in force in other countries (e.g. in Poland: Journal of Laws, 1984, No. 39, items 201 and 202). There were numerous amendments made, but it was finally modified by two codes in 1995. It comprises a list of minimal requirements in terms of theoretical and practical knowledge and skills, certificates and qualifications required to perform tasks at the particular workstations on a vessel – the Author's note.

requirements for navigation along the Rhine river, there are additional specific documents required there: the Rhine Boatmaster's Certificates. There are four types of such certificates, namely:

- the Rhine Boatmaster's Certificate Class 1 – allows its holder to navigate any type of vessels along a particular section of the Rhine,
- the Rhine Boatmaster's Certificate Class 2 – allows its holder to navigate along the Rhine where maritime regulations are applicable (as on lagoon waters),
- the Recreational Boatmaster's Certificate – allows its holder to navigate sports and recreational vessels (up to 12 crew members on board),
- the Administration Boatmaster's Certificate.

Regulations which refer to the limitation of GHG (greenhouse gases) emission generated by maritime transport come as the last fundamental component in the package of solutions related to safety in water (maritime and inland waterway) transport.

Limiting GHG (including CO_2) emission from all its potential sources, including GHG emission generated by maritime and inland waterway navigation, into the atmosphere closely follows a global tendency to counteract the phenomenon of general climate warming on Earth. Therefore, the IMO implemented relevant modifications to the MARPOL Convention (stated in the VI Annex of 15th July 2001). The above-mentioned modifications refer to the necessity of applying some specified means which limit GHG emission generated by vessel engines in international maritime navigation into the atmosphere.

New regulations by the IMO automatically implement an obligation of applying the Energy Efficiency Design Indices (EEDI) and the Ship Energy Efficiency Management Plan (SEEMP) which refer to energy saving on newly constructed vessels. The application of the EEDI refers to all vessels above 400 GT, and it has been obligatory in maritime navigation since 1st January 2013.[58] In accordance with the above-mentioned Annex to the MARPOL Convention – Chapter 4, the process of limiting GHG emission by oceangoing vessels has been divided into several time stages.[59] Table 2-2 presents the

---

[58] Jurdziński M., *Ocena metod i możliwości zmniejszenia zużycia paliwa oraz emisji gazów na statkach morskich*, Prace Wydziału Nawigacyjnego Akademii Morskiej w Gdyni no. 26, Gdynia 2013, p. 23.

[59] Ibidem, p. 23.

range and deadlines for the implementation of tools applied to lower CO_2 emission during the years 2015–2050.

Table 2-2. The schedule for the implementation of regulations applied to reduce CO_2 emission in maritime navigation (according to the IMO)

Scheduled dates	Reduction in CO_2 emission generated by oceangoing vessels
By 2015	Regulations applicable to approximately 90% of vessels in the global fleet come into force
By 2015	Meeting requirements referring to the EEDIs by new vessels
By 2020	New vessels must reduce CO_2 emission by 20%
By 2025	Newly constructed vessels must prove lowered CO_2 emission by another 20%
By 2030	Obligatory reduction in GHG emission by another 20%
By 2050	Reduction in CO_2 emission by 50% per tonne/kn

Source: adapted from M. Jurdziński, *Innowacje technologiczne na statkach morskich w celu redukcji zużycia energii i emisji CO_2*, Zeszyty Naukowe Akademii Morskiej no. 77, Gdynia 2012, p. 24.

Since the introduction of the EEDIs and the SEEMP, a number of solutions and procedures that facilitate the implementation, collection and full use of information obtained in this field have been developed in order to achieve the assumed outcome (monitoring, reporting and verification – MRV of CO_2).[60]

Pro-ecological regulations in maritime navigation have recently started to appear, especially in the areas where possibilities of maintaining homeostasis are limited.[61] Such an approach has resulted in the establishment of

[60] Considered by ship-owners as very strict, the new regulations stated by the IMO came into force on 1st January 2018. Their main objective is to provide the EU legal framework for collecting and publishing verified annual data on CO_2 emission from all large vessels (over 5 000 GT) which call at the EU ports, regardless of their registration place. Ship-owners and operators are obliged to monitor emission and to provide reports on the levels of verified amounts of CO_2 emitted by their ships during all their voyages in the EU waters (to, from and between the EU ports). Operators are also obliged to provide data which are required for the calculation of the ship energy efficiency, and they are allowed to choose a proper monitoring method. The system is ultimately intended to become an element of extensive global solutions. The presented regulations are aimed at gradual (it is going to be a long-term process) definition of global standards of energy efficiency for operated vessels – the Author's note.

[61] *BSR Innoship*, Project seminar *Ograniczanie emisji toksycznych związków spalania paliw ze statków w rejonie M. Bałtyckiego i portów*, non-published materials of the Polish Register of Shipping, Gdańsk 2012, p. 14.

Particularly Sensitive Sea Areas (PSSA), Emission Control Areas (ECA) and Sulphur Emission Control Areas (SECA).[62] The subsequent IMO regulations (Tier I – III standards) refer to the limitation of SO_x, NO_x and dust (soot) emission from vessels.[63] The minimisation of the amounts of these substances can be achieved with the use of facilities for exhaust gas treatment in the process of selective catalytic reduction (SCR) or with the use of alternative drives and fuels (e.g. LNG).[64]

In order to implement the directives of the IMO and the EU, numerous operators implement an extensive pro-ecological programme under the SEEMP on their vessels (along with its telematics equipment) to lower GHG emission generated by exhaust fumes. In the programme, two fundamental groups of factors are considered:[65]

- optimisation of a voyage (optimal vessel speed which comes as a compromise between the expectations of the cargo supplier and the amount of consumed fuel – see Figure 2-13, the choice of a route which provides more favourable weather conditions, functionality of the autopilot and the steering gear, a proper trim, that is namely the inclination angle of the bow with regard to the stern, the use of the ECDIS and INS/IBS, etc.);
- the use of non-standard technical solutions (e.g. covering the hull with silicone anti-fouling paint, polishing the propeller – anti-cavitation, using special additional substances to fuel, etc.);

---

[62] On 1st January 2015, in accordance with IMO MARPOL Annex VI, the limitations in the amounts of sulphur in the ECAs (Batlic Sea region) to 0,1%, were implemented. It generates some unusually radical changes in the economy of vessel operation. The previous EU Directive 2005/33/WE which had been applicable since 1st January 2010, limited the amount of sulphur in fuel to 0,1%, but only during the moorage and manoeuvering in the EU ports and canals – the Author's note.

[63] See more in: R.K. Miler, A. Bujak, *Implementacja systemów monitoringu zużycia paliw i efektywności energetycznej statku jako pochodna dyrektyw proekologicznych UE w transporcie morskim*, "Logistyka" no. 2/2014, p. 56; *LNG nowe paliwo w żegludze bliskiego zasięgu*, Det Norske Veritas Poland, on-line publication, p. 10, www.dnv.pl [accessed on 26th July 2016].

[64] The LNG drive provides a well-proven and efficient solution to the problem of SOx and other pollution, however, considering the high costs of its implementation and the lack of distribution infrastructure, it has not triggered a lot of interest among ship-owners and operators yet – the Author's note.

[65] The presented problems are discussed further in the following online publications: http://morzaioceany.pl/zegluga-morska/pzm/1070-mniej-gazów-cieplarnianych-ze-statków-pżm.html; http://www.polsteam.com/mniej-gazow-cieplarnianych-ze-statkow-pzm [accessed on 8th February 2017]– the Author's note.

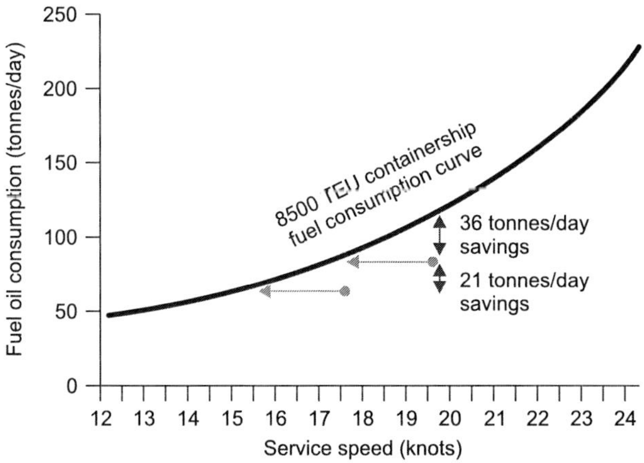

Figure 2-13. The relation between fuel consumption and speed developed by a container vessel [8,5 T TEU in tonnes per 24 h]

Source: adapted from H. Igliński, *Wybrane sposoby ograniczenia zużycia paliwa w transporcie morskim*, "Logistyka" 5/2012, p. 459.

The implementation of the Directive of the European Commission on limitation of GHG emission has resulted in the fact that most vessels have been already equipped with some special systems of fuel consumption monitoring, related to the drive and energy generation (the main engine, auxiliary engines, the boiler). The data obtained from such devices are transmitted to the bridge and to the engine room.[66] In modern telematics systems (condition-based monitoring CBM), the data can be also accessed by relevant onshore units of the ship-owner in the real time. In this way ship-owners can also decide about the most advantageous options in order to limit fuel consumption and, at the same time to limit GHG emission.[67]

The element which is closely related to excessive GHG emission in maritime and inland waterway transport is congestion. The phenomenon can be of the following nature:[68]

[66] http://www.polsteam.com/mniej-gazow-cieplarnianych-ze-statkow-pzm [accessed on 26th July 2016].

[67] See more in: R.K. Miler, A. Bujak, *Implementacja systemów monitoringu zużycia paliw i efektywności energetycznej statku jako pochodna dyrektyw proekologicznych UE w transporcie morskim*, op. cit., p. 57 – the Author's note.

[68] https://unicen.smu.edu.sg/sites/unicen.smu.edu.sg/files/Intelligent-Maritime-Traffic-Prediction-Coordination.pdf [accessed on 26th July 2016].

- primary – generated by the modes of water transport; it can be manifested by the overcrowding of navigable areas, including roadsteads, approach fairways to seaports as well as rivers (congestion of water transport routes);
- primary – generated by the modes of land transport (road and rail); it can be manifested by the overcrowding of roads and railways (congestion of hinterland);
- secondary – which is characterised by considerable delays in the operation of ports and cargo handling terminals in the transshipment function (congestion of ports).

Hence, various interested parties are looking for conceptual solutions in the field of intelligent telematics systems which will be able to identify, to analyse and to counteract congestion in the area of sea waterways and approach fairways to ports. A model of such an analytical and decisional solution is presented in Figure 2-14.

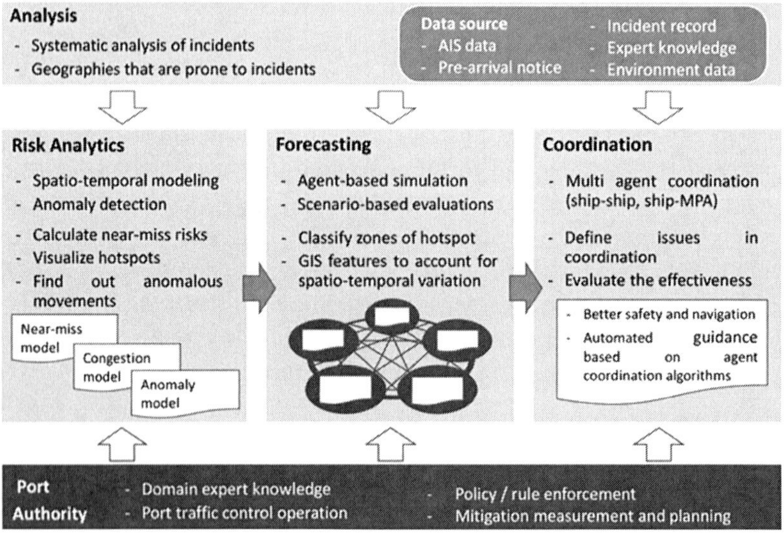

Figure 2-14. A model of an intelligent telematics solution in the field of analysis and counteraction of congestion in maritime and inland waterway transport

Source: adapted from https://unicen.smu.edu.sg/sites/unicen.smu.edu.sg/files/Intelligent-Maritime-Traffic-Prediction-Coordination.pdf [accessed on 14th May 2018].

Such a solution can be integrated with the next generation of e-navigation, and it can be commonly used for counteracting congestion on water transport routes. Handling a ship in congested or high traffic areas is not an easy task. Congested waters are high-density traffic areas where a vessel is likely to collide with another vessel if ship navigation is not carried out in the right manner. Factors such as the state of sea, weather, proximity of vessels, status of the ship (underway or making way, laden or under ballast condition, etc.), interaction with other vessels causing differences in pressure at the bow or the stern, proximity to the land, steering gear effectiveness, effectiveness of navigation aids, rudder movements, effect of shallows, are among many things that are to be adhered to. All of the aforementioned elements are to be applied in reference to COLREGs to ensure that the vessel is safely navigated into and out of the congested area. Rule 18 of COLREGs explicitly states as to which vessel is to be kept out of the way of another, which basically means that a vessel cannot bypass preexisting rules with regard to vessel responsibility.[69]

In accordance with the European Conference of Ministers of Transport (2007),[70] there is some common understanding of the notion of congestion, although its main meaning in such a context refers to the secondary character of congestion (port congestion as a result of primary congestion). Port congestion [...] *arises when demand levels approach the capacity of a facility and the time required to use it (or travel through it) increases well above the average under low demand conditions.*[71]

In this context, demand levels could be, for example, voyage time, berthing window, moves/hour, truck throughput time, or handling time. Capacity can be broken down into capacity of port access, berth, crane, personnel, equipment and, for example, yard handling. The accessibility from the sea refers to the shipping route, a river channel, the berth, the terminal (quay, yard, hinterland hand-over); and the average time is either fixed in contracts or undefined/experienced.[72]

---

[69] https://www.marineinsight.com/marine-navigation/how-to-handle-a-ship-in-congested-high-traffic-waters/ [accessed on 26th March 2016].

[70] European Conference of Ministers of Transport (2007): *Congestion: A Global Challenge the Extend of and Outlook for Congestion in Inland*, Maritime and Air Transport, Brussels 2007; http://www.internationaltransportforum.org/IntOrg/ecmt/cm/pdf/ITF200706e.pdf [accessed on 17th July 2015].

[71] Dios Ortúzar, J. de, Willumsen, L.G., *Modelling transport*, Chichester Wiley 2011, p. 5.

[72] P.W. de Langen, 7th Framework Programme SST.2013.6-2. *Towards a competitive and resource efficient port transport system* Collaborative Project, Deliverable 4.2 Connectivity, costs and congestion indicators, DOCUMENT ID PORTOPIA|D|4.2.2|DT|2014.03.20|V |01, 2014-03-31, p. 50–55.

Today, port congestion has been growing worldwide, as an increase in container traffic over the past year and ongoing consolidation in the shipping industry compound other problems, such as inefficient transfer to inland transportation, customs delays, and episodic weather- and labor-related slowdowns.[73] The acceptance and unification of a wide range of solutions described above considerably contributes to the improvement of the safety level in maritime navigation, it lowers ecological risk and, in the same way, considerably restricts the level of threats against safety in maritime transport. Unfortunately, it is not the only one risk field in broadly understood water (maritime and inland waterway) transport.

2.2.2. The field of *security* in maritime and inland waterway transport (incidents and their consequences)

Another field of threats in the domain of maritime transport includes security risks which refer to maritime security of the state. Considering the general characteristics of global security of maritime transport over the last decade, the most frequently occurring threats of anthropogenic nature (caused by purposeful and negative activities of people – individuals, groups or organisations) in the field of maritime transport security have involved the following:[74]

- maritime terrorism[75] aimed at facilities of the nodal infrastructure of maritime transport (ports, terminals, lock systems on canals, etc.) as well as at superstructure (vessels and even warships e.g. the USS Cole[76]);

[73] https://www.americanexpress.com/us/content/foreign-exchange/articles/port-congestion-an-international-trade-challenge/ [accessed on 17th July 2015].

[74] Cf.: R. Miler, *Bezpieczeństwo transportu morskiego...*, op. cit., pp. 88–89.

[75] It is believed that the first terrorist attack in history was the arson of the Temple of Artemis committed by Herostratus of Ephesus in 356 BC. In 1090 an association of assassins was established in Persia; it affiliated secret killers who at first used to murder pagans, but later on they also killed people of their own religion who became cumbersome henchmen, or those for whom they had been paid more. At present, terrorism – as we understand it – appeared in the middle of the 19th century, introduced by revolutionists and anarchists who wanted to change the social order (as it was during the French Revolution). At that time, Europe and the world witnessed numerous assassinations of rulers and representatives of government establishment. The meaning of the *terror* term has been evolving, and now it is a synonym of fight led against countries, nations and political systems. During the first twenty years of the 20th century, the most notorious terrorist attack was the assassination of Archduke Franz Ferdinand of Austria on 28th June 1914. It was the beginning of the First World War which cost over 37 million lives on both sides of the conflict – the Author's note.

[76] The suicidal terrorist attack on the USS *Cole* was committed by the al-Qaeda terrorists in 2000. They used the tactics of the Trojan horse: two terrorists hit the American

- piracy;[77]
- asymmetric activities aimed at ports, port facilities and terminals (including cybercrime – attacks against IT/telematics infrastructure).

The attacks on the World Trade Centre came as a specific catalyser for the absorption of modern understanding of the *terrorism* notion in the international community. It transpired that such actions, even if they had not been new in the history of our civilisation, were redefined at the beginning of the 21st century, and now they come as a real threat to international security.[78] The scale and drastic nature (suicidal attacks) and global consequences of such actions (a global effect of the communicating vessels) are highly alarming too. According to most frequently cited experts, maritime terrorism involves *[...] a planned and organised violent attack aimed at people, vessels, port facilities and installations at sea.*[79] It is intended to force state authorities, societies or individuals to act in a particular way, to make concessions or to provide financial means. Figure 2-15 presents trends in the fluctuations referring to the main motives of terrorist attacks.

destroyer manoeuvering at the roadstead of the Port of Yemen in the Gulf of Aden with a boat resembling a port service vessel, which was carrying approximately 270 kg of C4 explosives. Apart from two "sea shahids"-suiciders, 19 American seamen were killed and 39 of them were injured. Apart from the loss of human lives and undeniable loss of prestige in the US Navy, the attack resulted in a very severe economic loss with regard to the renovation of the warship (renovation work done on the USS *Cole* in a dry dock was estimated for USD 250 million); additionally, the destroyer was out of duty for 14 months – the Author's note.

[77] The notion of *piracy* is over 3000 years old, and it was probably used for the first time by a Roman historian, Polybius, in 140 BC. It comes from the word *pierato*, which refers to those who attack illegally. It can be stated without any doubts that the Golden Era of piracy occurred in the 16th–17th centuries, when seas and oceans were dominated by the Spanish fleet which transported all the means indispensable for the functioning of the state and further conquests from the New World. The atmosphere of national and colonial mutual antagonisms fostered the boom of piracy. On one hand, given their letters of marque, English buccaneers robbed Spanish vessels, and on the other hand, pirates of North Africa were allowed to rob English ships. Pirates of Madagascar were also granted independence of a specific kind, robbing vessels for France – the Author's note.

[78] The difference between terrorist attacks and pirate attacks is merely subjective, and it most often depends on the author of a particular publication; for example, K. Kubiak assumes that a factor which differentiates a terrorist attack from a pirate attack at sea is an attitude of the attackers towards the problem of maintaining their anonymity – the Author's note.

[79] See more in: K. Kubiak, *Piractwo i terroryzm morski. Nowe wyzwania dla bezpieczeństwa międzynarodowego*, http://www.dsw.edu.pl/fileadmin/user_upload/wszechnica/04.pdf, p. 23 [accessed on 13th August 2016].

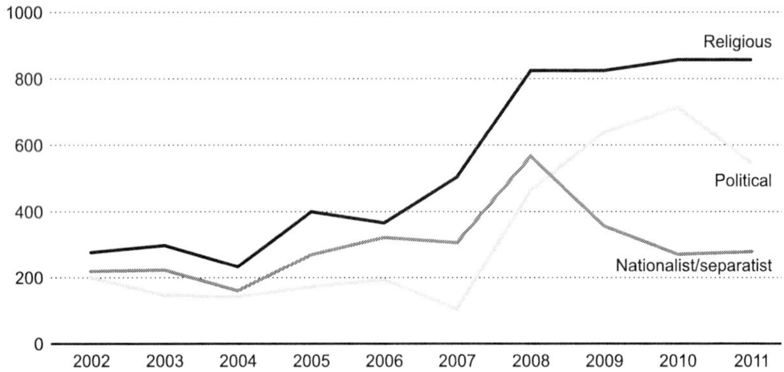

Figure 2-15. Changes of terrorists and assassinators' motives during the years 2002–2012

Source: adapted from http://www.washingtonpost.com [accessed on 22nd January 2015].

Over the last several decades, terrorism has evolved and has been transformed into a specific total war. Terrorists have changed not only their forms and methods of action, but they also have improved their weapon, equipment and, first of all, their determination. The fight against terrorism has been translocated into the international space with regard to the fact that the consequences of terrorism are often of global nature.[80]

Considering the significance of maritime transport (maritime navigation and ports) in global economy, terrorist attacks on maritime facilities can guarantee the achievement of terrorists' most desirable objectives: loss of human lives, economic loss incurred by the victim countries and global economy, media coverage which will intensify fear and panic.[81] The critical aim of terrorist attacks involves not only navigation itself but also seaports and cargo handling terminals which are the hubs in the network of land-sea logistical chains. In the discussed environment, we refer to a gigantic range (in terms of its scale, involved activities and time pressure) of logistical work (cargo handling operations, temporary storage, internal transport, changes of the means of transport, etc.), and terrorist threats result in a necessity of applying additional procedures which provide and increase the level of security (e.g. container scanning), but which also generate disproportionally

[80] Cz. Dyrcz, *Wybrane problemy terroryzmu XXI w. (2)*, "Przegląd Morski" no. 10/2005, p. 7.
[81] S. Kalitowski, *Terroryzm morski*, op. cit., http://www.terroryzm.com/terroryzm-morski/ [accessed on 17th July 2015].

high costs, not mentioning the delays in the implementation of basic processes (congestion of ports/terminals and transshipment processes).[82]

Paradoxically, terrorist attacks aimed at maritime facilities account only for 2% of all terrorist incidents that have been recorded so far in the world over the last 30 years; however, the scale of their impact is immense.[83]

The "balance" of pirate attacks is quite similar – between the years 2008–2013 the scale of pirate attacks was high, but due to determination of the international community, the phenomenon has been reduced over the last several years. Figure 2-16 presents pirate attacks in the area of the Somalian coast.

For the requirements of this publication, it has been assumed that piracy is understood as any *[...] lawless act of violence against a moving or moored/anchored vessel, that is not subject to the jurisdiction of any state, in order to achieve private gain.*[84] Another source that should be referred to for a definition of piracy is the United Nations Convention on the Law of the Sea – UNCLOS.[85] It states that piracy includes any illegal act of violence, hijacking or theft of a private vessel of an aircraft, aimed[86]

a) against any other oceangoing vessel at sea or an aircraft,
b) against people or property on the board of an oceangoing vessel or an aircraft,
c) against any oceangoing vessel or an aircraft, people or property which are not subject to the jurisdiction of any state.

---

[82] R. Miler, *Bezpieczeństwo transportu morskiego...*, op. cit., p. 112.

[83] The statistics are increased every month by numerous terrorist acts on land – in Iran, Iraq, Afghanistan, Syria and in other countries struggling with destabilisation – the Author's note.

[84] See more in: K. Szczyrbowski, op. cit., p. 23.

[85] The article 101 of UNCLOS defines piracy in a broader way, stating that it also includes:
a) a lawless act of violence, hijacking or theft done for personal purposes by the crew or passengers of a private vessel, an aircraft, and aimed against:
  - any other oceangoing vessel at sea or an aircraft;
  - people or property on the board of a seagoing vessel or an aircraft, people or property at a place which is not subject to the jurisdiction of any state;
b) an act of voluntary participation in an operation undertaken by an ocean of an oceangoing vessel or an aircraft, with the knowledge about the facts which give the vessel or the aircraft pirate attributes;
c) incitement to or purposeful facilitation of acts stated in item a and b.
The United Nations Convention on the Law of the Sea, Art. 101 http://www.vilp.de/Plpdf/p120.pdf [accessed on 14th July 2014] – the Author's note.

[86] Cf.: K. Wardin, *Współczesne piractwo morskie zagrożeniem dla międzynarodowego transportu morskiego*, ZN AMW no. 3(178), Gdynia 2009, p. 91.

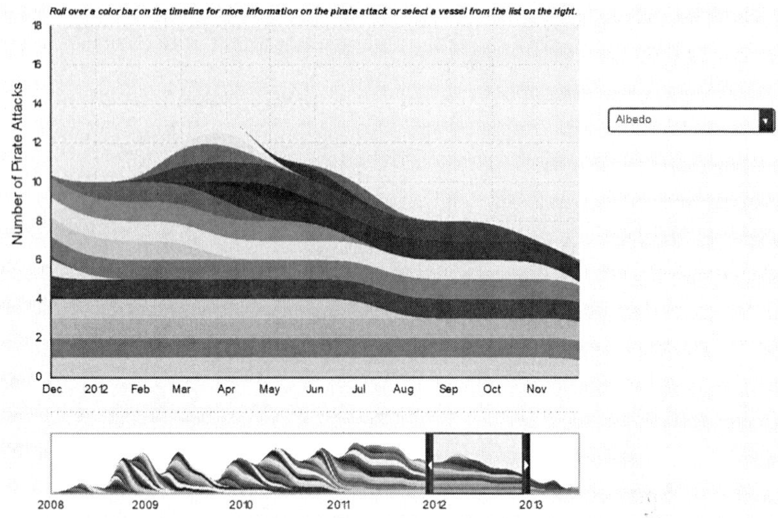

Each incident is marked with a different colour, the length of the line represents the period of time in which a vessel was hijacked, the tag (gray area) provides the details of the vessel and the hijacking characteristics; in this example these are the details of the m/v *Albedo* (more information – see the footnote).[87]

Figure 2-16. Piracy incidents at the Somalian coast in 2008–2013

Source: adapted from http://oceansbeyondpiracy.org/attack-timeline/somalia [accessed on 13th May 2016]. [c]

A good example of successful international (civil-military) fight against piracy at sea in the regions which are crucial for maritime transport, with particular consideration of the Horn of Africa (HOA), are operations undertaken with the participation of the NATO navy forces and the EU, namely: the Ocean Shield and the Atalanta. Due to the forces and means deployed in these operations, it was possible to minimise possibilities of activities undertaken by pirate groups and, in this way, to decrease the risk.

[87] The history of hijacking m/v *Albedo* comes as a perfect illustration of the scale of the problem. The vessel under the Malaysian flag was hijacked in the Gulf of Aden in November 2010. 23 crew members included citizens of Iran, Sri Lanka, Bangladesh and Pakistan. The ransom was quickly paid and the Pakistani crew members were released, but the crew members of other nationalities were still kept hostage. The families of 6 citizens of Sri Lanka were running a long-lasting (several years long) campaign to collect money for the ransom. Eventually the remaining 11 crew members were released on 8th June 2014 sic! – the Author's note.

Another sensitive area refers to threats related to cybercrime.[88] The progressing dependence of operational activities in water (mainly maritime) transport on IT/ITC systems and broadly understood telematics of transport results in the fact that the IT assistance for transport processes, informational supply, e-navigation and analysis in the field of decision making become targets for hackers' attacks and implementation of malware (bugs, Trojans, viruses, etc.).

Table 2-3 presents proven cyberattacks on informational and IT infrastructure of maritime transport (considering its current weak development and its immanent features, IT infrastructure of inland waterway transport has not been exposed to such strong pressure of cyber threats).

Table 2-3. Cyberattacks on informational and IT infrastructure of maritime transport

Date	Victim	Consequence
NOV 2017	Clarksons	Perpetrators gained unauthorised access to computer systems, accessing confidential information and threatening to release information unless ransom payment is made. Company share prices decreased by 2.71%.
JUN 2017	Ships in Novorossiysk, Russia	At least 20 ships in the Black Sea were reporting false data was being transmitted, indicating the ships were 32 km inland of their actual position. It is now believed to have been as a result of a GNSS spoofing attack.
JUN 2017	A.P. Moller Maersk	NotPetya also known as ExPetr ransomware led to outages on A.P. Moller Maersk computer systems impacting both oil and gas production and port operations. Following the incident, Maersk claimed to have changed its IT systems to prevent similar incidents from occurring in the future. The incident resulted in an estimated USD 300 million of losses.
APR 2016	South Korea	280 ships were forced to return to port due to problems on their navigation systems. The issue was largely blamed on North Korea however this remains unconfirmed.
2012–14	Danish Port Authority	An email virus spread through the port network that was likely initiated through an infected pdf document. The virus spread and successfully reached other Danish government institutions.

[88] https://www.jltspecialty.com/our-insights/publications/cyber-decoder/nis-directive--explained [accessed on 17th March 2018].

Table 2-3. cont.

Date	Victim	Consequence
2012	Australian Customs and Border Protection Service agency	Cargo systems controlled by customs and border protection were hacked in order to determine which shipping containers were suspected by authorities.
2011–13	Port of Antwerp	The port had been a victim of an APT attack since 2011 commissioned by a drug cartel. The attack targeted terminal systems which were subsequently compromised by hackers and used to release containers without port authorities becoming aware. Illicit drugs and contraband worth approximately EUR 307 million (USD 365 million), firearms and EUR 1.3 million (USD 1.5 million) were seized when authorities finally became aware.
AUG 2011	Iranian Shipping Line (IRISL)	The servers were hacked resulting in damage to data relating to rates, loading, delivery and location. Consequently, the location of many cargo containers remained unidentified and an undisclosed amount of financial losses were incurred as a result.

Source: adapted from Andrew Huxley – TT Club cyber threat to shipping (non-published materials) https://www.ttclub.com [accessed on 5th May 2018].

In 2016, BIMCO / Fairplay Survey revealed a true picture of threats: over 21% of respondents from the water transport sector confirmed that they had been victims of attacks aimed at the informational and IT infrastructure of their companies at least once. At the beginning of 2017 Seaintel informed that 44% of top 50 maritime carriers had very weak systems of protection and insufficient remedy procedures to be applied during a cyberattack.

Hence, the level of cyberattack threat and associated risk have been explicitly defined as highly probable. As a result, some documents have been developed in order to obligatorily impose particular levels of protection and methods of counteracting threats in IT networks. These are namely:

- IMO new regulation guidelines on cyber risk management (1st June 2016),
- Resolution MSC.428(98) (7th June 2017) which requires administrations […] *to take the necessary steps to incorporate cyber threat considerations appropriately through safety management systems* […] (by 1 January 2021),

- The EU General Data Protection Regulation (GDPR) (EU) 2016/679 – is a regulation in the EU law on data protection and privacy for all individuals in the European Union (GDPR effective May 25, 2018),
- Security of Network and Information Systems Directive (NIS Directive)[89] (NIS effective May 9, 2018).

Considering their character and often unbelievable precision in their preparation, the above-mentioned threats to security (maritime terrorism, maritime piracy and asymmetric threats to seaports – cybersecurity hazards) belong to a group of threats which are feared most in modern maritime transport. Most of these activities may be secretly prepared, in a way which makes preemptive detection and prevention very difficult or sometimes even impossible.

Over the last few decades a number of solutions and tools for providing security have been implemented into the pragmatics of maritime transport processes. The most important ones in the field of security are the following:
- regulations of the International Ship and Port Security (ISPS),
- a civil-military concept of Maritime Security Operations (MSO) and Maritime Situational Awareness (MSA),
- implementation of analytical and decisional systems based on the monitoring systems of maritime navigation (based on obligatory reporting).

The regulations implemented by the ISPS Code have become highly significant in the protection of seaports and vessels performing port operations.

In accordance with the accepted ISPS regulations, port facilities and the processes which take place there are particularly protected in order to minimise risk and to provide a desirable security level of the operational (transport and logistics) processes implemented there.[90] Figure 2-17 presents protection-related processes.

---

[89] The Security of Network and Information Systems Directive (NIS Directive) will significantly expand the scope of cyber security regulations in the EU. Essentially, the legislation aims at raising levels of security and resilience of critical infrastructure, such as power and utilities, but also digital infrastructure such as cloud service providers. It came into force in August 2016 and the EU member states have had to apply the Directive by May 9 2018 – the Author's note.

[90] The IMO (*International Maritime Organization*) defines minimal requirements for providing security to vessels and ports with a set of regulations and guidelines which are aimed at increasing security of navigation and protection against terrorism. They are stated in the International Ship and Port Facility Security Code (ISPS Code). ISPS Code

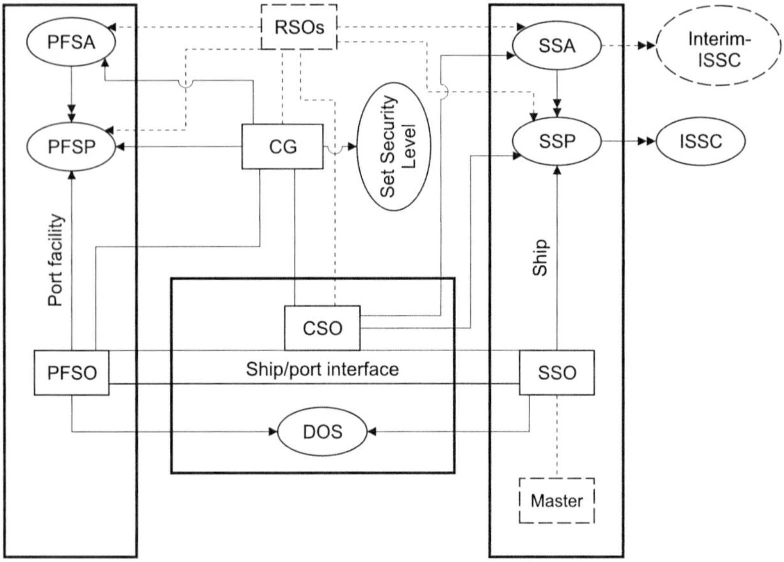

PFSA – Port Facility Security Assessment, PFSP – Port Facility Security Plan, PFSO – Port Facility Security Officer(s), RSO – Recognized Security Organization, CG – Contracting Government, CSO – Company Security Officer, DOS – Declaration of Security, SSA – Ship Security Assessment, SSP – Ship Security Plan, SSO – Ship Security Officer, ISSC – International Ship Security Certificate, Master – captain on board (skipper)

Figure 2-17. The functional scheme of ISPS

Source: adapted from A. Mazaheri, *How the ISPS code affects port and port activities,* University College of Borås, Borås 2008, p. 17.

Modern port facilities are more exposed to asymmetric operations than to attacks during an open armed conflict. Providing security to port facilities, as well as to vessels which are moored there, largely depends on the tightness and efficiency of port security systems which protect them from water and from land (e.g. telematics systems of the GIS type).[91]

---

has been introduced in all maritime countries all over the world, e.g in Poland as the Act on Protection of Navigation and Seaports (Journal of Laws, 2008, No. 171, item 1055). Apart from being compliant with the ISPS Code, it also meets the requirements of the Ordinance of the European Community on Increased Protection of Vessels and Port Facilities (Ordinance of EC, 2004) – the Author's note.

[91] J. Zboina (ed.), *Bezpieczeństwo na lądzie, morzu i w powietrzu w XXI wieku,* Wydawnictwo CNBOP-PIB Józefów 2014 p. 206.

Security systems implemented at each port facility are mainly of preventive nature. They are supposed to prevent events/threats against the security of port facilities. There are three levels of security constituted (Figure 2-18).

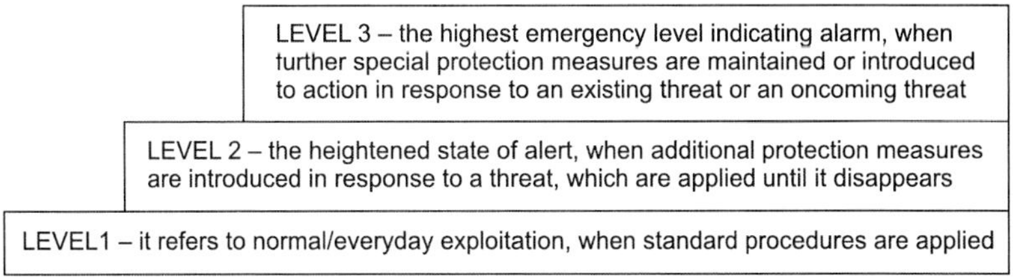

Figure 2-18. Levels of security in the ISPS regulations

Source: adapted from J. Zboina (ed.), *Bezpieczeństwo na lądzie, morzu i w powietrzu w XXI wieku...*, p. 207.

The foundation of prevention against threats comes with obligatory development of security plans for each port facility. They are developed on the basis of an individual evaluation of the protection status of a port facility, which comes as a fundamental part of the process referring to the acknowledgement of such a security plan (Port Facility Security Plan – PFSP, Ship Security Plan – SSP).

In practice, the Geoinformatics Port Security System (GPSS)[92] may be responsible for the implementation of the ISPS code. The system applies geoinformatics technology to provide security to operational activities related to monitoring and protection of ports from the sea-side and shore-side, in the restricted access areas (RAA).

Another aspect related to the package of tools implemented to improve security of maritime transport (it does not refer to inland waterway transport) includes systemic solutions which come along with the MSO (Maritime Security Operations) and MSA (Maritime Situational Awareness) concepts. During the NATO summit in Riga, an idea of a pan-European, civil and military approach

---

[92] The scientific research team of the Institute of Geoinformatics at the Maritime University of Szczecin, Poland has embarked on development and pilot implementation of such a system for a selected part of Szczecin and Świnoujście Seaports, within the framework of a research and development project: *Geoinformatics security system for operational activities related to protection of ports from the sea-side.* The project is financed by the National Centre for Research and Development – the Author's note.

towards the MSO problems was presented.[93] There were strategic courses of action identified, including:[94]

- a political line, which assumes increased activities oriented towards attracting the particular NATO countries to the participation in the concept;
- civil and military cooperation, which assumes intensification of national and international operations oriented towards the unification of terms and standards and the cooperation of NATO with international organisations, such as the IMO, the EMSA, Paris MoU, etc.;
- early warning which indicates the necessity of developing a unified system for navigation data display with a possibility to provide an analysis of anomalies that could result in some hazard, and with a system of rapid response (an experimental BRITE system);
- layered security of navigation interests, in which the development of a display and an early warning system is strongly emphasized. Such a system would display and provide early warning about incidents in the World Ocean, not only in territorial waters or exclusive economic zones of particular countries;
- implementation of security elements into everyday commercial practice of navigation agencies, where strong emphasis is laid onto the monitoring of vessels carrying hazardous cargo, large passenger vessels, in relations: a safe port of loading, a safe carrier and a safe vessel, a safe port of unloading (as a reference to the ISPS Convention)[95];
- a legal line, which indicates areas of national and international jurisdiction requiring some additional specification or a complete definition in order to obtain a formal and legal basis for the implementation processes of the MSO concept.

Understood in such a way, Maritime Situational Awareness (MSA) becomes a development in the field of integration referring to information exchange. It is often identified with an initiative of the USA – Maritime Domain Awareness (MDA). According to the NATO doctrine, MSA is […] *understanding (or providing a picture) of military and non-military events,*

---

[93] M. Grzybowski, R.K. Miler, *Bezpieczeństwo na europejskim i globalnym rynku transportu morskiego*, ZN AMW no. 1, Gdynia 2008, p. 34.

[94] M. Grzybowski, R. Miler, *MSA as the Aspect of European Maritime Security*, "The Romanian Economic Journal" no. 25, Bucharest 2007, p. 125.

[95] Adopted on 13th December 2002 as the Resolution No. 2 at the Conference of Contracting Governments – the parties of the International Convention for the Safety of Life at Sea (SOLAS, 1974) (Journal of Laws 2005 No. 120, item 1016) – the Author's note.

operations and circumstances (connected with maritime environment) which are related to current or future NATO operations and exercises. Maritime environment refers to oceans, seas, gulfs, estuaries, waterways, coastal regions and ports.[96]

Considering the scope, the character and the complexity of collected information, it becomes rationally justified to maintain a three-level process of providing an MSA picture (a higher level provides a more complete picture), where:[97]

- Level I (Maritime Administration) – it should provide data from civil systems of acquiring information, based on the AIS signal and from the vessel traffic surveillance system (VTS) at the level of local authorities and maritime administration (the level of Maritime Offices and their counterparts); having been collected and processed at a local Maritime Security Centre, the data should be transmitted to an international server (development of an Integrated Maritime Picture – IMAP);
- Level II (Coast Guard) – information obtained from Level I (only a display) should be supplemented with additional information from other "stakeholders" of the system, including counterparts of the Coast Guard (a system based on IR/TV cameras and radars);
- Level III (Navy) – it should include data obtained from Level I and II and the own sources of the Navy (the Navy surveillance and detection systems, data provided by warships and air force, data obtained from the NATO systems – including MCCIS); on such a basis a recognised picture of a maritime situation shall be developed (Recognised Maritime Picture – RMP).

Figure 2-19 presents the path and the method of developing information in the MSA system.

The key advantage of such a system is the fact that it has intelligent in-built algorithms of data evaluation, their comparison to reliable data or to a specifically required status (so called smart agents), with a possibility of automation of a decision-making process within the framework of effector operations during the implementation of the particular (standard) procedures when threats appear (Contingency and Emergency Plans – CEP).

[96] M. Kościelski, R. Miler, M. Zieliński, *Maritime Situational Awareness (MSA)*, ZN AMW no. 4 (171), Gdynia 2007, pp. 79–80; http://tide.act.nato.int/mediawiki/index.php_30_Mar_07 [accessed on 19th July 2008].

[97] Cf.: M. Dura, *Koniec wolności pływania...*, pp. 5–25

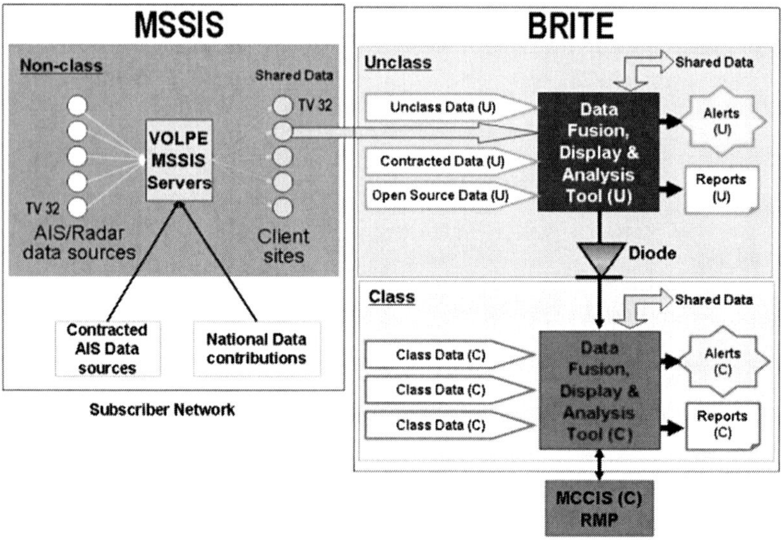

Figure 2-19. The scheme (architecture) of acquiring information in the MSA system with the application of an experimental BRITE tool

Source: adapted from http://tide.act.nato.int/mediawiki/index.php/NATO MSA Architecture [accessed on 11th September 2007].

The implementation of the above-mentioned concepts in the field of operational activities aimed at the improvement of maritime transport security assumes providing tools for the process of acquiring information in the field of maritime navigation and some elements of the process of maritime transport, and also data collection, storage and transmission for further analytical processes.

One of the EU Directives[98] imposes an obligation on the member countries which refers to the implementation and maintenance of systems designed for the exchange of information obtained from vessel monitoring. The scope of information exchange refers to all the data acquired due to the application of the regulations provided in the Directive, Articles 4, 5, 8, 13, 15, 16 and 17. The specified ranges of information which should be acquired can be presented as follows:[99]

---

[98] Directive 2002/59/ EC of 27th June 2002 establishing the EC monitoring and information system of vessel traffic and repealing the Directive of the Council 93/75/EC – the Author's note.

[99] R. Krystek (ed.), *Zintegrowany System Bezpieczeństwa Transportu…*, op. cit., pp. 339–340.

1. information which must be submitted, in accordance with Art. 4 (general identification information):
   a. details of the vessel (name, call signal, IMO number or MMSI number),
   b. destination port,
   c. estimated time of arrival at the destination port (ETA) and estimated time of departure from that port (ETD),
   d. the total number of people on board;
2. information which must be submitted, in accordance with Art. 13:
   a. details of the vessel (name, call signal, IMO number or MMSI number),
   b. destination port,
   c. for outbound vessels leaving a port of the EU member country: estimated time of departure from the port or pilot disembarkation, estimated time of arrival to a destination port,
   d. for inbound vessels arriving to a port of the EU member country from any port located outside the European Community: estimated time of arrival at a destination port or pilot embarkation, in accordance with the requirements of relevant authorities,
   e. the total number of people on board,
   f. data referring to the cargo, particularly: proper technical names of hazardous cargo, pollutants, their UN numbers if applicable, an IMO class of cargo, in accordance with the IMDG, IBC and IGC Codes and a class of the vessel, if applicable, in accordance with the INF Code, the quantity of such cargo, its location on the vessel and the identification number of a loading unit if the cargo is transported in loading units other than containers,
   g. a confirmation that there is a cargo manifest, a loading declaration or a stowage plan/bay plan with the details of hazardous or pollutant cargo which is transported and its location on the vessel,
   h. the address where detailed information about the cargo can be obtained,
   i. in accordance with Art. 17, the following information must be stated: identification of the vessel, its position, the latest port of call, the port of destination, the address where information about hazardous cargo and pollutant cargo transported by the vessel can be obtained, the number of people on board, a detailed description of any incidents and any other information defined in Resolution A.851(20) IMO.

Despite great determination of the EU, a number of European countries have not fully implemented the uniformed standards defined in the Directive 2002/59/EC. Some national systems which have been currently applied still do not allow their users to transmit full information to the SafeSeaNet (SSN)[100] – an international IT network developed to streamline data exchange (mainly because of technical reasons). Some problems also occur at the ultimate connection of databases of the currently operated SIReNaC[101] system and a database for passenger, ro-ro and HSC[102] vessels.

Outlined in this way, the problems referring to safety and security in the domain of maritime and inland waterway transport lead to a necessity of indicating some tools for risk assessment, which may become a foundation of analytical and decisional sub-systems in telematics systems of water transport.

2.3. Risk management in maritime and inland waterway transport (including emission control)

2.3.1. The structure of security and safety system in maritime and inland waterway transport

The essence of the process of management in maritime and inland waterway safety/security must involve not only some specific properties of the process itself but also the complicated nature of the environment in which the process takes place. Such immanent features of the process of manage-

[100] The SSN system (discussed further in this publication) connects national IT systems which have their separate systems dedicated for storing and accessing information. The system incorporates various databases, and it allows its users to follow information with the use of a central catalogue. The central catalogue is not a database, it works as a set of references to locations of the specified data in the resources of the particular countries. All the information considered to be significant should be stored and made accessible by the EU member countries. They should provide permanent access to such data 24/7 – the Author's note.

[101] SIReNaC is a database used for navigation safety monitoring, where the results of vessel inspections performed by the PSC inspectors are recorded. The EQUASIS system is more easily accessible (login, password, on-line access); it provides access to important data about numerous oceangoing vessels. The data for this system are loaded from the SIReNaC system and also from the databases of Tokyo MoU and the USCG; see more in: R. Krystek (ed.), *Zintegrowany System Bezpieczeństwa Transportu…*, op. cit., pp. 339–340.

[102] HSC *High Speed Craft* – a high speed passenger vessel equipped with water jet propulsion systems – the Author's note.

ment in maritime and inland waterway safety/security result from the following reasons and premises:[103]

- the aim of such management is not a total elimination of risks (because it is practically impossible, considering the stochastic character of nautical phenomena), but it comes down to maintaining basic threats to people, vessels, transported cargo, port facilities, maritime and fluvial environment at "an acceptable level"[104];
- formulated for the management of maritime and inland waterway navigation safety, the postulated aims are achieved through the use of integrated safety systems of the local, regional and global character. However, when inland waterway transport is considered, such systems are of a regional character at their most (e.g. the EU level);
- a system must include the informational and IT potential of all its stakeholders (the necessity of integrating systems with regard to compatibility of data streams).

Having recognised the need for relevant organisation of maritime safety management and maritime environment protection, the IMO has approved a resolution in which the International Safety Management Code is presented (the ISM Code).[105] Its aim is to implement universal international standards and procedures related to safety. Following the resolution is obligatory for ship-owners (ship-owners, other institutions and legal persons operating vessels), based on the implementation of the regulations stated in the 9th Chapter of the SOLAS 1974 Convention. Ship-owners who operate special purpose vessels, such as passenger ships, bulk carriers or tankers have been obliged to implement the ISM Code since 1998, whereas any other types of oceangoing vessels have been obliged to do so since 2002.[106]

---

[103] R. Miler, *Bezpieczeństwo transportu morskiego*..., op. cit., pp. 93–94

[104] The level of risk is defined in the regulations of the Formal Safety Assessment (FSA), as amended. MSC and MEPC/IMO, 2002 – the Author's note.

[105] The International Safety Management Code approved by the IMO with its Resolution A 741(18) came into force with the implementation of the 9th Chapter of the SOLAS Convention referring to the management of safe vessel operation (1st July 1998). The ISM Code presents an international standard which refers to safe management and operation of vessels. It also prevents pollution. In accordance with the assumptions, the implementation of the ISM Code should foster and motivate the development of safety culture in navigation – the Author's note.

[106] See more in: R. Krystek (ed.), *Zintegrowany System Bezpieczeństwa Transportu*..., op. cit., p. 332

The adoption of the ISM Code took place after the recognition of the fact that – apart from the design, construction and equipment provided to safer and safer seagoing vessels – the most important means to prevent accidents and sea pollution is proper operation of vessels (most often understood as a combination of factors, such as maintenance of vessels at the level which guarantees their operational serviceability, current operation identified with the capabilities of being maneuvered by a properly qualified crew in dynamic nautical conditions). The ISM Code directly refers to ship-owners and vessels they operate, and also to the cooperation among ship-owner – vessel – crew.

The regulator responsible for safety processes in inland waterway navigation in the EU territory is the European Parliament with its commissions of inland waterway transport (The European Commission Directorate-General for Mobility and Transport/Inland Waterway) and Central Commission for the Navigation of the Rhine (CCNR) specific for the Rhine basin. The most important legal regulations are the following:

- Directive (EU) 2017/2397 of the European Parliament and the European Council of 12 December 2017 on the recognition of professional qualifications in inland navigation and repealing Council Directives 91/672/EEC and 96/50/EC and Directive 96/50/EC on the harmonisation of the conditions for granting national Boatmaster's Certificates for the carriage of goods and passengers by inland waterway in the Community;
- Directive (EU) 2016/1629 of the European Parliament and the European Council of 14th September 2016, stating technical requirements for inland waterway vessels, amending Directive 2009/100/EC and repealing Directive 2006/87/EC;
- Regulation (EU) 2016/1628 of the European Parliament and the European Council of 14th September 2016, on requirements relating to gaseous and particulate pollutant emission limits and type-approval for internal combustion engines for non-road mobile machinery, amending Regulations (EU) No 1024/2012 and (EU) No 167/2013, and amending and repealing Directive 97/68/EC;
- Directive 2005/44/EC on harmonised river information services (RIS) on inland waterways in the Community.

Considering the specific features of both modes of water transport discussed above, it should be also indicated that maritime transport is exposed to much more serious risks, and the scale of transport operations is incomparably larger; hence the tools for risk assessment and systemic solutions in the field of risk management for maritime transport are much more complex than for inland waterway transport. Therefore, this sub-chapter is focused on the aspects referring to the structure of safety management in maritime transport, however, provided that similar (but much less complicated) systemic structures can be applied in inland waterway transport as well.

The term maritime safety management is naturally broader than a definition stated in the ISM Code. Considering its broader meaning, maritime safety management is a constant process of affecting the level of threats which occur at sea in order to maintain the level of such threats, that is the level of risk, below the assumed acceptable level.[107]

The ISM Code includes the most important – but unfortunately not all of them – elements of maritime transport which are related to its safety. Considering the macro-scale, it does not include safety management in the field of linear and nodal infrastructure of maritime transport, which should be implemented by the maritime administration of a coastal country (for inland waterway transport there is a relevant administration body responsible for inland waterway navigation).[108]

The European Maritime Safety Agency (EMSA)[109] has been making attempts at the coordination of operations undertaken by national units of maritime administration for maritime transport safety in the field of their standardisation; however, it is still possible to observe prevailing strong local particularism and national "customary" solutions. The fundamental documentation of the Safety Management System (SMS) is stated in *Vessel Safety Management Manual* and *Shore Safety Management Manual*. The entities responsible for the development of the above-mentioned documents

[107] Cf.: R. Miler, *Bezpieczeństwo transportu morskiego*..., op. cit., pp. 77–78

[108] Activities undertaken by the maritime administration of a coastal country are directly translated into maritime transport safety. Implementation of active and passive means and systems of maritime transport safety (mapping navigation routes, systems for vessel traffic separation, systems of reporting and surveillance of vessel traffic, systems of navigation and navigation marking) comes as a key element of the safety system – the Author's note.

[109] The European Maritime Safety Agency (EMSA) was established after the catastrophy of *the Erika* tanker. The tasks performed by the agency include improving safety at sea, decreasing the number of maritime catastrophes and environment pollution generated by vessels. These aims are implemented in the cooperation with the European Commission and renowned classification associations in the field of, among others, technical requirements which must be met by vessels – the Author's note.

are ship-owners. The approval and issuance of documents which certify the implementation of the system is the responsibility of state administration of the flag country or of institutions authorised by such a country (e.g. classification associations).[110]

Assuming that the above-mentioned particular limitations are to be taken into consideration, the recommendations provide only general definitions and aims.[111] Each ship-owner should develop, implement and maintain a system of safety management which obligatorily should state functional requirements in the field of safety policy, environmental protection and procedures of safe vessel operation, in accordance with relevant legal regulations applied internationally or in the country of the flag.[112] The system should also state the scope of authorisation and communication channels for crew members, onshore and vessel staff, procedures for reporting accidents, emergency operations, internal audits, including reviews of the safety management system.[113]

The required level of maritime safety and navigation security is most often achieved by operation of an integrated System of Maritime Safety and Navigation Security. Figure 2-20 presents the structure of such a system.

The morphology of the components in the System of Maritime Safety and Navigation Security, understood in such a way, is adequate to the tasks; however, the notion referring to the environment of maritime navigation requires some more specification. It involves physical, geographical, hydrological, political and legal elements which are defined by the regulations of

---

[110] See more in: R. Krystek (ed.), *Zintegrowany System Bezpieczeństwa Transportu*, op. cit., pp. 333–334.
[111] Ibidem., pp. 333–334.
[112] Most vessels operated by European ship-owners, including Polish ones, have been registered in the Flag of Convenience (FOC) countries. There countries are: Antigua and Barbuda, Bahamas, Barbados, Belize, Bermudas (UK), Bolivia, Burma/Myanmar, Cambodia, Cayman Islands (UK), Comoros, Cyprus, Equatorial Guinea, German International Ship Register – GIS, Gibraltar (UK), Honduras, Jamaica, Lebanon, Liberia, Malta, Marshall Islands (USA), Mauritius, Netherlands Antilles, Panama, São Tomé and Príncipe, St. Vincent and the Grenadines, Sri Lanka, Tonga, Vanuatu. See more in: *Tanie bandery*, http://mminakowski.republika.pl/tanie_bandery_i_itf.htm, 28.12.2008 [accessed on 17th July 2015] – the Author's note.
[113] Inspections are carried out by maritime administration inspectors (Port State Control – PSC) on vessels under foreign flags to check whether they conform with safety requirements and are serviceable. Similar inspections are carried out by inspectors of the flag country on vessels under their own national flag (Flag State Control – FSC) – the Author's note.

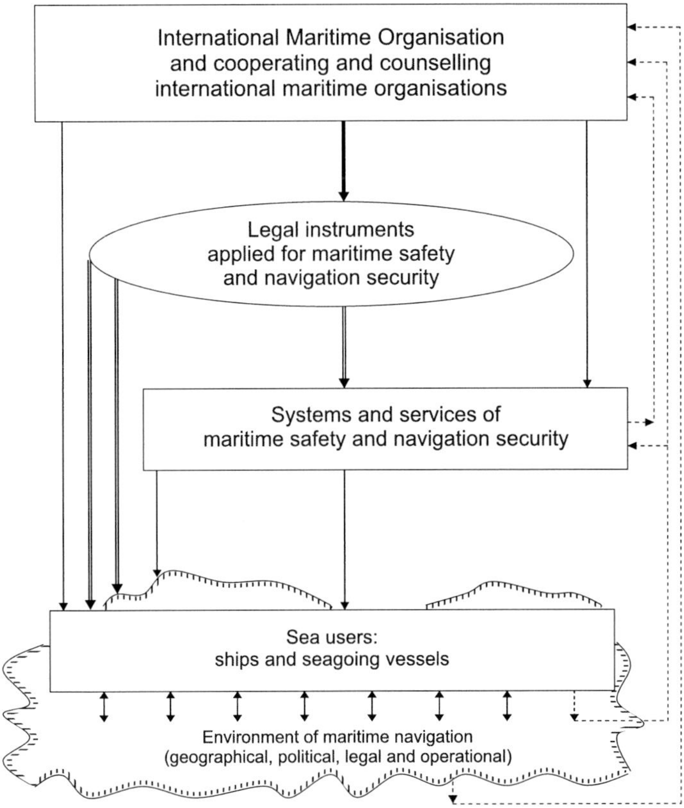

Figure 2-20. The structure of the System of Maritime Safety and Navigation Security

Source: adapted from J. Urbański, W. Morgaś, Z. Kopacz, *Bezpieczeństwo morskie i ochrona żeglugi oraz zarządzanie nimi; próba identyfikacji i specyfikacji przedmiotu…*, p. 20.

the UN Convention on the maritime law and by operational regulations for navigation, that is namely, by a set of current principles and operational procedures which refer to safe navigation stated in the IMO regulations.[114]

Considering the subsequent element, namely maritime systems and services, it should be emphasized that it involves a wide range of systems, starting from maritime navigation systems, such as the IALA, and ending

[114] J. Urbański, W. Morgaś, Z. Kopacz, *Bezpieczeństwo morskie i ochrona żeglugi oraz zarządzanie nimi; próba identyfikacji i specyfikacji przedmiotu…*, op. cit., p. 21

with satellite navigation or communication systems, such as INMARSAT. Services of maritime navigation safety and security are presented (institutionally) in Table 2-4.

Table 2-4. State maritime services responsible for maritime safety/security

No.	Type of maritime services
1	Maritime Administration (Port State Control, Flag State Control, certification, crew qualifications, etc.)
2	Navigation Infrastructure Service
3	Hydrographic Service
4	Maritime Safety Information and Alerts at Sea
5	Search and Rescue Service
6	Service for Response to Maritime Environment Pollution
7	Vessel Traffic Management Systems (Services)
8	Vessel Traffic Monitoring and Information Systems (Services)
9	Services for Technical Supervision and Vessel Inspection (Register of Shipping)
10	Ship Salvage Organisation and others

Source: adapted from J. Urbański, W. Morgaś, Z. Kopacz, *Bezpieczeństwo morskie i ochrona żeglugi oraz zarządzanie nimi; próba identyfikacji i specyfikacji przedmiotu…*, op. cit., p. 17.

Coming as a general concept of all the regional and national security systems which have been actually implemented, the System of Maritime Safety and Navigation Security is a complex organisational and technical system, which is managed at its various levels. Table 2-5 presents the levels of maritime security management, their scope and characteristic attributes.

Table 2-5. Characteristics of the levels of maritime safety management and navigation security

No.	Management level	Scope and field of management	Basic attributes pertaining to the particular level of management
1	International (global) level of management	The whole area of the World Ocean	Defining legal instruments applied for management in the whole area of the World Ocean and providing management with the use of such instruments
2	National (domestic) level of management	State territorial sea areas	It comes as a fundamental level for the management of operational services of maritime safety and navigation security

No.	Management level	Scope and field of management	Basic attributes pertaining to the particular level of management
3	Sub-regional level of management (e.g. the level of management in the area of the Baltic Sea	Sea areas of a particular sub-region (e.g. the Baltic Sea, the Mediterranean Sea, etc.)	The basic aim at this level of management is mainly to increase maritime environment safety, particularly in such sensitive areas as landlocked seas etc., and their protection against pollution generated by vessels
4	Regional level of management (e.g. the level of management in the area of the European Union	Sea areas of a particular region (e.g. sea areas of the European Union, etc.)	The basic aim at this level of management is mainly to increase the level of maritime safety, particularly maritime environment safety, its protection against pollution, and to increase anti-terrorist protection of navigation in a particular region

Source: adapted from J. Urbański, W. Morgaś, Z. Kopacz, *Bezpieczeństwo morskie i ochrona żeglugi oraz zarządzanie nimi; próba identyfikacji i specyfikacji przedmiotu...*, op. cit., p. 25.

In order to function properly, the system of safety management in maritime transport, which has been defined in such a way, must apply adequate systemic instruments and algorithms of risk analysis. As a result, it is possible to state that management of maritime safety and navigation security is characterised by three fundamental features:[115]

- management involves functions of planning, organisation, management, coordination and monitoring with the use of adequate instruments and methods;
- management is implemented with the use of a complex organisational and technical system, such as the System of Maritime Safety and Navigation Security;
- management is implemented at several levels, where the basic levels are international and national ones; however, it is also possible to discuss strategical, tactical and operational levels (in an instrumental approach – the functionality of telematics systems).

To sum up the considerations presented above, it is possible to state that management of maritime transport safety is implemented at three levels.

[115] J. Urbański, W. Morgaś, Z. Kopacz, *Bezpieczeństwo morskie i ochrona żeglugi oraz zarządzanie nimi; próba identyfikacji i specyfikacji przedmiotu...*, op. cit., p. 28.

The leading role in international maritime navigation and its safety processes is played by the IMO. In terms of international requirements, maritime navigation in sea areas of the EU is regulated by the European Maritime Safety Agency (EMSA). The third, national level is regulated by the maritime administration of a coastal country, and inland waterway navigation is regulated by the inland waterway administration (of all the levels).[116] In terms of applicable instruments, the tasks related to management of maritime safety, navigation security and inland waterway transport are implemented with the use of relevant legal regulations and systems applied for acquiring, collecting, processing and transmitting information (related to the fields of safety and security) in a way which provides identification of all (or at least most critical) contemporary threats and risks in maritime transport and inland waterway navigation.

2.3.2. The instruments of the security and safety system and risk analysis in maritime and inland waterway transport

Risk is inseparably connected with all activities undertaken by individuals, groups and societies. A basic method recommended for risk assessment, which results from some general requirements referring to the measurement and assessment of risk factors, is the adoption of a three-level scale, where a determined level of risk is referred to a statistically defined value, so called the volume, which defines acceptable exposure (exposure understood as the level of risk exposure). The general rules of risk assessment (and acceptability) (*Ri*) for such a three-level scale are characterised by the relations presented below:[117]

- (Ri_l) low risk (negligibly low), where $P \leq 0{,}5\ P_{acc}$,
- (Ri_m) medium risk (acceptable), where $0{,}5P_{acc} < P \leq P_{acc}$,
- (Ri_h) high risk (inacceptable), where $P > P_{dop}$.

where:

P – measured or expected value of the volume of risk exposure,

P_{acc} – acceptable value of the volume characterising the risk exposure.

---

[116] Cf.: R. Miler, *Bezpieczeństwo transportu morskiego…*, op. cit., pp. 89–90.

[117] T. Abramowicz-Gerigk, Z. Burciu, P. Kamiński, *Kryteria akceptowalności ryzyka w żegludze morskiej*, Prace Naukowe Politechniki Warszawskiej – Transport (96), Warszawa 2013, p. 9.

Considering their complexity levels, methods and techniques of risk analysis resulting from operational and environmental factors can be divided into three types, namely:[118]

- qualitative methods and techniques applied mainly to identify threats,
- quantitative methods and techniques (tree-based methods and techniques) applied mainly to assess risk (the level of a threat),
- methods of risk analysis of dynamic systems in maritime transport applied to provide risk assessment analysis for unmanned engine rooms and inside vessel security systems, etc.[119]

Considering the above-mentioned statements referring to the ontology and morphology of maritime and inland waterway transport, it should be assumed that risk in water transport may occur in the following areas:

- *safety*:
  - for the means of transport, that is namely, vessels (theories of reliability),
  - for a human factor in the relations among navigators, vessels and their environment (theories on man-technology-environment relations, theories of human reliability),
  - navigation tools (systems) (stochastic theories of navigation system errors),
  - natural environment – waters of the World Ocean and navigable inland waterways – rivers, canals and lakes (statistical and probability theories of risk distribution);
- *security*:
  - for the implementation of the power function of the state (theory of defence sufficiency and the Cline formula),
  - for piracy threats (theories of decision-making based on a decision-making tree),
  - for terrorism threats (anticipation theories),
  - for smuggling threats and others (anticipation theories).

Safety of a vessel (a seagoing vessel and a river barge) during the transportation process (usually understood as the lack of accidents and reliability) has become a subject of numerous research studies and model analyses. One of the measures related to the accident rate is the probability of a vessel failure

[118] Z. Kopacz, W. Morgaś, J. Urbański, *Formalne i nieformalne oceny bezpieczeństwa morskiego*, Zeszyty Naukowe AMW no. 3 (162), Gdynia 2005, p. 54.

[119] See more in: *Guidelines for Formal Safety Assessment (FSO) for use in IMO rule-making process*, MSC/Circ. 1023 and MPEC /Circ. 392, 2002.

or malfunction which comes as a threat to its own safety, to safety of any cooperating units, to safety of natural environment or human life. The occurrence of failures allows interested parties to realise that vessels and their systems (including navigation and telematics systems) can be in various conditions which can be divided according to the criteria listed below:[120]

- considering safety, a vessel having or not having a damage/claim;
- considering reliability, serviceability or unserviceability of a vessel (Figure 2-21).

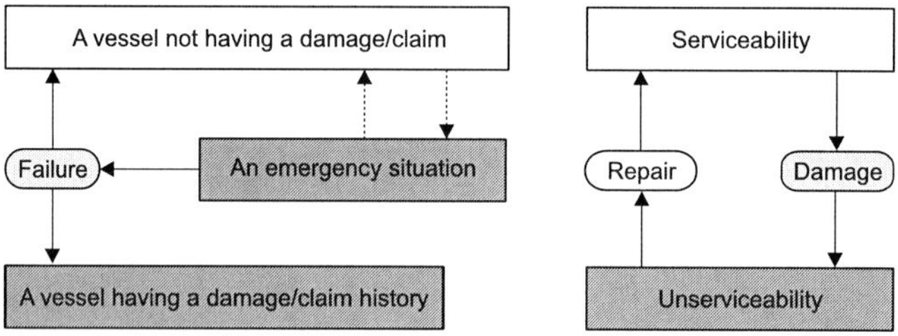

Figure 2-21. Conditions of vessels and vessel systems with regard of systemic safety

Source: adapted from R. Krystek (ed.), *Zintegrowany System Bezpieczeństwa Transportu*..., op. cit., p. 311.

The transition between the particular statuses takes place as a result of a failure or a damage. The status of unserviceability can be reversible, but it can also turn out to be irreversible. It is characterised by the partial or full loss of the capability to function efficiently. Such a situation can be reversed, that is namely, the system can reach its full functioning capabilities after a repair. One of the most significant problems referring to the assessment and acceptability of risk in maritime traffic engineering[121] is a problem which is related to the identification of failure probability in a particular configuration of the determined factors – a man (a navigator equipped with proper knowledge and experience), a water area which determines the range (scope) of the impact of the system (sub-system) and a number of objects (vessels) in a particular area, within the framework of a holistic system of risk assessment. In such a case, we have to consider the following:

[120] Ibidem, p. 310.

[121] Maritime traffic engineering is a branch of general science which refers to the problem of safety in maritime traffic and vessel traffic flows in the restricted areas. The system of maritime traffic engineering may be defined as a purposive system which is operated in stochastic conditions; most findings also refer to inland waterway traffic – the Author's note.

- a micro-system of maritime traffic engineering which refers to one vessel, a water area and external conditions (Figure 2-22);
- a macro-system of maritime traffic engineering which refers to a group of vessels and the external conditions of a water area (Figure 2-23);

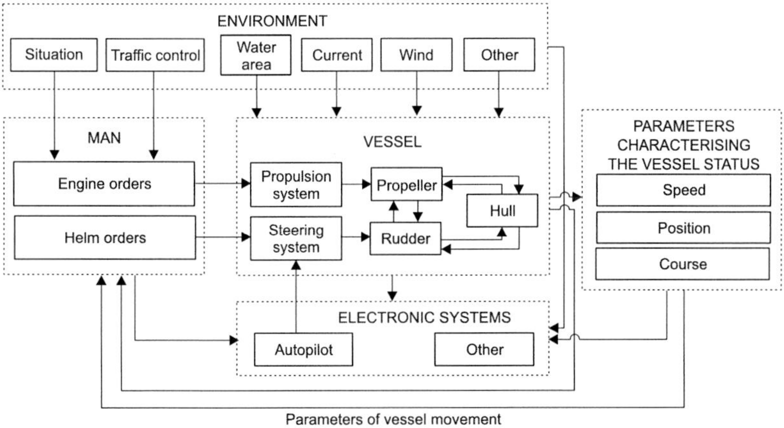

Figure 2-22. A simplified MTE model of a system of maritime traffic engineering (a micro-system)

Source: adapted from R. Krystek (ed.), *Zintegrowany System Bezpieczeństwa Transportu…*, op. cit., p. 312.

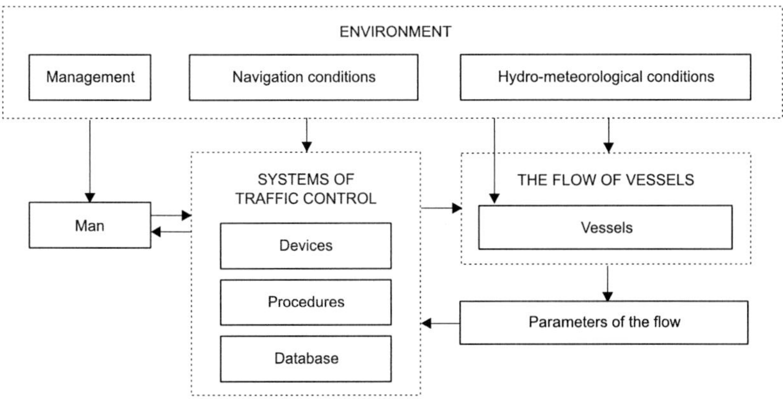

Figure 2-23. A simplified MTE model for many vessels (a macro-system)

Source: adapted from R. Krystek (ed.), *Zintegrowany System Bezpieczeństwa Transportu…*, op. cit., p. 312.

In order to determine the level of navigation safety and the related risk on current and modernised waterways, it is required to assume a number of numeric indicators which will be used as the criteria of navigation safety in the particular area.[122]

Indicators referring to the safety of navigation systems (broadly understood as the AtoN system – Aid to Navigation) include all functional or numerical characteristics which conventionally determine the status of the system safety. These can be probabilistic characteristics of a random variable, such as the time of the functioning of the system since the moment of its transition into the loss status, and also characteristics which refer to the obsolescence of the system, that is namely to the time in which the system can function in the conditions of limited unserviceability.

In statistical reporting, the criteria of navigation safety can be divided into qualitative and quantitative ones. The qualitative criteria of safety allow us to determine causes of accidents, factors threatening navigation safety and means which can improve safety. The criteria can be used as guidelines in accident investigation. The quantitative criteria of navigation safety can be divided into statistical (absolute and relative criteria developed on the basis of data referring to accidents) and probabilistic criteria (calculated as a result of estimation and extrapolation).[123] Figure 2-24 presents the criteria of navigation safety.

In the most recent audits, a parametric assessment of navigation systems comes as a fundamental method of their qualitative classification, because – along with technical development and in response to growing requirements of the navigation process – the comparative criteria of navigation systems have been undergoing a process of reassessment with the simultaneous change in their ranks.[124] The analysis of the criteria allows us to identify

---

[122] At present, in some EU countries there are not any explicit and specific legal regulations on the criteria referring to the safety of vessels manoeuvering in restricted areas; regulations do not state precisely which institution, using which methods, should determine the level of safety on waterways and which criteria should be applied to provide its assessment; at present, various methods are applied, e.g. the Monte Carlo method – the Author's note.

[123] The categories of systemic safety are more broadly discussed in the following publication: R. Krystek (ed.), *Zintegrowany System Bezpieczeństwa Transportu…*, op. cit., pp. 313–315.

[124] C. Specht, *Availability, Reliability and Continuity Model of Differentia GPS Transmission*, 'Annual of Navigation', 2003, No 5, p. 16.

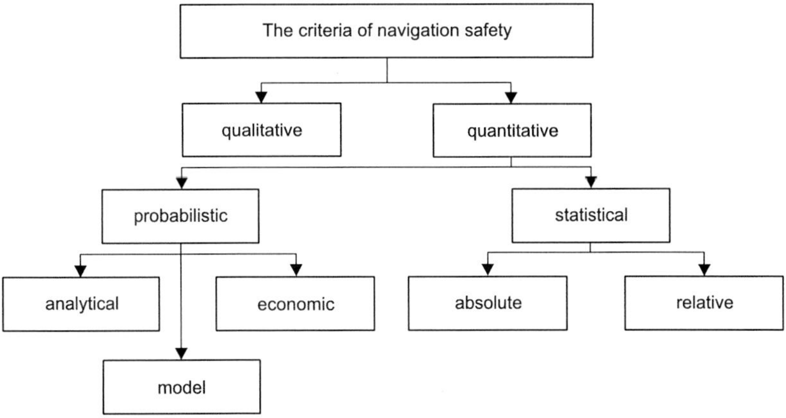

Figure 2-24. Taxonomy of the navigation safety criteria

Source: adapted from R. Krystek (ed.), *Zintegrowany System Bezpieczeństwa Transportu...*, op. cit., p. 313.

three groups which have been corresponding to the particular stages in the development of the system over the recent years:[125]

- position criteria – they characterise the system in terms of the quality referring to the positioning accuracy, with the use of measures pertaining to accuracy, frequency, explicitness and measurability of coordinates, with consideration of the system capacity;
- reliability criteria – include a group of indicators which refer to the operational characteristics of the systems (reliability, accessibility and consistency);
- operational safety criteria – they provide users with current information about the quality of the work performed by the system which is characterised by its reliability (described by such variables as time after which the alarm is initiated, probability of a false alarm, etc.).

An increase in the system integration of navigation processes has brought an idea of e-navigation,[126] which definitely extends the current understanding of the field in which traditional navigation takes place. Figure 2-25 presents

[125] Cz. Dyrcz, *Wpływ informacji nautycznej na bezpieczeństwo morskie*, Zeszyty Naukowe AMW no. 4 (183) Gdynia 2010, pp. 52–76.

[126] IALA Vision of e-navigation, Nordic Institute of Navigation, e-navigation, Oct 16th 2007, Oslo, p. 47.

the environment of e-navigation that incorporates on-board systems, nautical publications, digital charts, satellite positioning systems, integrated systems of traffic management and many others.[127]

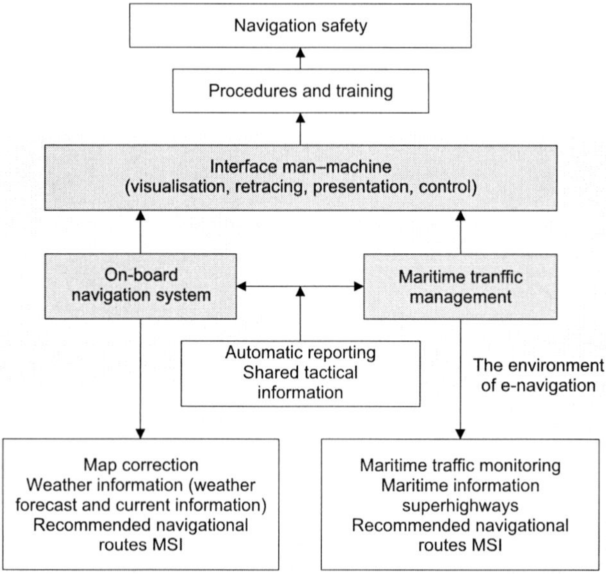

Figure 2-25. The flow of nautical information in the environment of e-navigation

Source: adapted from T. Abramowicz-Gerigk, Z. Burciu, *Systemy bezpieczeństwa i obsługi autostrady morskiej,* Prace Naukowe Politechniki Warszawskiej, Z. 70 "Transport", Warszawa 2009, p. 12.

The identified system is based on broadly understood nautical information[128] which is provided to the maritime safety system along with other information (meteorological reports, SAR reports, etc.).[129] Nautical infor-

[127] T. Abramowicz-Gerigk, Z. Burciu, *Systemy bezpieczeństwa i obsługi autostrady...*, op. cit., p. 12

[128] Nautical information comes as a result of the operations performed by hydrographical services; it takes the form of paper and digital products, nautical publications (reports for sailors, navigation warnings, navigation course books, lists of navigation lights, lists of radio signals, paper and digital maritime navigation charts, specialist newsletters and databases for navigation systems) – the Author's note.

[129] Cz. Dyrcz, *Model służby hydrograficznej na obszarach morskich RP*, Wydawnictwo AMW, Gdynia 2007, p. 53.

mation is now more often referred to as maritime geospatial information, which is related to the development of modern geospatial information systems (GIS). Apart from meteorological data, geospatial information also includes a wide range of oceanographic data.[130] Nautical information significantly affects the provision of conditions which are indispensable for maritime safety through maintaining the parameters of the navigation system, such as its accessibility, reliability, consistency and trustworthiness at the proper level (discussed in detail in Chapter 4 – the Author's note).[131] Considering the model approach, the system of the navigation information flow becomes an ordered set:

$$(S_1, S_2, \ldots, S_n, S, \psi) \tag{2.1}$$

where:

$S_1, S_2, \ldots, S_n$ – sets of the statuses referring to the reliability of elements,
S – the status of the system,
ψ – the function of the system structure which assigns the status of the system to the status of its elements.

In the considerations on the system of the navigation information flow, which has been defined in the above-mentioned way, it is possible to assign its elements and sub-systems to two binary statuses – (B) which are related to their use:[132]

- "0" – means the unserviceability status of an element in the structure or in the navigation system;
- "1" – means the status of its serviceability

Hence, it is possible to provide a following equation:

$$S_1 = \ldots = S_n = S = B = \{0, 1\} \tag{2.2}$$

where:

B – the binary status of the system, on which the level of the parameters depends (accessibility, reliability, consistency and trustworthiness), along with the risk related to the use of the navigation system.

---

130 K. Kubicki, *Rola morskiej informacji geoprzestrzennej dla zabezpieczenia działań morskich*, "Przegląd Hydrograficzny", 2007, no. 3, pp. 67–68.

131 C. Dyrcz, *Model służby hydrograficznej na obszarach morskich RP...*, op. cit., p. 54.

132 C. Specht, *Availability, Reliability and Continuity in Navigation and Hydrography – terminology discussion*, 'Annual of Navigation', 2002, no. 4, p. 98.

Nautical information, its trustworthiness and the reliability of the systems of its application come as the significant elements (sub-components) of efficient solutions in the field of maritime safety assessment. In 2002, the IMO (the MSC and the MEPC) developed and published Formal Safety Assessment – FSA) for the requirements of the decision-making process taking place at the International Maritime Organisation.[133]

The methodology of formal maritime safety assessment (FSA) incorporates five stages (steps):[134]

- identification of types and sources of threats,
- risk analysis,
- options for risk control,
- assessment (the balance) of costs and benefits,
- recommendations for the decision-making process.

Figure 2-26 presents the methodology applied in the process of formal safety assessment (FSA).

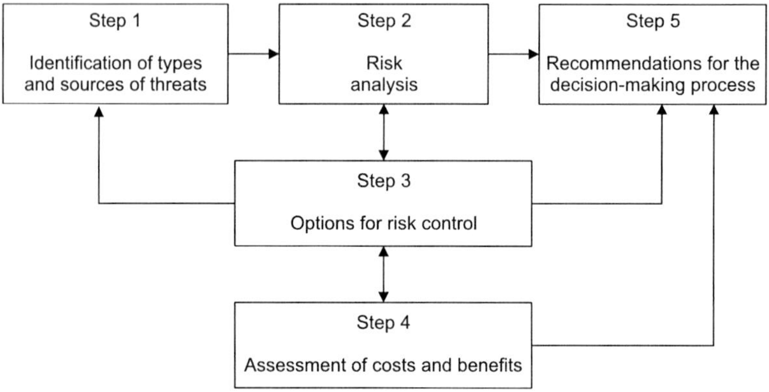

Figure 2-26. A model of the process of formal safety assessment (FSA)

Source: adapted from Z. Kopacz, W. Morgaś, J. Urbański, *Formalne i nieformalne oceny bezpieczeństwa morskiego*, Zeszyty Naukowe AMW no. 3 (162), Gdynia 2005, p. 54.

---

[133] J. Urbański, W. Morgaś, C. Specht, *Bezpieczeństwo morskie – ocena i kontrola ryzyka*, Zeszyty Naukowe AMW no. 2 (173), Gdynia 2008, p. 54.

[134] See more in: *Guidelines for Formal Safety Assessment (FSA) for Use in the IMO Rule – making Process*, MSC/Circ. 1023, MEPC/Circ. 392, April 5, 2002; Z. Kopacz, W. Morgaś, J. Urbański, *Formalne i nieformalne oceny bezpieczeństwa morskiego*, Zeszyty Naukowe AMW no. 3 (162), Gdynia 2005, p. 52.

Taking the above-mentioned methodology into account, Table 2-6 presents identification of the particular stages of the FSA process, tasks implemented in these stages and expectations towards the outcomes of the risk analysis (the equivalent names for the stages of the FSA process applied in the IALA literature are provided in brackets).[135]

Table 2-6. The stages of the FSA process (with the reference to the IALA) and the place of the risk (assessment) analysis

No.	The name of the stage in the FSA process	The tasks implemented during the particular stage	Notes
1	Identification of types and sources of threats (risk identification)	Identification of threats and their probable course or probable scenarios of maritime accidents, their probable causes and consequences	Risk analysis (risk assessment)
2	Risk assessment (quantitative risk assessment)	Indication of risk coefficients, that is the frequency of maritime accidents (F), their consequences (C) and the level of risk (R)	
3	Options for risk control	Indication of ways in which the level of risk can be lowered, i.e. how to control risk	
4	Assessment of costs and benefits	Identification of costs, benefits and cost efficiency for each risk control option; hierarchisation of risk control option based on the above-mentioned operations	
5	Recommendations for the decision-making process	Recommendations for the implementation of the risk control options which have been identified by the experts as the most efficient ones in terms of their operational and economic aspects.	

Source: adapted from J. Urbański, W. Morgaś, C. Specht, *Bezpieczeństwo morskie – ocena i kontrola ryzyka*, Zeszyty Naukowe AMW no. 2 (173), Gdynia 2008, p. 56.

An algorithmic form of the steps in the FSA process is presented in Figure 2-27.

Generally, considering the methodological approach defined in the above-mentioned way, risk management comes down to assessment and indication of an acceptable level of risk (risk assessment). The acceptable level

[135] See more in: IALA NAVGUIDE. 5th Edition, 2006; IALA Recommendation 0-134 on the IALA Risk Management. Tools for Ports and Restricted Waterways, May 2006 [in:] J. Urbański, W. Morgaś, C. Specht, *Bezpieczeństwo morskie – ocena i kontrola ryzyka...*, op. cit., p. 56

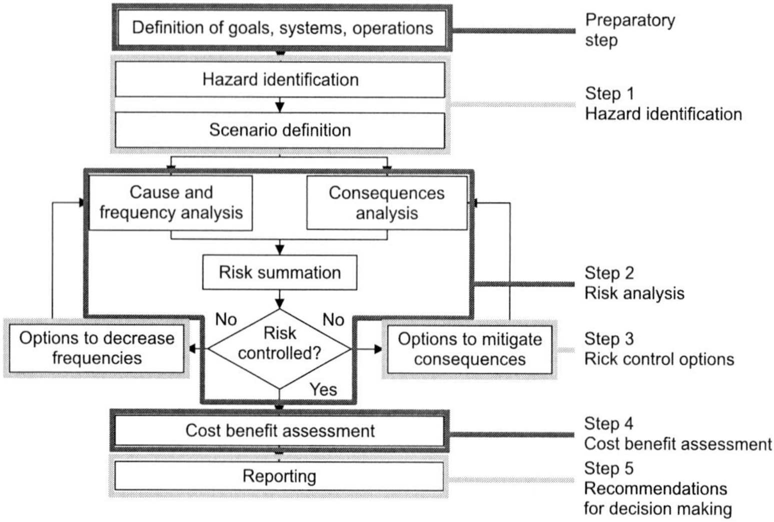

Figure 2-27. The algorithm of proceedings for risk assessment provided with the use of the FSA method

Source: adapted from International Association of Classification Societies, 2002 Document No. W_WP6_4_01 Document Access: *Public Methods to Quantify Maritime Accidents for Risk-based decision-making* Deliverable No. W WP6 4 01 [accessed on 24th January 2012].

of risk, or more precisely the range of risk, is commonly referred to as the ALARP level (As Low As Reasonably Practicable).[136] Table 2-7 presents acceptability criteria for individual risk, as recommended by the IMO.[137]

In the ALARP methodology, the social risk is the risk of a fatal accident for *n* people, where *n* ranges from several to a dozen people. The level of social risk largely depends on the type, size, seaworthiness, technical conditions of a vessel carrying people (its crew and passengers) and also on their number at the time of the accident (obviously, the differences between a commercial vessel and a passenger vessel are very significant in terms of the number of people – the Author's note). Figure 2-28 presents the relation

---

[136] It should be mentioned that the level of risk (R) is expressed also in some other measures, such as the number of Disabled Adjourned Life Years (DALY), or Potential Loss of Live (PLL). However, these are not universal risk indicators – the Author's note.

[137] Cf.: T. Rosquist, R. Tuominen, *Qualification of Formal Safety Assessment. An Explamatory Study*, "Safety Sciences" 2003, pp. 14–15.

Table 2-7. Acceptability criteria for individual risk ALARP, according to the IMO

Decisional parameters		Acceptability criteria	
		The bottom level of acceptability (ALARP)	The top level of acceptability (ALARP)
		The negiligible, that is generally acceptable level of risk, expressed with a number of fatal accidents in a year	The maximal acceptable number of fatal accidents in a year
Individual risk	for crew members	10^{-5}	10^{-3}
	for passengers	10^{-5}	10^{-4}
	for third parties, i.e. people ashore, etc.	10^{-5}	10^{-4}
	target volumes for new vessels	10^{-5}	The above-listed values should be reduced by one rank of their magnitude

Source: adapted from J. Urbański, W. Morgaś, C. Specht, *Bezpieczeństwo morskie – ocena i kontrola ryzyka*, Zeszyty Naukowe AMW no. 2 (173), Gdynia 2008, p. 56.

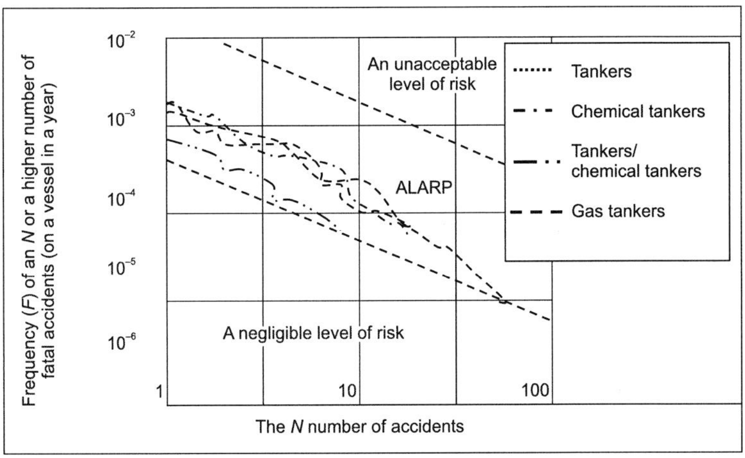

Figure 2-28. The *F*/*N* chart. The relation between the frequency of fatal accidents for *n* people (*F*) and the general number of fatal accidents at sea (*N*)

Source: adapted from J. Urbański, W. Morgaś, C. Specht, *Bezpieczeństwo morskie – ocena i kontrola ryzyka...*, op. cit., p. 56.

between accident frequency in the field of social risk ALARP and a general number of fatal accidents (*F*/*N*).[138]

Table 2-8 presents the research methods and techniques applied for identification of all types of threats, including threats to environment and risk assessment – referring to life and property only.

Table 2-8. Methods and techniques of risk analysis applied in the FSA process

Methods and techniques applied for the analysis of risk caused by operational and environmental conditions and factors	**Methods and techniques applied for the analysis and assessment of the impact exerted by the human factor, i.e. the analysis and assessment of human reliability**
I. Qualitative methods and techniques	1. Techniques applied to predict the frequency of the human error (THERP – Technique for Human Error Rate Prediction) 2. A technique applied to reduce human error assessment (HEART – Human Error Assessment Reduction Technique)
1. A method of preliminary risk assessment (HAZID – Hazard Identification) 2. A method applied for the studies of threats and operability (HAZOP – Hazard and Operability Studies) 3. A method applied for the analysis of failure and effect (FMEA – Failure Mode and Effect Analysis); a research technique (What if)	
II. Quantitative methods and techniques i.e. tree-based methods and techniques	
1. An analysis of an error tree 2. An analysis of an accident tree 3. A method of risk contribution tree 4. A technique applied for the assessment referring to the organisation of safety management (Safety Management Organisation Review Technique)	

Source: adapted from J. Urbański, W. Morgaś, C. Specht, *Bezpieczeństwo morskie – ocena i kontrola ryzyka*, Zeszyty Naukowe AMW no. 2 (173), Gdynia 2008, p. 60.

Apart from the IMO, there is a number of other international maritime organisations which focus their interest and operations on maritime safety. The most significant one among them is the IALA[139] (with regard to the subject of

[138] Proceedings of the 83rd Session of the Maritime Safety Committee (MSC 83(2007)) on the FSA [in:] J. Urbański, W. Morgaś, C. Specht, *Bezpieczeństwo morskie – ocena i kontrola ryzyka…*, p. 56

[139] International Association of Lighthouse Authorities, the abbreviation is the IALA, and its full name is International Association of Marine Aids to Navigation and Lighthouse

maritime navigation safety). In 2005, the IALA published and recommended two methods for risk assessment and risk control. These are namely:[140]

- the PAWSA method (Ports and Waterways Safety Assessment) considered as a qualitative method applied for the analysis, assessment and control of risk at ports, along fairways and navigable inland waterways;
- the IWRAP method (IALA Waterway Risk Assessment Programme) considered as a quantitative method applied for the analysis, assessment and control of risk along fairways and on other restricted water areas.

It is necessary to emphasize one more important problem related to the identification of threats and risk assessment for maritime environment referring to the use of the FSA methodology. It is a universal methodology because it refers to all types of maritime accidents and to all types of seagoing vessels, including tankers. It also presents methods and techniques indispensable for risk assessment, that is namely frequency of the particular types of maritime accidents and their consequences; however its methods and techniques allow us to identify such consequences with great precision with regard to people and property. Applying the FSA methods and techniques, it is impossible to assess the consequences of maritime accidents with regard to natural maritime environment in a highly probable way.[141] Therefore, in order to assess the risk for maritime environment, the IMO recommends application of *the methodology for establishing the criteria of risk acceptability for maritime environment,*[142] which are stated in the Appendix C[143] of the NORSOK

Authorities. It was established in 1957 as an international organisation for association of national organisations responsible for navigation marking, producers, distributors and consultants in the field of navigation aids recommended by the authorities of a particular state. The headquarters of the organisation is based in Saint-Germain-en-Laye in France. The most significant achievement of the IALA was the implementation of almost a uniform international system of maritime navigation aid marking (the A and B system), which has become more and more popular also in inland waters – the Author's note.

[140] See more in: IALA Recommendation 0-134 on the IALA Risk Management. *Tools for Ports and Restricted Waterways*, May 2006; Ports and Waterways Safety Assessment (PAWSA), Workshop *Guide Office of Vessel Traffic Management*, USCG, July 2005 [in:] J. Urbański, W. Morgaś, C. Specht, *Bezpieczeństwo morskie – ocena i kontrola ryzyka*, Zeszyty Naukowe AMW no. 2 (173), Gdynia 2008, p. 55.

[141] R. Miler, *Bezpieczeństwo transportu morskiego…*, op. cit., p. 125

[142] Proceedings of the 83rd Session of the Maritime Safety Committee (MSC 83(2007)) on the FSA – the Author's note.

[143] See more in: NORSOK Standards. *Risk and Emergency Preparedness Analysis*, Z-013, 1998; and in: T. Rosquist, R. Tuominen, *Qualification of Formal Safety Assessment. An Explamatory Study*, "Safety Sciences", 2003; Z. Kopacz, W. Morgaś, J. Urbański, *Formalna*

standards, developed by the Norwegian Office of Technical Standards.[144] The level of risk for maritime environment is often defined on the basis of the time estimated for environment to recover to its normal state. The recovery time depends on the magnitude and type of a spill, hydro-meteorological conditions and capabilities of marine or water ecosystems to achieve homeostasis.[145] Table 2-9 presents recovery time required by maritime environment, according to the NORSOK standards, with reference to the main types of risk.

Table 2-9. Types of maritime environment pollution, according to the NORSOK standards

The magnitude of maritime environment pollution	Recovery time	Acceptable frequency	Acceptable annual value of frequency
Low	½ of a year	< 1 accident in 10 years	0.1
Medium	2 years	< 1 accident in 40 years	$2.5 \cdot 10^{-2}$
High	5 years	< 1 accident in 100 years	$1 \cdot 10^{-2}$
Very high	20 years	< 1 accident in 400 years	$2.5 \cdot 10^{-3}$

Source: adapted from J. Urbański, W. Morgaś, C. Specht, *Bezpieczeństwo morskie – ocena i kontrola ryzyka*, Zeszyty Naukowe AMW no. 2 (173), Gdynia 2008, p. 63.

In the field of security, a seagoing vessel or a barge with its legal and administration system (its ship-owner, maritime/inland operator, insurer), its cargo but also a port, a terminal and navigation infrastructure become subjects of security processes and assessment of risk coming from the sources indicated as potential threats in this aspect of maritime transport security (e.g. piracy, terrorism, cyber-crime). On one hand, such risks come as derivatives of the geopolitical situation in the world, which is mapped onto the pragmatics of processes taking place in global and international maritime transport, and more locally – in inland waterway transport; on the other hand, they are also derivatives of phenomena taking place in particular countries, and they reflect their strength.[146] These are the particular countries which participate

ocena bezpieczeństwa morskiego, AMW, Gdynia 2006; Z. Kopacz, W. Morgaś, J. Urbański, *Próba zilustrowania zasad zastosowania Formalnej Oceny Bezpieczeństwa Morskiego (FSA)*, "Zeszyty Naukowe" AMW, 2006, no. 4.

[144] Z. Kopacz, W. Morgaś, J. Urbański, *Próba zilustrowania zasad zastosowania Formalnej Oceny Bezpieczeństwa Morskiego (FSA)*..., op. cit., p. 23.

[145] See more in: R. Miler, *Bezpieczeństwo transportu morskiego*..., op. cit., p. 145.

[146] R. Miler, *Bezpieczeństwo transportu morskiego*..., op. cit., p. 134

in trade, using international (maritime and inland) waterways and waters of the World Ocean, rivers, channels and lakes for transport.[147]

Therefore, it is possible to assume that at present the key to security of maritime (and inland waterway) economy in particular countries, and at the same time, the key to security of water transport in their territories, is the strength of these countries. Hence, it is commonly expected that countries should strive for reinforcement of their position in order to implement their objectives that would result in maintaining the acceptable level of maritime and inland waterway security and rolling over potential risks. The equation presented below (the Cline formula) perfectly displays the essence of the state power:[148]

$$S_P = (M + E + P) \cdot (S + W_s) \tag{2.3}$$

where:

- S_P – power of the state,
- M – critical mass (demographic factors: population, territory, geopolitical location),
- E – economic potential (the size of economic macro-aggregates, with particular consideration of foreign trade implemented with the use of maritime transport),
- P – military potential (quantitative and qualitative levels of military resources, their allocation and possibilities of deployment, the size of military and industrial facilities, including the potential of the Navy),

[147] Ibidem p. 122.

[148] See more in: P. Sienkiewicz (ed.), *Wystarczalność obronna*, Bellona, Warszawa 1996, pp. 56–57; there are various methods applied for practical measurement of the state power, depending on the extent to which the particular factors have been considered in them. According to one of the models, the GDP represents an economic factor, military expenditure and the number of soldiers on active duty represent a military factor. The military expenditure rate represents a moral factor, that is namely intensity of the will, and a strategic factor, which is the recognition of the role performed by the military factor; a demographic factor is represented by people (population) and a geographical factor is represented by the state surface area. The power of the country is measured in milimirs, assuming that the general world power equals 1 mir. The list of top 10 world powers which was calculated in accordance with the above-mentioned methodology in 2011 starts with the USA – 139,46 milimir, PRC – 119,02; Japan – 43,48 – http://geopolityka.net/wp-content/uploads/2013/07/BG_no._2.pdf *Globalny układ sił ujęcie ilościowe* [in:] Biuletyn Polskiego Towarzystwa Geopolitycznego IV–VI 2013, pp. 6–9 [accessed on 17th July 2015] – the Author's note.

S – strategy (the result of defensive policy, including the strategy of maritime safety and safety of maritime economy of the state),

W_s – the will to implement the strategy (adequate to the b safety indicator and the $1/K$ coefficient, where K is a threat of a military, economic, asymmetric or any other conflict in the domain of maritime economy, $b = W/K$ – an indicator of security taking the value $b > 1$ which means domination, $b = 1$ means balance, $b < 1$ = insufficiency; the $1/K$ indicates the inverse relation of the W_s determination and time which has elapsed since the last conflict).

Understood in such a way, the power of the state (if high) can be an efficient element of discouragement for the above-mentioned "symmetric" threats, because it "naturally" decreases the risk of aggression and non-standard actions undertaken by the neighbouring countries, whereas the same volume of state power can become a "catalyzer" for asymmetric processes (such as terrorism, maritime piracy and cyberattacks against informational and IT infrastructure and telematics systems) of the state. It happens because numerous terrorist, extremist and orthodox organisations by assumption aim their attacks at strong countries which play a key role in the processes taking place in their region or in the world. The USA comes as an excellent example of a country with a super power status which is one of the most frequently selected targets for asymmetric actions. Hence, the level of risk in the field of security which depends on the power of the country will be decreased with the growth of such power with regard to symmetric risk factors, and it will be increased with the growth of that power with regard to asymmetric risk factors.[149] Unfortunately, as it can be observed in history, it is not easy to find an optimum for such conflicting relations. Considering the above-mentioned findings, it is possible to describe the level of risk in the field of maritime security as the following function:

$$R_{security} = f\,(S_P, K_{sym}, K_{asym}{}^{-1}) \tag{2.4}$$

where:

S_P – the power of the state,

K_{sym} – a threat of a conflict based on "symmetric" factors,

K_{asym} – a threat of a conflict based on "asymmetric" factors.

[149] R. Miler, *Bezpieczeństwo transportu morskiego…*, op. cit., p. 124.

To sum up, it should be assumed that the general (aggregated) volume of risk resulting from various types of accidents, failures and incidents on (maritime and inland) waterways and risk factors related to pirate, terrorist or asymmetric threats (R_s) comes as a sum of volumes of partial risk resulting from the factors of safety and security nature. Such a relation may be described with a following equation:

$$R_S = \sum_{i=1} R_i \tag{2.5}$$

where:

R_i – the volume of risk and i^{th} type of threat (all types of threats in the field of safety and security can occur here).

Assumed in the sub-chapter presented above, identification of contemporary threats and risks (along with the methods of their estimation) subsequently allows us to indicate systemic mechanisms implemented in the processes of maritime and inland waterway transport, which are applied to counteract current and potential threats, and which take forms of telematics systems of maritime and inland waterway transport.

2.4. The origin and significance of telematics systems in maritime and inland waterway transport (introduction of transport intelligence)

2.4.1. The systemic approach towards processes of maritime and inland waterway transport

Its introduction to the language of science and its present popularity make the notion of a *system* one of the most significant phenomena in contemporary transformations taking place in science and technology.[150] The application of a systemic approach is a function of some considerable complication of processes and an attempt at replacing the current method of cognitive procedures (referred to as a reductionist approach)[151] which have already stopped being an efficient tool and a scientific pattern to follow because real-

[150] T. Nowakowski, *Niezawodność systemów logistycznych*, Oficyna Wydawnicza Politechniki Wrocławskiej, Wrocław 2011, p. 36.

[151] This way of approaching reality has greatly succeeded in the research on simple objects of low complexity (having a small number of elements and their mutual relations), and it is still applied in the analysis of such objects. The reductionist approach is often referred

ity has become much more difficult in its cognitive aspect.[152] A fundamental principle for the reductionist approach towards reality involves a division of an analysed object or problem into components and the analysis of the attributes pertaining to these parts through some subsequent simplifications, in order to apply the synthesis of the results and to draw conclusions referring to the functioning of the whole object or problem. Contrary to the reductionist approach, the primary principle in the theory of systems[153] involves a holistic approach towards the perceived reality. The systemic approach has provided a possibility of finding a justification for the synergy effect which occurs in the holistic approach and in which *the entirety is more than a sum of particular elements.*[154]

Identified in the theory of systems, the axiomatic and formal line assumes development of an axiomatic theory of systems in which derivative statements[155] result from the smallest number of basic definitions and statements which particularly refer to the notion of a system. In this theory the way in which information is conveyed and the achievement of aims set for the

to as a mechanistic approach. See more in: E. Laszlo, *Systemowy obraz świata*, PIW, Warszawa 1978 – the Author's note.

152 See more in: L. von Bertalanffy, *Ogólna teoria systemów*, PWN, Warszawa 1984.

153 The theory of systems is a hypothetical and the most general "theory of everything" which assumes that the attributes of systemic structures are of a universal character. Developed by Ludwig von Bertalanffy, it originates from biology, and it has been further developed and completed by cyberneticists and social cyberneticists. It has been adopted in engineering as engineering of systems, and it is now being extended onto socio-economic and cognitive sciences, e.g. socio-cognitive engineering. The general theory of systems refers to all the objects which can be recognised as intelligent, e.g. human organisations and complex systems, such as man-technology-environment (MTE) – the Author's note.

154 It is considered that the cognition of our surrounding reality – from general to specific – is a logical order which is proper to interpret the acquired truth. The method of acquiring knowledge about our surrounding reality – from specific to general – is a proper way for the human mind. See more in: T. Nowakowski, *Niezawodność systemów logistycznych…*, pp. 36–37; C. Cempel, *Teoria i inżynieria systemów – zasady i zastosowania myślenia systemowego*, Instytut Technologii Eksploatacji, Radom 2008 – the Author's note.

155 An attempt at developing such a theory is presented by M. Mesarović who applies the theory of sets and defines a system as a set of relations taking place among the features of the system which can be observed. Mesarović assumes that the theory of systems is focused on the explanation of phenomena or conceptual structures in the categories of information processing and the process of decision-making. These problems are also discussed by E. Michlowicz in *Systemy logistyczne a teoria systemów*, 13th Conference of Applied Logistics "Total Logistic Management", TLM 2009, Zakopane 2009 – the Author's note.

system are particularly significant. In order to provide a description of an analysed system, the theory refers to the method of formalisation, in which two stages can be distinguish:[156]

- formulation of a definition describing a system which is consistent with an intuitive meaning of that system in relevant fields of application;
- an axiomatic definition of the notion of a system recognised in such a way, with the use of a minimal mathematic structure.

The advantage of such an approach is a possibility of analysing very complex systems as mutually related sub-systems. It should be emphasized the systemic approach characterises the following transitions:[157]

- from an element to entirety with consideration of the role performed by the element in that entirety;
- from a systemic structure to the processes which take place in the system;
- for objective (absolute) science to epistemic science depending on the reference system that leads to contextual statements;
- from certain knowledge to approximate knowledge which comes as the next stage in approximation of reality;
- from a linear model of cognition and implementation in the following sequence: fundamental sciences – applied sciences – new (process and product) technologies to a network model of interactions taking place at each level with other levels.

The word *system* comes from Ancient Greek (σύστημα *systema – a complex thing*), and at present, it is used in several meanings:[158]

- an internally integrated system of elements, which has a specific structure of operation (e.g. a production system composed of various machine tools);
- a set of operational methods applied to perform complicated activities (e.g. a system applied to design some elements of a vehicle);
- a set of organisational rules, a set of norms, principles and standards which are obligatory in a particular field (e.g. a financial system of an enterprise);

---

156 T. Nowakowski, *Niezawodność systemów logistycznych...*, op. cit., p. 38.

157 C. Cempel, *Teoria i inżynieria systemów – zasady i zastosowania myślenia systemowego*, Instytut Technologii Eksploatacji, Radom 2008, p. 58.

158 J.F. Espiritu, D.W. Coit, U. Prakash, *Component criticality importance measures for the power industry*, "Electric Power Systems Research", no. 77 (2007), p. 47 [in:] T. Nowakowski, *Niezawodność systemów logistycznych*, Oficyna Wydawnicza Politechniki Wrocławskiej, Wrocław 2011, p. 39.

- a holistic and ordered set of tasks which are mutually connected by some specific logical relations (considering such a meaning, a system is each theory which is methodologically correct and which refers to a sufficiently large fragment of reality).

Considering the above-mentioned findings, a system should be understood as a certain entirety which is incorporated into a set of larger entireties and which is composed of parts (which come as smaller entireties possible to be distinguished), related in such a way that gives it a particular structure. It is distinguished from reality with regard to some specific functions that are performed by such an entirety.[159] The above-mentioned distinguishing of a system, which is synonymous with its identification, consists in the consideration of only those components and relations in its structure which undergo analysis[160] because of the purpose (the reason for its existence, structure) of that system. A schematic structure of the notion of a system is presented in Figure 2-29.

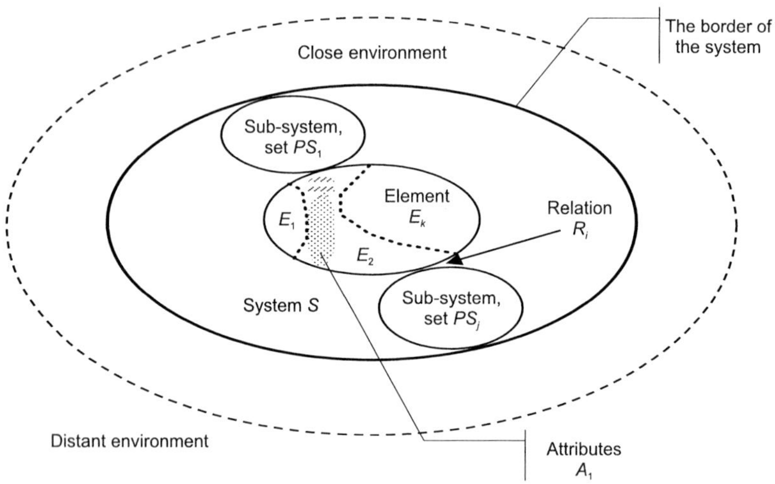

Figure 2-29. A schematic notation of the *system* term

Source: adapted from A. Villemeur, *Reliability, Availability, Maintainability and Safety Assessment*, Wiley and Sons 1992 [in:] T. Nowakowski, *Niezawodność systemów logistycznych…*, op. cit., p. 41.

[159] C. Cempel, *Teoria i inżynieria systemów – zasady i zastosowania myślenia systemowego*, op. cit., p. 59.
[160] S. Oziemski, *Entropia układów logistycznych*, Proceedings of the 10th Conference of Applied Logistics "Total Logistic Management", Zakopane 2006, p. 157.

In the S system (entirety) some sub-systems are distinguished (smaller entireties): S_1, S_2, …, S_j. Each of the S_j sub-systems can be divided further into some subsequent components (parts), until the E_k elements are ultimately defined. An element is a component of a system which cannot undergo any further division into smaller parts. There are some connections between sub-systems and elements, which are referred to as R_i relations. Everything outside the analysed system (distinguished from reality) is in its environment. Depending on the position of the elements in a system, it is possible to distinguish internal elements of a system and boundary elements which are connected with the elements which do not belong to the analysed system (elements of the environment, boundary elements of some other systems); these are input and output elements of the systems (inputs and outputs).[161]

To simplify this definition, it is possible to assume that a system is determined by a set of elements characterised by some specific attributes and relations, which come as a purposeful entirety. If a set of the E elements of the S system has the A attributes (qualities, properties) which are connected by R relations, it is possible to describe a system as the following mathematical equation:[162]

$$S = (E, A, R), E = [E_1, \ldots, E_n], A = [A_1, \ldots, A_{m \cdot n}], R = [R_1, \ldots, R_{r \cdot n-1}] \quad (2.6)$$

According to the (2.6) equation, the system is composed of n elements which can have $m \cdot n$ attributes, and they can be connected by $r \cdot n-1$ relations.[163] As stated in Wintgen's set-theoretic definition, a simpler description of the S system can be applied:[164]

$$S = (E, R) \quad (2.7)$$

where:

$E = [E_i; i = 1, \ldots, I]$ – a set of elements,

$R = [R_j; j = 1, \ldots, J]$ – a set of system-forming relations between the elements.

---

161 T. Nowakowski, *Niezawodność systemów logistycznych…*, op. cit., p. 40.

162 C. Cempel, *Teoria i inżynieria systemów – zasady i zastosowania myślenia systemowego…*, op. cit., p. 61.

163 L. Powierża, *Elementy inżynierii systemów*, Oficyna Wydawnicza Politechniki Warszawskiej, Warszawa 1997, p. 149.

164 G. Wintgen, *System cybernetyczny w świetle teorii mnogości*, "Problemy Organizacji", 2 (1972), p. 21.

For some applications, it would be advisable to refer to a definition, according to which a system is described with three following sets:[165]

$$S = (X, Y, R) \tag{2.8}$$

where:

X – a set of inputs,
Y – a set of outputs,
R – a set of relations between X and Y.

The (2.8) notation is an ultimate, causative description, referred to in literature as a method of input-output description, which is understood as the clarification of Mesarivić's[166] formal description. Another approach suggested in expert literature is the method of description by pursued objectives, also referred to as a teleological description of decision-making. In such a case, it is assumed that:[167]

X – a set of inputs,
Y – a set of outputs,
D – a decisional object,
W – an object of valuation;

and:

$P: X \cdot D \rightarrow Y$ – the function of the result (of a process),
$G: M \cdot Y \rightarrow W$ – the function of the characteristics (of a process) or the function of a purpose.

Hence, the system $S \subset X \cdot Y$ can be defined in the following way:[168]

a $(x, y) \in S$ pair for $x \in X_{iy} \in Y \Leftrightarrow$ when there is $m_x \in M$, in which for each $m \in M$

$$G\big(m_x, P(x, m_x)\big) \leq G\big(m, P(x, m)\big) \tag{2.9}$$

and:

$$y = P(x, m_x) \tag{2.10}$$

[165] L. Powierża, *Elementy inżynierii systemów…*, op. cit., p. 152.
[166] E. Michlowicz, *Systemy logistyczne a teoria systemów…*, op. cit., p. 103.
[167] T. Nowakowski, *Niezawodność systemów logistycznych…*, op. cit., p. 44.
[168] Ibidem, p. 43.

that is:

for each $x \in X$ input the $y \in Y$ output is attributed in such a way that an adequate G function of the characteristics of the process should reach a minimum value, simultaneously complying with the limitations defined by the functions of the *P* result.

In system engineering, a system is treated as a performer of instrumentalised actions which are handled by an operator. Then, a system can be defined as:[169]

$$S = (O, T, E, Z, F) \tag{2.11}$$

where:

$O = [O_i; i = 1, \ldots, I]$ – a set of operators,
$T = [T_k; k = 1, \ldots, K]$ – a set of technical objects,
$E = [E_n; n = 1, \ldots, N]$ – a set of the elements of environment,
$Z = [Z_l; l = 1, \ldots, L]$ – a set of tasks,
$F = [F_j; j = 1, \ldots, J]$ – a set of technologies.

An operator (a user) is understood as a person who handles devices (technical objects) and uses them as instruments for action.[170] Technical objects are indispensable to perform such action. A task is most often understood as a set of activities (procedures) which must be performed in a specified sequence (order) defined according to:[171]

- time,
- methods of implementation,
- types of a technical object,
- other criteria (relevant to a particular system).

The sequence of tasks connected by relations constitutes technology – a particular procedure required to achieve the state which is the purpose of the operation of the particular system.

The most important point in the description of a system is the analysis of relations between its elements (components). These relations define the essence of the identification of the structure and the functioning of the analysed system. The set of relations observed between the elements, which are indispensable for the implementation of the particular objectives by the

[169] L. Powierża, *Elementy inżynierii systemów...*, op. cit., p. 152.
[170] Ibidem, p. 153.
[171] T. Nowakowski, *Niezawodność systemów logistycznych...*, op. cit., p. 41.

system, is referred to as the structure of a system. Formally, the structure can be formulated in the following way:[172]

$$R = [R_j; j = 1, \ldots, J] \qquad (2.12)$$

where:

R_j – a system-forming relation, which can be of a material, energetic, informational or temporal character.

Relations can have one or many parameters. Various relations between elements can form various structures. The type of a structure mainly depends on:[173]

- the number of elements,
- the number of relations between elements,
- diversification of elements,
- types of relations.

It is possible to distinguish numerous structures of a system. Considering the aspect of system reliability, the following structures can be mentioned: serial, parallel, mixed, bridge, threshold or complex types.[174]

Apart from its structure, a system is characterised by its behaviour and attributes. The behaviour of a system refers to the way it responds to stimulants. The attributes of a system are defined as functions defined in the set of system statuses, which reveal similarities in transitions of the system between its statuses. The status of a system is defined by a set of values reached by the variables describing the system in terms of quality and quantity.[175]

The elements of the environment are connected with the elements of a system by relations of various significance for the analysed problem. The part of environment the impact of which cannot be ignored is referred to as close environment. The impact exerted on a system and its behaviour by the part of environment which can be ignored is referred to as distant environment. The actual geometric distance does not define environment as close or distant. It is defined by the strength of the relation with the function of the purpose implemented by the system.[176]

---

172 L. Powierża, *Elementy inżynierii systemów…*, op. cit., p. 153.

173 T. Nowakowski, *Niezawodność systemów logistycznych…*, op. cit., p. 41.

174 L. Powierża, *Elementy inżynierii systemów…*, op. cit., p. 154.

175 Ibidem, p. 154.

176 These problems are discussed in more detail in the following publications: W. Zamojski, *Niezawodność i eksploatacja systemów*, Wyd. PWr., Wrocław 1981 and S. Ziemba,

Hence, identification (description) of a system requires a definition of:[177]

- an operational purpose of a system:
  - the main purpose,
  - the mission,
  - minor purposes which cannot be ignored;
  - the level of significance and importance of such purposes and functions;
- the structure of a system:
  - components of a system (sub-systems, sets, elements),
  - the role and characteristics of components (properties, attributes),
  - relations, connections between components and between attributes and properties of such components,
  - location of the components in the structure of a system;
- the environment of a system:
  - location of other analysed and cooperating systems,
  - conditions of system operation.

To sum up, each analysed system can be described with a few attributes. The most important features of a system are the following:[178]

- extensiveness – a feature which assumes that a system comes as an entirety with strictly defined attributes and behaviour; extensiveness depends neither on the size and the complexity of a system, nor on the distinctiveness of its physical features and the functions of the components;
- ability of a system to be deconstructed – ability of a system to be divided (deconstructed) into smaller, mutually related components (sub-systems, sets and elements) which can be separately analysed (as sub-systems for which the other components of the analysed system become their environment);[179]
- isolation – a possibility of isolating an object from its environment and analysing it as a closed system;[180]

W. Jarominek, R. Staniszewski, *Problemy teorii systemów*, Zakład Narodowy im. Ossolińskich – Wydawnictwo, Wrocław 1980 – the Author's note.

[177] S. Werbińska-Wojciechowska, *Zagadnienie utrzymania gotowości systemu wsparcia logistycznego z zależnością czasową* [in:] Metody utrzymania gotowości systemów, Instytut Technologii Eksploatacji, Radom 2008, p. 234.

[178] M. Jacyna, *Modelowanie i ocena systemów transportowych*, Oficyna Wydawnicza Politechniki Warszawskiej, Warszawa 2009, p. 149.

[179] An action opposite to deconstruction is aggregation of components into one system – the Author's note.

[180] This attribute states that there are not any elements of the system which are common for the system and its environment. The feature is in particular related to the principles of accuracy, consistency, absoluteness and separation. The principle of accuracy requires

- ability of a system to be identified – a possibility of distinguishing each mapped element of a system, with a simultaneous possibility of differentiating, classifying and characterising particular components of a system;
- diversity – each element of a system can be characterised by a unique way of operation and a unique status which is different from the status and way of operation applied by other elements;
- indeterminacy – a feature assuming that it is impossible to define explicitly all the attributes of the components and relations which take place between them; hence the analysis of a system requires an analysis of uncertainty and sensitiveness;
- adaptability – an attribute which enables a system to respond in a required way (achievement of an aim) to changes in the system status and to changes in the environment status;
- interdependency – a feature which depends on the number and the significance of relations taking place between those elements of a system the status of which will trigger a response of other elements if it is changed.

Listed by numerous authors, other elements of a system also include reliability, which is generally understood as an ability to indicate the required and previously defined behaviour in a process.[181]

Considering research studies on organised systems, it is impossible to ignore the cybernetic paradigm. Cybernetics defines a system as a set of elements which cooperate in order to achieve a particular aim. There are two types of systems identified:[182]

---

explicit determination of elements which belong to the system and elements which belong to its environment. The principle of consistency does not allow the elements of the system to change their affiliation (to the system or to its environment) during the analysis. Absoluteness means that all the elements of the system have their specific affiliation. Separation results in the fact that none of the elements can belong to more than one sub-system at the same time – the Author's note.

[181] The problems are discussed in more detail in the following publications: J. Konieczny, *Inżynieria systemów działania*, WNT, Warszawa 1983, J. Migdalski (ed.), *Poradnik niezawodności. Inżynieria niezawodności*, Wydawnictwo Przemysłu Maszynowego "WEMA", Warszawa 1992, W. Zamojski, *Niezawodność i eksploatacja systemów*, Wyd. PWr., Wrocław 1981, F. Grabski, *Semimarkowskie modele niezawodności i eksploatacji*, Polska Akademia Nauk, Badania Systemowe, vol. 30, Warszawa 2002 – the Author's note.

[182] Z. Biniek, *Elementy teorii systemów modelowania i symulacji*, Wydawnictwo Internetowe INFOPLAN, p. 6.

- closed systems (which do not exchange their factors with their environment; hence, they achieve their maximal entropy very fast, vanishing and leaving no further interest in them;
- open systems which are relatively separated from their environment, which exchange their informational, energetic and material resources with their environment; such systems strive to maintain the status of their dynamic equilibrium, despite the fact that the environment intrudes on the system, trying to achieve its maximal entropy – that is namely, to destroy the system.

An open system initiates its protecting mechanisms against the destructive influence of the environment. A system which is able to protect itself against the destructive influence of the environment (entropy) is a system which has achieved its dynamic equilibrium. The achievement of dynamic equilibrium is the aim of the existence of an open system.[183]

Considering its cybernetic approach, a system can be described by the notions of a variable and a status. Sometimes a notion of a variable of the system status is applied. The variable describes a specific attribute of a system which pertains to a defined potential element of the system. A status variable is often related to the function of time, hence, it defines the volatility of the potential of a particular element in a specific period of time (t_0, t). The status of a system is described with a vector, the components of which correspond with the particular values of a particular variable:[184]

$$X = [x_1, x_2, \ldots, x_n] \tag{2.13}$$

and in dynamic systems:

$$x(t) = [x_1(t), x_2(t), \ldots, x_n(t)] \tag{2.14}$$

Hence, in each system which is analysed in a cybernetic way the following can be distinguished:[185]

- the input-output factors that handle the processes of the exchange with the environment,
- a sub-system, referred to as an effector, which is responsible for the influence exerted by a particular system on the environment,

183 S. Oziemski, *Entropia układów logistycznych*, op. cit., p. 160.

184 Z. Biniek, *Elementy teorii systemów modelowania...*, op. cit., p. 7.

185 C. Cempel, *Teoria i inżynieria systemów – zasady i zastosowania myślenia systemowego...*, op. cit., p. 63.

- a sub-system, referred to as a receptor (input), which is responsible for receiving incoming (informational and feeding) stimulants.

Considering the cybernetic approach based on the theory of feedback, the behaviour of a system is distinguished from its functioning. Originating from functionalism, the functioning of a system is analysed with the use of analytical techniques different from systemic analysis, which is focused on studying the behaviour of a system. In cybernetics the following systems are distinguished:[186]

- feedback systems (composed of many elements),
- steerable systems (of purposive behaviour),
- dynamic systems (undergoing changes in time),
- open systems (relatively separated from their environment),
- specifically complex systems (implementing multiplied purposes),
- ultra-stable systems (having an ability to maintain their homeostasis),
- structurally diversified systems.

As a subject of cybernetic analysis, economic systems are much more difficult than technical systems. The processes which take place inside them are far more complicated, and they are expressed by multilateral dependency of the system elements and by the lack of temporal and spatial unity of cause and effect. Interactive and combined dependencies between the elements of a system do not occur openly. Hence, it is difficult to define all the significant factors which affect the final result of a system. Applying cybernetics for the analysis of economic systems allows us to develop a new, effective approach towards the modelling of such systems. Typical economic models developed in the field of econometrics or operational research describe relations between physical parameters (including financial ones) and they are oriented towards the search for explicit relations occurring between such parameters. The cybernetic approach, by assumption, combines a possibility of forming explicit relations between physical parameters with searching for other relations through the construction of an informational and decisional network. The behaviour of a system is explained when a decisional structure, which is based on the acquired and processed information, can be introduced to the model.[187]

Information decreases the level of uncertainty and risk; however, informational processes themselves undergo various disturbances, similarly to material

[186] Z. Biniek, *Elementy teorii systemów modelowania...*, op. cit., p. 8.

[187] Hence, a real structure is modelled, with consideration of the influence exerted by the decision on the course of the actual processes – the Author's note.

processes. Therefore, information processing must also include selection, collection and transmission of information, and the criteria for the selection of information become a derivative of the purpose and tasks of a particular system. Depending on the level of centralisation and decentralisation, there are two categories of information and information flow that can be distinguished:[188]

- information in a centralised system; a decisional administrative level receives information and passes it to other administrative levels without any processing,
- information in a decentralised organisation system: a decisional administrative level receives information and sends it to other administrative levels after processing.

Cybernetic analysis of a system involves a regulative aspect, which is essentially aimed at organising a process of purpose and task correction, depending on the changes in the operational conditions of a system. Figure 2-30 presents the correction of the aim and tasks in a form of a simple feedback system.

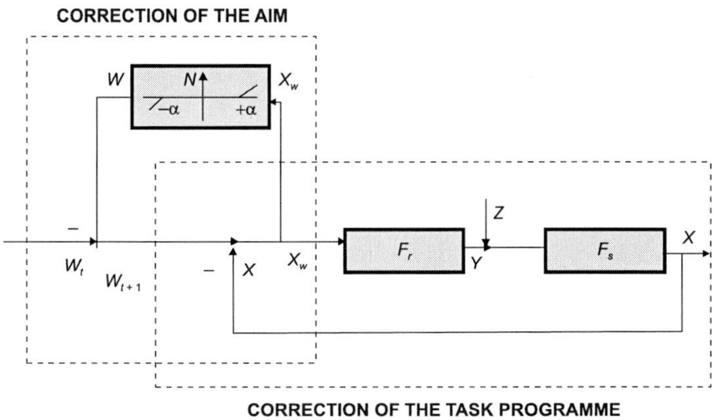

W_t – the magnitude of the aim at intervals (the status: "should be"); X – a regulative variable (the status: "is"); X_w – deviation (a difference between "is" and "should be"); Y – activities; Z – disturbances (deviation elements); F_r – frequency of the regulator (the object of planning); F_s – frequency of the regulated system (the object of planning); N – a non-linear function of description (sensitivity of the object of planning)

Figure 2-30. A regulative system in a cybernetic management system (correlator/homeostat)

Source: adapted from Z. Biniek, *Elementy teorii systemów modelowania...*, op. cit., p. 14.

[188] Z. Biniek, *Elementy teorii systemów modelowania...*, op. cit., p. 13.

The system works in accordance with the assumed sequence of activities; at first, the current status of the object of planning is each time compared with the magnitude of the aim within the system which is defined as a homeostat. In a case where there are some deviations, a decision-making administrative level (correlator/regulator) should make specific decisions (*Y*) that will aim (*X*) at the target (W_t), with consideration of the dispersion of deviations, where the correction of the aim (+ or –) should be made. The process of aim and task programme correction should be correlated (and supported) by the organisational structure of decision-making administrative levels. Decision-making administrative levels should act in accordance with the outlined competence programme, which is usually included in the description of the functional structure.[189]

The behaviour of such a system results from the operation of connected regulative systems which are controlled by positive and negative feedback. The level of system dependency is measured by the number of relations between its elements. According to the interpretation provided by J.W. Forrester,[190] a system is particularly complex if it indicates organisation of the 4th or 5th degree. The degree of systemisation results from the scale of integration referring to the status variables at a particular level of the system description. The level of description depends of an observer, who defines it by setting limits to the analysed level.[191]

Hence, the analysis of a system consists in observation of its transformation process and in the analysis of the behaviour of an imbalanced system. There are three basic types of system response to disturbances:[192]

---

[189] Z. Biniek, *Elementy teorii systemów modelowania…*, op. cit., p. 14.

[190] J.W. Forrester, *Grundzuge einer Systemtheorie*, Gabler, Wiesbaden 1978 [in:] Z. Biniek, *Elementy teorii systemów modelowania…*, op. cit., p. 14.

[191] Cybernetic interpretation juxtaposes a particularly complex system with a simple system; it also juxtaposes the behaviour of a system with its functioning. A system which is simple in terms of its behaviour may be complex in terms of its functioning. The behaviour of a simple system is reflected exclusively by the operation of negative feedback loops which define a regulative mechanism of the system. Hence, it is possible to state that the complexity of a system is not identical with its complication. The level of system complication results from the principles of its functioning, whereas system complexity results from the principles of its behaviour. The functioning is expressed by more technical aspects of systemic dynamics, and it should express the cooperation taking place between the particular elements of the system. The behaviour of a system indicates possible consequences of the cooperation between the elements of the system in time. See more in: L. Kondratowicz, *Modelowanie symulacyjne systemów*, WN-T, Warszawa 1978 – the Author's note.

[192] Z. Biniek, *Elementy teorii systemów modelowania…*, op. cit., p. 23

- a disturbance is compensated by the system, and the system returns to its primary state (a). The system and its balance are stable;
- a disturbance is not compensated, the system pursues its balance for a longer time. The system is stable and its balance (a) is unstable with regard to the disturbance;
- a disturbance is not compensated and the system undergoes a transformation which results in some manoeuvering in the system of balance volatility. The balance is unstable and the stability of the system is maintained only within some limits. The balance can be maintained but only within some other newly established limits, after the full process of transformation has been completed.

Figure 2-31 presents a static interpretation of stability.

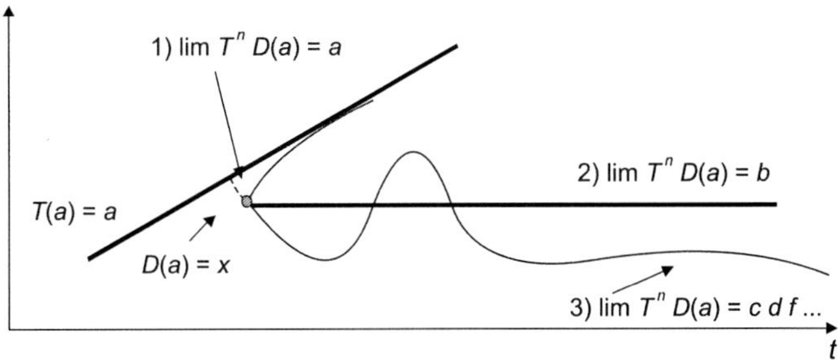

Figure 2-31. The response of a system to the element which disturbs the balance

Source: adapted from Z. Biniek, *Elementy teorii systemów modelowania…*, op. cit., p. 23.

The above-mentioned division indicates the differences between the stability of the balance and the stability of the system.

The achievement of aims assumed by the system requires the implementation of several coordinated actions, that is namely the implementation of a certain process. The notion of a process is defined in several ways, such as the following:[193]

- a phenomenon referring to a fragment of reality and indicating a transformation of the state of this particular fragment of reality;
- sequences of elementary changes in the state of a fragment of reality which take place in time;

[193] L. Powierża, *Elementy inżynierii systemów…*, op. cit., p. 154.

- the sequence of causatively related changes which come as stages or steps of changes in reality;
- the course, development and transformation of something.

In the most general approach, the notion of a process can be defined as some sequences of mutually related states of reality which occur in time:[194]

$$P(t_i) = \left(S(t_i), R(t_i)\right) \tag{2.15}$$

$$S(t_i)=\{S_l(t_i);\ l = 1, \ldots, L\}\ t_p < t_i < t_k \tag{2.16}$$

where:

$S(t_i)$ – the state of the system at the t_i moment,
$S_l(t_i)$ – the momentary value of the S_l variable,
$R(t_i)$ – the relations between $S(t_i)$ and $S(t_{i-1})$, where the state of the system is defined by a set of values which describe the system in quantitative or qualitative terms.

Hence, the analysed systems are assigned with a task of implementing some particular processes which are oriented towards the achievement of a particular aim. During the implementation of the process, it is possible to distinguish some key sub-processes, such as (Figure 2-32):[195]

- working $A(t)$,
- disrupting $Z(t)$,
- supporting $W(t)$,
- controlling $S(t)$,

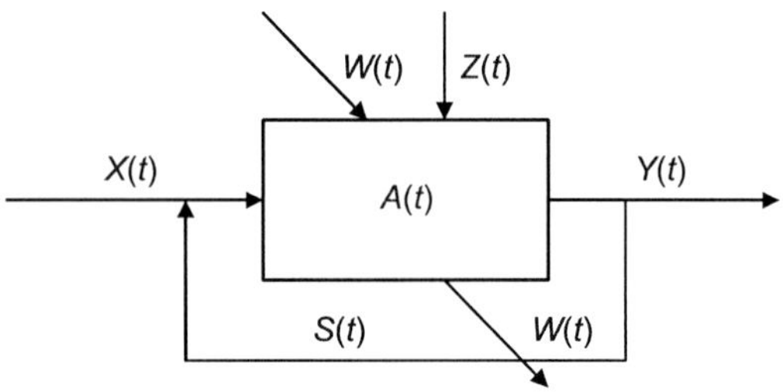

Figure 2-32. Elements of the process (on the example of a technological process)

Source: adapted from T. Nowakowski, *Niezawodność systemów logistycznych…*, op. cit., p. 44; L. Powierża, *Elementy inżynierii systemów…*, op. cit., p. 156.

[194] T. Nowakowski, *Niezawodność systemów logistycznych…*, op. cit., pp. 43–44.
[195] Ibidem, p. 44.

Disrupting processes involve processes generated by the environment of the system which implements a technological process. The destructive impact of disrupting processes makes it impossible for the system to implement that technological process. Such processes disrupt its course at an unacceptable level. In order to limit the destructing impact of the environment, it is necessary to recognise the nature and strength of the impact exerted by such processes and the level of system sensitivity to external extortion.[196]

Supporting processes affect the efficiency and even the possibility of implementing working processes. They involve the following processes:[197]

- acquisition,
- storage,
- preparation, production, processing,
- power supply,
- information supply,
- service,
- establishing contact with the environment,
- recycling (e.g. of waste products),
- disposal (e.g. of waste products),
- prevention of destructive consequences of disrupting processes.

Considering the aspect of a process approach, it is possible to divide systems by the key processes which regulate a system. There are three categories:[198]

- configurational systems – systemising, relational, static;
- operational systems – working, processing, dynamic;
- special systems – among which the most significant ones are the following:
  - a *P* process system, which provides the whole system (a super-system) with a capability to function properly and to achieve its aims;
  - a *C* controlling system, which is applied to control and to coordinate the functioning of the remaining systems which implement the aims of the whole system;
  - a *J* information system, which is applied to organise processing and distribution of information flows inside the system (super-system) as well as outside the system;

[196] L. Powierża, *Elementy inżynierii systemów...*, op. cit., p. 157.

[197] E. Michlowicz, *Systemy logistyczne a teoria systemów*, “Total Logistics Management”, Zakopane 2009, p. 49.

[198] S. Ziemba, W. Jarominek, R. Staniszewski, *Problemy teorii systemów...*, op. cit., p. 112.

- an *L* logistical system, which is applied to maintain and to supply other special systems (and the system itself) in order to provide the proper functioning of the whole system.

Understood in such a way, the approach towards the system and the process is also reflected in telematics systems of water (maritime and inland waterway) transport processes.

2.4.2. A notion of telematics systems in maritime and inland waterway transport

Maintaining the acceptable level of risk in operations of maritime and inland waterway transport is inseparably connected with a dynamic development of the potential and promulgation of information/communication technology in the field of telematics systems of maritime transport, in order to monitor and to provide information supply to intelligent systems of analysis and decision-making. The notion of *telematics*[199] is a term in the field of information technology and, considering the current state-of-the-art, it is mainly applied in the following solutions:[200]

- structural, in which electronic communication, electronic acquisition and processing of information come as integral elements of the system, constructed in accordance with the requirements of that system;
- technical, which use universal telecommunication and information systems in an integrating way.

Therefore, it is possible to assume that telematics refers to telecommunication, informatics and informational solutions and also to solutions applied for automatic control, adjusted to the requirements of physical systems which are handled – resulting from their tasks, infrastructure, organisation, processes related to maintenance and management – and integrated with those systems. The notion of *physical systems* refers to installations which are developed in order to perform particular actions – along with their ad-

---

[199] Some sources state that the ontological precursors of telematics are the notions of telecommunication and informatics (from which teleinformatics and ultimately telematics originate); the term is usually preceded by an adjective or a noun which defines the field of application, e.g. transport telematics, industrial telematics, etc. – the Author's note.

[200] K. Wydro, *Telematyka – znaczenie i definicje terminu*, "Telekomunikacja i Techniki Informacyjne" 1–2/2005, pp. 116–118.

ministration, operators, users and environmental conditions, which include natural, economic, formal and legal environment.[201]

Frequently, attributes of telematics systems, which are understood in the above-mentioned way, are the following:[202]

- a possibility of application in physical systems which are spatially dispersed and which consist of numerous elements;
- a possibility of application in cases in which the role of communication with users and the environment is significant when considered from the perspective of risk management and threat reduction (prevention);
- integration of various types of environment and electronic systems (including cooperation of various devices and software);
- a role of an integrator of (sub-)systems in the superior system (supersystem);
- a possibility of immediate response to changes in operational conditions;
- a possibility of transmitting, collecting and processing large amounts of various data (including the function of monitoring, analysis, anomaly detection, systemic response to threats – effector elements);
- high reliability in terms of user safety;
- a possibility of constant extension through addition of new elements and functions (open architecture).

The most important functions of telematics systems are functions related to information operation. It refers to acquisition, processing, distribution, transmission and application of information in various decision-making processes. These are processes implemented in a determined way (e.g. automatic control) as well as processes resulting from ad hoc discrete situations (decisions made by administrators, operators and also independent users of particular infrastructure). Hence, telematics systems and applications are developed for some specific processes. Another important feature of telematics applications is their ability to combine operation of various sub-systems efficiently and to implement them to a coordinated mode of functioning.[203]

Considering the context of this publication, a particularly interesting example of telematics application is modern transport. In its broadest approach,

---

[201] Ibidem, p. 117.

[202] K. Wydro, *Telematyka – znaczenie i definicje terminu…*, op. cit., p. 118.

[203] A. Bujak, M. Smolarek, A. Orzeł, *Telematic Innovations in Transport Safety* [in:] J. Mikulski (ed.), *Activities of Transport Telematics*, Springer, Berlin Heidelberg 2013, p. 216.

(intelligent) transport is supported by the use of integrated measurement systems (with proper detectors, sensors, etc.) and also telecommunication, information and informational systems with their indispensable automatics. This part of intelligent transport which comes as dedicated measurement, telecommunication, information and informational systems with their equipment in particular (software – relevant IT applications) and services implemented by such equipment, is referred to as transport telematics.[204]

Telematics solutions for transport can be applied to its particular modes (maritime or inland waterway transport) and cover a selected geographical area (e.g. a selected water area used for maritime transport or a particular waterway in inland navigation). However, they can be also used for integration and coordination of a continental or even global system of transport (in the context of this publication, it refers exclusively to solutions applied in global maritime transport).[205] They are usually of open architecture, and they can be extended, completed and modernised according to the requirements (they are scalable). Their fundamental aim is to provide a possibility of cooperation to the particular elements of a system (sub-systems) and interaction with users (stakeholders). It provides a significant increase in transport safety and in reliability of a particular transport mode. It also provides better use of infrastructure; hence, better economic results that can be achieved. Moreover, it makes it possible to limit ecological risk and degradation of natural environment.[206]

In telematics applications for transport, the processes of acquisition, transmission and processing of megadata (an ample set of information) adjusted to the requirements specified by the recipients – users of such information take place. They also involve provision of relevant information to particular places and generation or use of such information (geospatial – GIS) in proper time. Operations related to information supply in telematics applications are implemented automatically (most frequently by continuous monitoring) or interactively on the user's request.[207]

---

[204] K. Wydro, *Telematyka – znaczenie i definicje terminu...*, op. cit., p. 118.

[205] A. Bujak, A. Orzeł, R. Miler, *Telematics of Supply Chain – Areas, Opportunities, Challenges* [in:] J. Mikulski (ed.), *Telematics – Support for Transport*, Springer, Berlin Heidelberg 2014, p. 186.

[206] A. Wójcik, P. Banaś, Z. Pietrzykowski, *Schema of Inference Processes in a Preliminary Identification of Navigational Situation in Maritime Transport* [in:] J. Mikulski (ed.), *Telematics – Support for Transport*, Springer, Berlin Heidelberg 2014, p. 135.

[207] A. Bujak, M. Smolarek, A. Orzeł, *Telematic Innovations in Transport Safety...*, op. cit., p. 216

Monitoring consists in regular acquisition, processing and storage of current information on the analysed subject and its environment, in a systematic way, that is namely in a way that allows interested parties to obtain always updated, systematised and accessible information.

Monitoring comes as a sub-component of intelligent transport which must be efficiently designed to function well. It also requires consistent implementation of procedures which must be followed strictly. First of all, sources, ranges, methods and frequency (cyclicality) of acquiring information must be defined, along with the way of collecting, processing, systematisation and storage of information, the principles for the use of information (access levels and scopes). It is also necessary to define the way and the frequency of reports and their addressees.

Monitoring programmes in transport are most often applied to collect information on the number and quality of transport processes, with consideration of a possibility of using such information to make decisions about optimal economic solutions and to undertake relevant actions in the face of a risk which is too high and which involves human life, property or threats against natural environment. Such programmes also allow interested parties to undertake recovery actions or to remove consequences which have already occurred as a result of a congestion, an accident (failure) or an ecological disaster.[208]

Maritime transport (as a key element of the global economic system) and inland waterway transport should undergo analytical procedures which can increase the level of their safety, and in the same way, to increase the potential referring to further optimisation of transport processes and their derivatives (cargo holding operations, storage, etc.). The analytical and decisional process is impossible without relevant systems applied to acquire information about navigation, port operations and the environment[209] (generating, collecting and passing information), without analytical, decisional (processing information with the use of advanced intelligent algorithms for decision making) and executive elements (passing managerial information – a decision which

[208] A. Bujak, P. Zając, *Monitoring of Cargo in Logistic Systems of Transport and Storage* [in:] J. Mikulski (ed.), *Activities of Transport Telematics*, Springer, Berlin Heidelberg 2013, p. 362.

[209] Organisation of a measuring system and methods applied in bio-identification research might result from the European ICP/IM programme (Integrated Monitoring International Co-operative Programme on Integrated Monitoring on Air Pollution Effects) – the Author's note.

must be executed by the subordinated units of the system). It corresponds to the definition of monitoring presented above. Understood in such a way, a telematics system should consist of at least three modules:[210]

- a sensory module – to acquire information from accessible sources (e.g. from AIS, LRIT, SafeSeaNet, river monitoring and similar systems);
- an analytical and decisional module – to eliminate informational entropy and to process information into a decision (based on intelligent decisional algorithms, such as CPM, PERT, payoff matrices, game theory, data mining, smart agents and others with triggers installed in software which is to support analytical processes);
- an effector module – to pass a decision (best in an automatic way) for immediate execution (with consideration of preventive and optimising functions).

The important morphological elements of water transport (identified in the previous chapters) are sea and river ports. Entities operating in the area of sea and river ports are related to broadly understood logistical services provided to maritime and inland waterway transport processes. Such entities perform their operations in the regime of full cost efficiency. It is strictly related to relevant timing and sequences of particular operations (e.g. unloading, deconsolidation, customs procedures, short-term storage, the choice of transport means, road carrier, forwarding and insurance, etc.).[211] Figure 2-33 presents a concept of a land-sea logistical chain with consideration of operational, time and informational regimes.

In order to acquire all the required information in the sequence of logistical operations, as described above, that would allow interested parties to optimise the process in terms of time and safety, it is necessary to:[212]

- acquire information about the current position of a vessel and its estimated time of arrival (monitoring systems of maritime and inland waterway navigation);

[210] R. Miler, *Systemy zobrazowania żeglugi i aplikacje informatyczne transportu morskiego*,[in:] J. Dąbrowski (ed.), *Wyzwania i problemy transportu morskiego*, Fundacja Rozwoju Uniwersytetu Gdańskiego, Gdańsk 2011, p. 210.

[211] See more in: A.S. Grzelakowski, *Model zarządzania polskimi portami morskimi jako element systemu regulacji działalności sektora portowego* [in:] *Kierunki rozwoju polskiego transportu morskiego – programy i rzeczywistość*, H. Klimek, J. Dąbrowski (ed.), Fundacja Rozwoju Uniwersytetu Gdańskiego, Gdańsk 2010, pp. 12–17.

[212] R. Miler, *Systemy zobrazowania żeglugi i aplikacje informatyczne transportu morskiego…*, op. cit., p. 214.

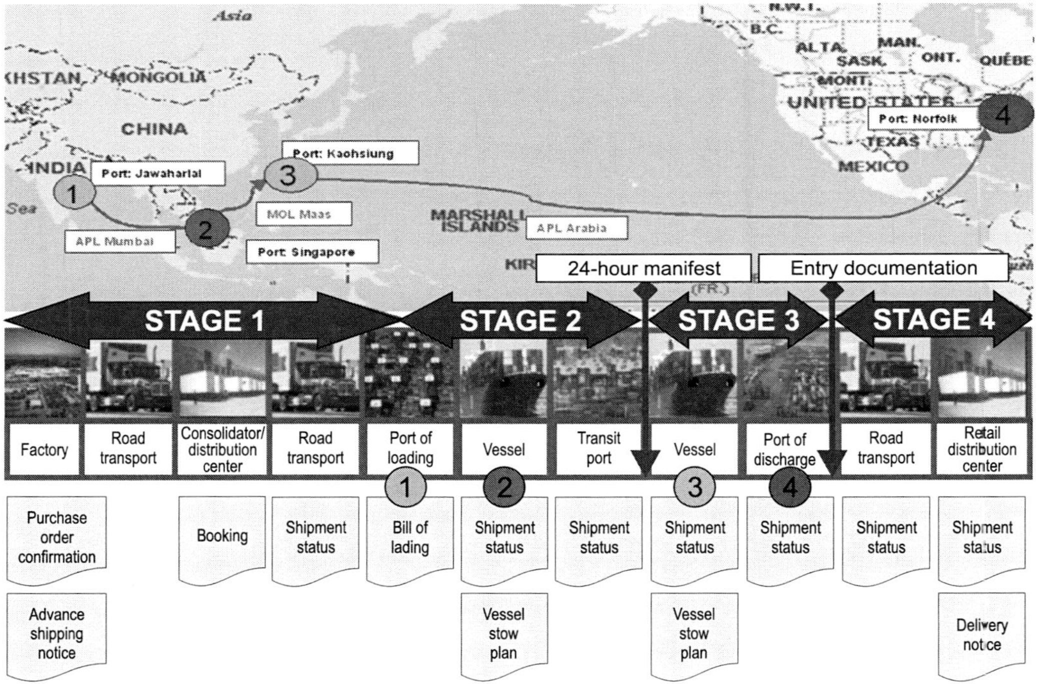

Figure 2-33. A concept of a land-sea logistical chain

Source: adapted from materials provided by NSC Northwood. [c]

- have a possibility of monitoring the route of the selected vessel in order to determine its actual time of arrival;
- acquire information about the cargo and cargo handling operations (e.g. the number of containers to be loaded/unloaded, systems of load capacity management, systems of cargo monitoring);
- have an IT tool to support terminal management and logistical operation of main processes (Terminal Operation Systems – TOS).

An additional stream of information, which would allow interested parties to optimise and to improve safety in the processes of maritime and inland waterway transport, should also consider threats against security. Therefore, information provided by the Border Guard (e.g. FRONTEX) becomes indispensable as well as information provided by the National Defence (informational and IT systems of the Navy and the NATO naval forces, e.g. the MCCIS) and by customs and financial services. Considering a large variety of such systems, their operational diversification and a various scope of application, it is necessary to identify the criteria and to present an original taxonomical division of the current systems operated in the field of telematics and monitoring of maritime and inland waterway security.

CHAPTER 3

TELEMATICS SYSTEMS IN INTELLIGENT MARITIME TRANSPORT

3.1. Classification of telematics and shipping monitoring systems in maritime transport

3.1.1. Tasks of telematics and shipping monitoring systems in maritime transport and their morphological elements with the focus on innovative technical solutions (automation and autonomy challenges)

Contemporary lines of maritime transport development are focused on spatial and temporal integration of all its stakeholders (seaports and navigation) on the regional and international (global) scale. However, the extent and the level of integrating solutions which have been implemented are not the same everywhere, considering the structural, technical and technological, legal and administrative, financial and organisational differentiation. It causes large regional disproportions and comes as a barrier to the development of integrated telematics systems dedicated to maritime transport. Considering its specificity, the implementation of various innovations into maritime transport mainly refers to design and construction of new transportation means (seagoing vessels) and to automation of cargo handling processes taking place at seaports. However, it is possible to observe that a trend of using the IT/ICT technologies and automation of processes related to cargo handling services been contributing to the development of numerous innovative solutions which are applied in maritime transport. They are based on telematics systems which go beyond the above-mentioned framework of technical and constructional innovations, reaching the level of integrated processes in transport management.[1]

As it has been indicated in the previous chapters, maritime transport comes as a significant element of the global economic system; therefore, it should undergo the procedures of monitoring, tracking and analysis in order to increase its safety/security. Information about such a broad field of activities (navigation and port operations) is acquired with the use of sensor modules; information processing is provided with the application of intelligent, advanced decision-making algorithms; decisions which are going to be ultimately implemented by sub-systems and by particular stakeholders are submitted with the use of effector modules. They all function within the framework of telematics systems dedicated to maritime transport.

The outcomes of the implementation of telematics systems in maritime transport are, first of all, of qualitative nature, and considering a holistic approach, they contribute to:

- an improvement in maritime transport safety and security,
- a decrease in energy consumption and anthropopressure of maritime transport (decreased fuel consumption and GHG emission),
- an improvement in the parameters of maritime transport economics,
- an increase in operational efficiency and productivity of seaport terminals,
- an improvement in the accessibility of maritime transport and the participation of this mode of transport in logistical supply chains,
- greater integration of maritime transport in intermodal and multimodal shipment.

Telematics systems, and first of all, information systems find their application in maritime transport in the field of:

- acquiring, processing, displaying and transmitting information (more and more often the access to information in real time becomes a stan-

[1] Incorporating maritime transport into integrated transport chains requires telematics solutions in order to optimise and to streamline processes related to transportation of passengers and cargo. Through the combination of IT and technical tools, such systems should efficiently use the existing linear and nodal infrastructure that is available at a particular moment and place. Hence, telematics systems are solutions which provide access to information resources indispensable for planning, implementing and having control over processes of transport and port operations (that is namely implementation of the main attributes of management) – the Author's note.

dard, e.g. data in the cloud, IoT – Internet of Things, blockchain technologies[2]);
- monitoring current positions of seagoing vessels and setting their optimal routes (including navigation in congestion conditions);
- providing support to maritime transport management and its infrastructure;
- providing support to the management of drive unit parameters (particularly in the context of decreased emission and external costs).

The key to the efficient functioning of the maritime transport system is finding a possibility of a systematic increase in its reliability. The particular elements of that system, such as maritime navigation, port operations, storage or order processing and handling, function on the basis of the most advanced technologies (including telematics) which are available to the particular entrepreneurs who form a logistical supply chain.[3] Apart from quite obvious benefits which result from streamlined supply chains, it is also possible to expect that they will contribute to the improvement in security of transport and its environment as well as to sustainable development.[4] The place of transport (including maritime transport which in the future will be based on autonomous vessels) in a new industrial and logistical ecosystem in the future is schematically presented in Figure 3-1.

Hence, considering such an understanding of the future essence of maritime transport, everything indicates that the competitiveness of its components and possibilities of their further development will be decided by only two groups of factors:
- digitalisation,
- protection of natural environment.

---

[2] Blockchain (or block chain) is a decentralised and dispersed database in an open source model in the Internet. It has a peer-to-peer (P2P) architecture, without any central computers and any central place dedicated for data storage. It is used for recording particular transactions, payments or ledger records. It is encoded with the use of cryptographic algorithms. In fact, blockchain is a specific decentralised and dispersed registry of transactions, or in other words, it is a decentralised platform for transactions in a dispersed network infrastructure. Most often, blockchain is a public and open register that is accessible for everyone. – the Author's note.

[3] Hence, the reserves of potential should be sought between the particular links of the chain. It is an area where intelligent transport systems (ITS) and intelligent logistics concepts can be applied – the Author's note.

[4] *The future of freight transport: integration with ITS and intelligent logistics*, www.connect.nl [accessed on 13th July 2016].

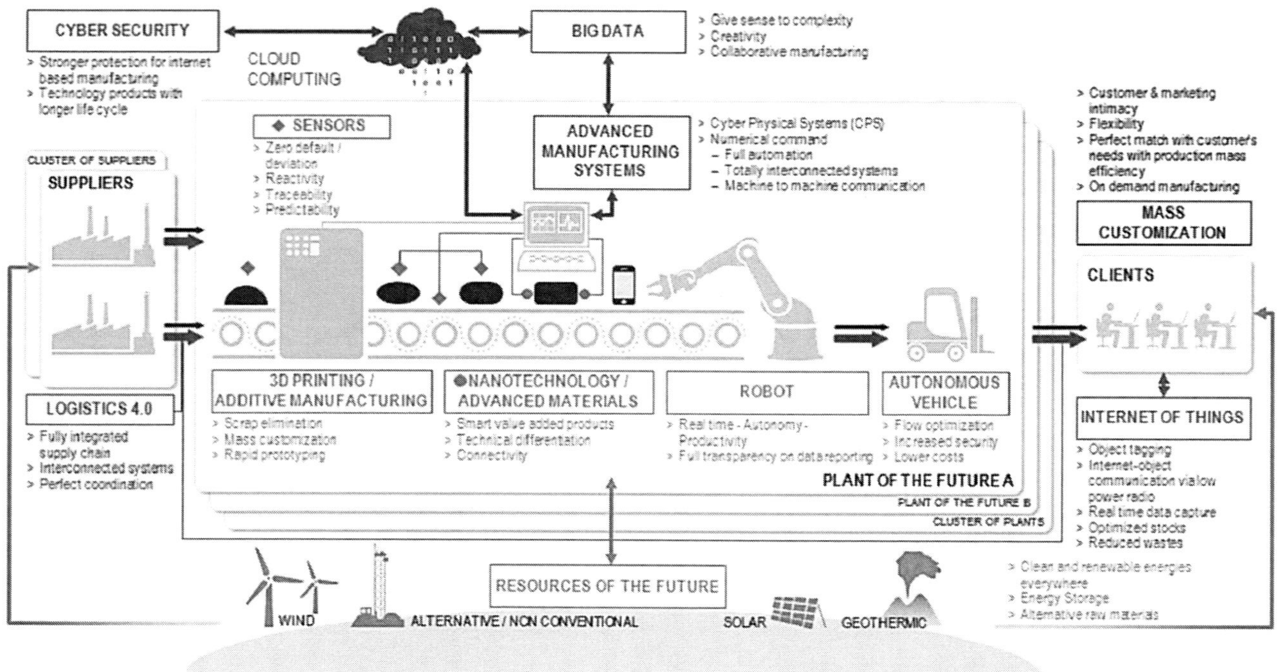

Figure 3-1. The industry (Logistics) 4.0 ecosystem

Source: adapted from https://supplychaingamechanger.com/industry-4-0-ecosystem-infographic/ [accessed on 23th May 2018].

Today, it is impossible to maintain a proper level of competitiveness in the world without innovative tools applied for the exchange of digitalised information indispensable for formation of efficient and safe supply chains. Protection of natural environment and sustainable development are an inseparable and multi-dimensional outcome of their application, regardless of green technologies which are specially dedicated to the sector of maritime transport.

One of telematics systems indicating a high level of aggregation (integration of a logistical chain which is also composed of the sub-systems of maritime transport) is the Port Community System (PCS).[5] The PSC is a tool which has been applied for over a dozen years at various ports in the world in order to improve their position in global maritime transport. It is a modern, neutral and safe platform which is not dominated by any other systems, but it is compatible with them, providing a proper, common and integrated format of information which flows smoothly to the selected recipients (the EDI – Electronic Data Interchange – standard is applied in logistical processes). Seaports also use the processes of automation in more and more applications. In Hamburg, an automated container terminal Altenwerder (CTA) has been operated for many years. It implements cargo handling operations without any direct participation of manpower. Operations carried out at the terminal are implemented with the use of driverless AGVs – Automated Guided Vehicles and double Rail Mounted Gantries – DRMGs. The International Container Terminal Services (ICTSI), a Philippine port group, has opened an autonomous container terminal Victoria in Melbourne in Australia. Having handled over 40 million TEU in 2017, the largest port of the world, Shanghai, opened a fully autonomous terminal, Yangshan Deep Water Port (YDWP) of an annual capacity of 6.3 million TEU.[6] More and more often digitalisation becomes involved into operational aspects of the functioning of maritime transport, particularly in the field of interactions between operators of maritime transport and their

---

[5] It functions at 33 ports in the world (as for the end of 2017). In Europe, in Rotterdam, it is referred to as *Portbase,* in Hamburg as *Dakosy,* in Antwerp it is the *Port of Antwerp* system, in Bremen and Bremerhaven it is a common *dbh* system. In fact, the authorities of all large seaports are thinking about the implementation of the PCS – the Author's note.

[6] It allows the interested parties to decrease human labour (and to lower the risk of error in the MTE relations), to lower costs and negative influence exerted by cargo handling operations on natural environment in a considerable way. In the case of YDWP, it is possible to lower the CO_2 emission by 10% – the Author's note.

customers (or more broadly: stakeholders of logistical supply chains), e.g. in the form of the blockchain technology.[7]

Apart from the above-mentioned processes, there have been innovative projects simultaneously implemented in the field of maritime transport telematics for protection of natural environment and autonomous transport (autonomous seagoing vessels with an electric drive). Maritime transport has been changing under the influence of the IT/ICT technologies. It can be more and more often observed that the value (and unreliability) of the human factor has been decreased through the use of augmented reality (AR), virtual reality (VR) and artificial intelligence (AI).[8]

The introduction of autonomous vessels is a solution which becomes a step into the future for the maritime sector.[9] A Norwegian company, Yara, which

---

[7] In 2018 the largest container ship-owner in the world, Maersk of Denmark, in cooperation with IBM, started work on the development of a system based on the blockchain technology which would enable the company and its customers to exchange all the information and documents (without their paper version) which are indispensable in the whole logistical chain in the real time. The system is going to be operational in 2020, and it is expected that it will generate savings at the level of USD 2 billion. The French ship-owner CMA CGM in cooperation with the largest e-commerce platform in the world, the Chinese Alibaba, has started some actions which are supposed to make it possible for the goods purchased by the Internet from One Touch to go directly to its container vessels operating between Asia and the Mediterranean Sea, omitting forwarding agents who have so far participated in that supply chain. It will shorten the time of transactions and lower their costs – the Author's note.

[8] http://www.gospodarkamorska.pl/Stocznie,Offshore/statki-autonomiczne-rozwiazaniem-przyszlosci.html [accessed on 11th May 2018].

[9] The idea of introducing autonomous vessels proves to have some massive potential. Additional reduction of costs related to vessel maintenance would emphasize the advantage of such vessels even more. It would significantly improve the competitiveness of short sea transport, the biggest competitors of which are rail and road modes of transport. The project of autonomous vessels fits perfectly into the rhetoric on the automation of the supply chain, and it should be developed as a future solution. However, it brings as many advantages as hazards. Therefore, they should be minimised by best possible endeavours. Economic values are undeniable; however, navigation of autonomous vessels at sea may turn out to be problematic. The COLReg would have to undergo some fundamental changes to make ship officers and software developers know how their vessels should respond to each other. The ARPA radar systems are able to inform navigators how to avoid a collision due to the use of calculating algorithms. Until now, people have been making the ultimate decisions about a change of the course, considering not only the radar observations but also all the external factors which can affect the movement of the vessel. However, the risk involved in navigation of autonomous vessels, which is based exclusively on radars, seems

produces artificial fertilisers, and a producer of technological and maritime telematics solutions, Kongsberg, have joined their efforts in the implementation of an exceptional project: a feeder line from a factory to the local port will be provided and handled in a fully autonomous way as soon as in 2021. The distance between a factory in Porsgrunn and a port in Larvik is 14 km by sea and 26 km by land (including an inland waterway). The first autonomous vessel, *Yara Birkenfeld*, a small container ship of the capacity of 150 TEU, equipped with an electric drive, will enter into service most probably at the end of 2020. Figure 3-2 presents a design/concept of the very first autonomous unmanned feeder vessel.

Figure 3-2. A project of an autonomous vessel, Yara Birkenland, of an electric drive

Source: adapted from https://www.km.kongsberg.com [accessed on 15th March 2018].

---

to be considerably high. Maritime administration should consider such possible situations, and it should develop relevant regulations of the maritime law and regulations on maritime traffic. Joining the commercial fleet, autonomous vessels will become a massive challenge for classification associations which will have to provide new safety and technological standards. The minimal levels of safety are defined by contemporary vessels. In order to enter into service, autonomous vessels will have to be as safe as those which operate at sea at present. The IMO precisely defines regulations referring to the equipment and reliability of vessels. Apart from the aspects related to navigation safety and technology, autonomous vessels will have to face the problem of the Internet piracy. It will be the key problem for the producers of autonomous vessels to secure all the access to communication channels in order to provide economic and ecological security. Apart from technological development, it will be necessary to provide relevant standards and requirements in the field related to management of maritime transport safety – the Author's note.

The concept of an autonomous vessel assumes that through the use of advanced cameras and microphones, it will be possible to replace work of navigational watch officers, enhancing it additionally with work of infrared systems in order to obtain still better outcomes. The support provided to autonomous interpretation of data acquired from the equipment which has been so far operated by people, such as sonars, radars and LIDARs,[10] will be provided to the vessel only in the form of some occasional verification remotely done by onshore operators (operating/management centres). Satellite communication and GPS systems will allow the interested parties to solve communication issues with management centres, mutual location and communication between vessels in a fleet. The lack of crew will contribute to reduction of costs, because the infrastructure and rooms indispensable for people will be eliminated. There will be more space for cargo, and it will be translated into higher profits gained on the operation. Another advantage will be optimal adjustment of vessel construction to the loading/unloading procedures. As it has been estimated in the *Yara Birkenfeld* project, as a result of launching this new vessel, the distance between Porsgrunn and Larvik will be covered by 40 000 fewer trucks (tractors and container semi-trailers) annually.[11]

Other companies have already followed Yara and Kongsberg to join their efforts in developing projects that involve construction of autonomous vessels. MOL Line and Rolls-Royce, Deltamarin, Inmarsat, DNC GL, NAPA have defined the following steps in the development of unmanned vessels (which correspond to five stages of autonomation – starting from operator's control – stage 1, to a fully autonomous unit – stage 5):

- 2020 – a decreased number of crew members due to the implementation of remote assistance,
- 2025 – remote control, unmanned vessels in coastal navigation,
- 2030 – remote control, unmanned seagoing vessels,
- 2035 – fully autonomous unmanned seagoing vessels.

---

[10] LIDAR, Light Detection and Ranging – a device the operation of which is based on a similar principle of radar operation; however, it uses light instead of microwaves. The device is characterised by high resolution, and it is used in the procedures of numeric modelling of the terrain to provide a numeric discrete (nodal) display of topographic features of an area, along with an interpolation algorithm (also in 3D technology) – the Author's note.

[11] https://www.chip.pl/2017/05/autonomiczny-statek-towarowy-zasilany-energia-elektryczna/ [accessed on 13th May 2018].

One of the most advanced projects in the discussed field is a concept of an autonomous vessel developed by a Japanese container navigation company, MOL Line, and a British concern, Rolls-Royce Marine. The core of the project is artificial intelligence, a solution developed by Rolls-Royce, named the Intelligent Awareness System (IAS). It has been just installed on a MOL ferry which will soon start its sea trials.[12] During the first stage, the IAS was equipped with a set of sensors supporting automatic navigation which enabled the system to read radar measurements, including the position of the vessel and collision courses of other vessels and navigational obstacles in the route of the vessel. Figure 3-3 presents data acquired in this mode.

Figure 3-3. Display of navigational data in the IAS system

Source: Rolls Royce materials available at https://www.google.pl/search?q=Intelligent+Awareness+System+(IAS)&tbm [accessed on 24th March 2018].

The data are analysed by the central computer of the system, and they form a basis for decisions still made by a navigational watch officer, but in the future – by an autonomous and intelligent vessel control system. Soon, the system will be equipped with modules acquiring other data from the vessel board as well as from its environment, e.g. data about the weather, water currents or any

[12] Practical operation of the IAS ferry will be soon tested in the water area of Seto Naikai, that is in the internal sea, among the islands of Honsiu, Kiusiu and Sikoku. The area is one of the most congested sea areas in the world, sic! – the Author's note.

other conditions which determine maritime navigation. Ultimately, the system will provide full autonomy to the vessel in all the aspects which refer to execution of navigational processes – in accordance with the standards presented in Figure 3-4, fleet management will be performed through a land-based control center, known as the Future Operator Experience Concept or oX.

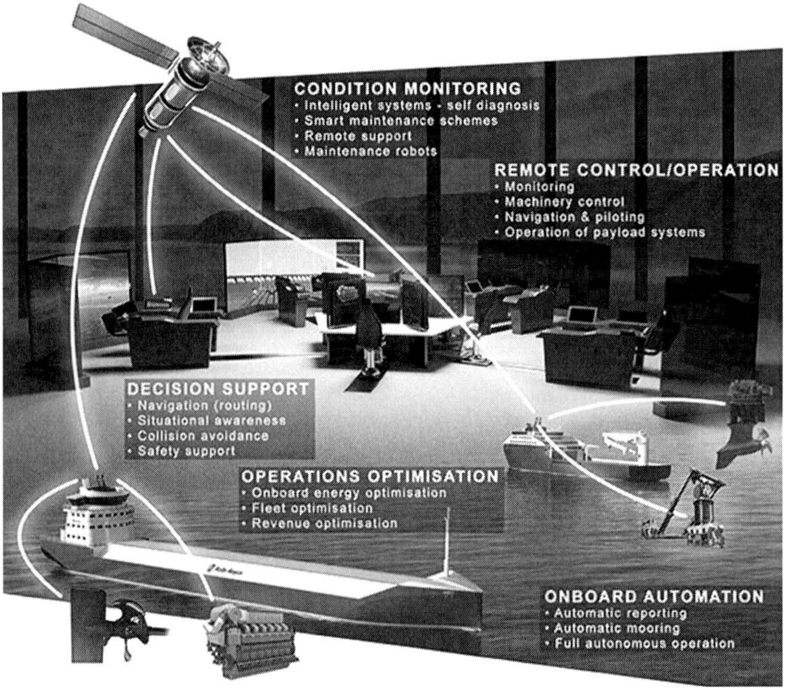

Figure 3-4. The areas referring to the full autonomy of a vessel in the technology of the Intelligent Awareness System – IAS (a concept by Rolls-Royce Marine)

Source: Rolls Royce materials available at http://gpsworld.com/rolls-royce-joins-partnership-to-develop-autonomous-ships/ [accessed on 24th March 2018].

Leaving aside telematics technologies applied in maritime transport of the future, it should be emphasized that at present currently implemented telematics systems are capable of acquiring a wide range of significant information from two basic fields of maritime transport (navigation and port operations). It provides the basis for the original taxonomic division of telematics systems and the monitoring of maritime transport safety.

3.1.2. Taxonomy of telematics and shipping monitoring systems in maritime transport

The demand for specific information from the field of maritime transport, which has been discussed in the previous sub-chapters, has resulted in the development and implementation of numerous systems used for acquiring, storing, processing and transmitting information in the transport practice. At first, the initial anchoring of all systems of navigation monitoring (and ultimately telematics systems) had its substantial justification in concepts of navigation safety, e.g. in the MSA.[13] However, the immense significance of such information for the optimisation of land-sea logistical chain processes was noticed relatively quickly, and its dynamic exploitation began in this field.

Considering the variety of systems and their main purpose in maritime transport, it is possible to distinguish the following:[14]

- applications (including telematics ones) used in regional and national systems of navigation monitoring by government institutions and agencies at the strategic level, mostly within the framework of maritime safety concepts;
- commercial applications based on the available AIS signal, which are dedicated to entities operating in maritime economy, including logistical operators, in order to streamline management at the integrated level;
- applications supporting processes taking place in land-sea logistical chains which optimise operational processes;
- ship-owner applications.

Considering the position taken by the particular systems, the above-mentioned division can be defined as a division according to a vertical criterion. Another division criterion is a horizontal criterion which indicates a significant diversification of current systems with regard to the distinguished homogenous factors. Based on such a division, it is possible to distinguish the following telematics systems of maritime transport:

---

[13] MSA – *Maritime Situational Awareness*, see more in: M. Grzybowski, R. Miler, *MSA as the Aspect of European Maritime Security*, The Romanian Economic Journal, Bucharest, 25/2007 and M. Kościelski, R. Miler, M. Zieliński, *Maritime Situational Awareness (MSA)*, Zeszyty Naukowe AMW, Gdynia, no. 4 / 2007, p. 2,– the Author's note.

[14] R. Miler, *Use of Vessel Monitoring Systems in the Logistics Chain of Supply within the EU Seas* [in:] J. Jaworski, A. Mytlewski (eds), *Funkcjonowanie Systemów Logistycznych,* Prace Naukowe WSB w Gdańsku, vol. 2 / 2009, pp. 3–5.

- considering the subject of the monitoring:
  - systems for monitoring vessel traffic (systems of navigation monitoring),
  - systems for monitoring loading units (e.g. containers),
  - systems for monitoring parameters which refer to the operation of vessel systems (including the drive system),
  - systems for monitoring maritime environment (spills in the waters of the World Ocean),
  - systems for monitoring fishing,
  - systems for monitoring port facilities;
- considering the object collecting and processing information provided by monitoring systems (data administrators):
  - civil and military systems,
  - national (domestic), sub-regional (e.g. HELCOM) and international systems (e.g. the EU systems),
  - public (of maritime administration) and private (commercial) systems;
- considering the sensory factor:
  - systems based on the AIS signal (including AIS-S),
  - systems based on the LRIT signal,
  - systems based on visual technologies (including infrared cameras),
  - systems of satellite photographs;
- considering the purpose of collected information:
  - systems displaying situation at sea (monitoring function only),
  - analytical and decisional systems (effector systems, telematics, automation and autonomation systems);
- considering the range (of the field of data collection):
  - local,
  - regional,
  - global;
- considering the accessibility of databases and the character of collected information:
  - public (with an unlimited access through the Internet) and requiring authorisation (user registration),
  - non-classified and classified,
  - confidential, with an access to commercial sensitive data provided only to the previously defined entities (e.g. ship-owner systems, systems for groups of users, national systems);

- considering the type of information and its significance for safety/security processes:
  - safety systems (navigational including nautical information, systems for monitoring vessel parameters and operational regimes, IBS/INS, monitoring systems of cargo, spills, etc., port terminal and ship-owner systems),
  - security systems (informational, analytical and decisional systems used by services responsible for state security in the maritime domain).

On the basis of the above-mentioned considerations, it is possible to present the following equation for monitoring systems of maritime transport safety:

$$SMTS_{def} = SMTS\ (DC, SC) \qquad (3.1)$$

where:

DC – division criterion,

SC – safety category in the systemic approach.

As a result of the above-mentioned equation, it is possible to present a horizontal classification of systems currently applied for monitoring safety of maritime transport. Table 3-1 presents a division of systems for monitoring safety of maritime transport safety according to the horizontal criterion in a synthetic (tabular) approach.

Table 3-1. Horizontal classification of systems for monitoring safety of maritime transport

Division criterion	Categories of systems for monitoring safety of maritime transport
the subject of the monitoring	• systems for monitoring navigation, • systems for monitoring loading units (e.g. containers), • systems for monitoring parameters which refer to the operation of vessel systems, • systems for monitoring maritime environment, • systems for monitoring fishing, • systems for monitoring port facilities
an entity collecting and processing information provided by the monitoring systems (data administrators)	• civil and military systems, • national (domestic), sub-regional and international systems, • maritime administration and commercial systems
the sensory factor	• systems based on the AIS signal, • systems based on the LRIT signal, • systems based on visual technologies

Table 3-1. cont.

Division criterion	Categories of systems for monitoring safety of maritime transport
the purpose of collected information	• systems displaying situation at sea, • analytical and decisional systems (mixed effector systems)
the range (the field of data collection	• local systems, • regional sytems, • global systems
the accessibility of databases and the character of collected information	• public (with an unlimited access through the Internet) and requiring authorisation (user registration), • non-classified and classified systems, • confidential systems with an access to commercial sensitive data provided only to the previously defined entities (e.g. ship-owner systems)
the type of information and its significance for safety/security processes	• safety systems (navigational including nautical information, systems for monitoring vessel parameters and operational regimes, IBS/INS, systems for monitoring cargo, spills, etc., port terminal and ship-owner systems), • security systems (informational, analytical and decisional systems used by services responsible for state security in the maritime domain, systems of seaport security)

Source: the Author's own study.

Rapid development of vessel traffic control systems has taken place as a result of the IMO regulations and development of sensory technologies used for collecting information referring to the whole process of maritime transport. After the initial period of uncoordinated development of various systems presenting different architecture, purposes and sensory equipment, some tendencies towards unification and coordination of undertakings in that field have started to appear, and as a result, three basic types of systems can be distinguished:[15]

- VTMIS (Vessel Traffic Monitoring and Information System) at the national and regional level (the EU);
- partially integrated (semi-integrated) tool systems applied in new concepts in maritime safety and protection of navigation (e.g. BRITE in a concept of the MSA in NATO, IMDatE in Europe);

[15] R. Miler, *Systemy zobrazowania żeglugi i aplikacje informatyczne transportu morskiego…*, op. cit., p. 216.

- commercial systems providing support to operators of maritime transport in land-sea logistical chains (as a derivative of the implementation of unified standards).

Considering a large variety of systems, it is possible to present their types, the scope of their influence and their specific features (functions) on the example of systems which have been implemented in the doctrine of state security of one country (however, with a provision that it is systemically connected with the solutions applied in its sub-region, and that it follows international standards).

Hence, the monitoring of vessel traffic is provided by several institutions which report to their superior units, namely various ministries. Maritime administration usually reports to the Ministry of Maritime Economy, Infrastructure or Maritime Transport, and it deals with the monitoring of vessel traffic with regard to nautical safety as well as to the aspect of security mentioned in the previous chapters. In order to perform such tasks, the VTS-class (Vessel Traffic Service) systems are usually applied. The Centre for Fishing Monitoring reports to the Ministry of Agriculture, Fishing or Maritime Economy – in accordance with the EU law, it applies the VMS (Vessel Monitoring System) to monitor all the fishing vessels, the length of which is over 15 meters. Maritime Units of the Border Guard report to the Ministry of Interior and Administration. The Border Guard is interested in vessel traffic with regard to the protection of maritime borders of the state, particularly in security of its territorial waters. Therefore, a dedicated system of the Coastal Integrated Monitoring Solution has been developed. The Navy is subject to the Ministry of National Defence, and it applies systems displaying the situation at sea for early identification of danger. It applies its own (national) systems as well as the ones applied by the allied parties (NATO). Additionally, while implementing the resolutions of the IMO Convention referring to PSC and FSC (Port State Control, Flag State Control), users of national systems receive an access to the data collected in the European system of information about vessels, such as (Paris MoU) – SIReNac (notified visits and scheduled inspections) and a list of banned ships (vessels under the prohibition of port entering).[16] Figure 3-5 presents the Paris MoU application.

In practice, there are numerous applications functioning, however such systems work in a selective way, or they cannot be (by definition) integrated

[16] R. Miler, *Bezpieczeństwo transportu morskiego…*, op. cit., p. 244.

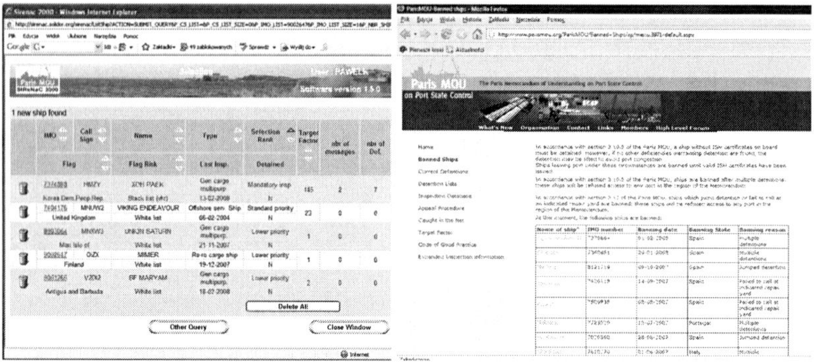

Figure 3-5. The Paris MoU – SIReNac application used for the implementation of the PSC and FSC functions

Source: adapted from A. Królikowski, *Shipping data collection and utilization…*, op. cit., pp. 6–8. [c]

with any other systems. It particularly refers to applications developed as auxiliary sub-systems for national systems of navigation display of the National SSN (National SafeSeaNet) class. However, information can be grouped and provided in the form of a database system: Global Integrated Shipping Information System (GISIS) which is a tool administered by the IMO. Considering the criteria of vertical and horizontal classification of telematics and systems for monitoring safety of maritime transport safety defined in the previous part, it should be emphasized that the systems of navigation monitoring have been the most developed ones.

3.2. Vessel traffic monitoring systems as an important element of maritime transport telematics

3.2.1. The VTMIS-class systems (Vessel Traffic Monitoring and Information System)

The VTIMIS (VTS + MIS) system, treated as a national system of vessel monitoring and maritime navigation management (that is namely the national VTMIS) is composed of:[17]

[17] The Vessel Traffic Management Information System (VTMIS) is an extension of the Vessel Traffic Service (VTS), in the form of Integrated Maritime Surveillance, which in-

- AIS – Automatic Identification System,
- VTS – Vessel Traffic Services,
- LRIT – Long Range Identification and Tracking,
- SRS – Ship Reporting Systems,
- MAS – Maritime Assistance Services,
- MIS – Management Information System,
- national computer centre of data exchange (national SafeSeaNet and CleanSeaNet systems).

Figure 3-6 presents the basic components of the VTMIS (at the sensory level). Figure 3-7 (a diagram below) illustrates the operation of the VTMIS.

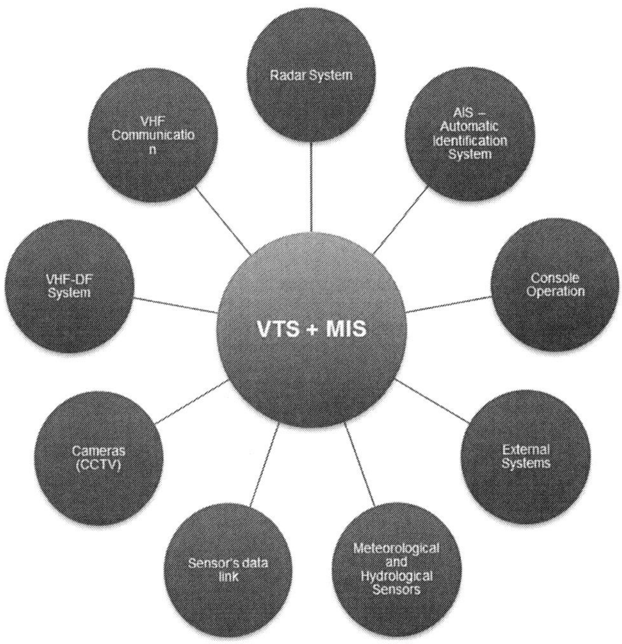

Figure 3-6. A schematic diagram of the input and output components and the sub-systems of the VTMIS

Source: adapted from www.sheltermar.com.br/en/vts/vtmis/ [accessed on 19th April 2018].

corporates other telematics resources to allow allied services and other interested agencies in the direct sharing of VTS data to get an access to certain sub-systems in order to increase the effectiveness of port or maritime activity operations as a whole, but that does not relate to the purpose of the VTS itself – the Author's note.

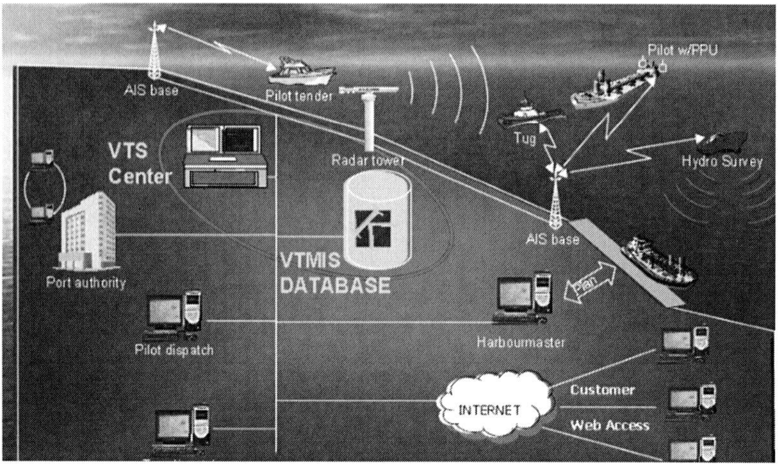

Figure 3-7. Basic operations of the VTMIS

Source: adapted from https://www.sheltermar.com.br/en/vts/vtmis/ [accessed on 24th March 2018].

The features of the VTMIS are:[18]

- port management systems (including information from Terminal Operation Systems – TOS),
- systems dedicated to port security (according to the ISPS standards),
- support systems and management of pilotage (if applicable),
- load management systems and overall property (including information from vessel planning systems),
- Docking Planning (including information from Terminal Operation Systems – TOS),
- systems for collecting port taxes (within the range of Port Community System),
- quarantine control, detection of ships (according to the Port State Control protocols),
- customs control (including e-Custom),
- support for Coast Guard operations such as repression of illicit acts on ships, smuggling, drug trafficking, illegal immigration, etc.

[18] https://www.sheltermar.com.br/en/vts/vtmis/ [accessed on 27th July 2017].

The role of the AIS system is to provide automatic exchange of data referring to a particular ship.[19] In accordance with the International Convention for the Safety of Life at Sea (SOLAS 1974), the AIS of a vessel should provide automatic transmission and reception of information stating identification of a vessel, including such elements as: vessel type, position, course, speed, navigational status and any other additional information related to safety. It should also monitor and record data which are automatically transmitted by the VHF bandwidth.[20] However, the potential capabilities of that system are much higher, and it is possible to have omnidirectional or dedicated transmission of over twenty various messages, such as:[21]

- the MMSI and IMO numbers, the name and the call sign of the vessel, its type and dimensions (length, beam and height) and its position,
- geographical position with the indication of its precision (precision above or below 10 meters),
- universal time coordinated (UTC), current values of the course over the ground and speed over the ground, gyrocompass course, navigational status of the vessel, angular velocity if the vessel is equipped with a relevant gauge,
- the current value of the maximal draft of the vessel,
- information about hazardous, harmful or pollutant cargo, required by the maritime administration of a coastal country, by the ship reporting system or by port authorities,
- destination port and estimated time of arrival (ETA),
- the scheduled route of the vessel (geographical positions of the subsequent turning points or a description of the route in words),
- the number of people on board, including the crew members,
- information referring to safety, given in 126 words at maximum.

In accordance with the IMO recommendations, the AIS system should be installed on all vessels of over 300 GT and on all passenger vessels. It means that is has been installed on over 40 000 vessels, and this number has been constantly growing, considering the policy of maritime administration and

---

[19] R.K. Miler, A. Bujak, *exactEarth-AIS as one of the most advanced shipping monitoring systems* [in:] J. Mikulski (ed.), *Activities of transport telematics*, Springer, Berlin – Hildeberg 2013, p. 330.

[20] M. Kościelski, R.K. Miler, M. Zieliński, *Automatic Identification System (AIS) as a Main Tool of NCAGS ADP Systems*, Zeszyty Naukowe AMW no. 3 (170), Gdynia 2007, p. 34.

[21] R. Wawruch, *Możliwości wykorzystania statkowych systemów automatycznej identyfikacji (AIS) do monitorowania ruchu morskiego*, Wyd. AM, Gdynia 2007, p. 43.

governments of numerous coastal (maritime) countries. Ship-owners who operate vessels of lower gross tonnage have been encouraged to use the system as well.[22]

It should be also remembered that the AIS is an additional device to transmit information concerning maritime safety, and its use in this field does not exclude the obligation of following the requirements defined in the Global Maritime Distress and Safety System (GMDSS). The architecture of the system and the flow of information is presented in Figure 3-8 (at present, the AIS A and B is operated for smaller vessels).

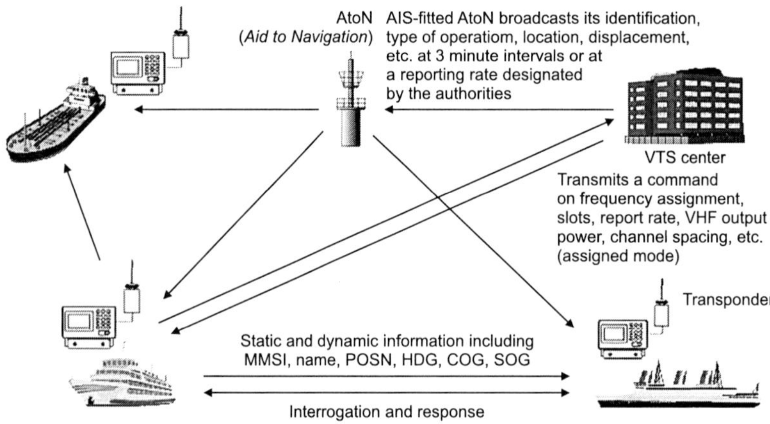

Figure 3-8. The architecture and the flow of information in the AIS

Source: adapted from M. Kościelski, R.K. Miler, M. Zieliński, *Automatic Identification System (AIS) as a Main Tool of NCAGS ADP Systems…*, op. cit., p. 14.

The first VTS (Vessel Traffic Services) system was established as early as in 1948, and it was applied in the port of Liverpool. VTS systems are implemented in order to improve safety and efficiency of vessel navigation.

During the 14th Session of the IMO General Assembly in 1985, the Resolution on Guidelines for Vessel Traffic Services (Res. IMO A 578(14) was approved.[23] These guidelines define functions which must be performed by

[22] W. Drozd, M. Dziewicki, M. Waraksa, *Status polskiego systemu automatycznej identyfikacji statków (AIS)*, A report provided by the Maritime Office in Gdynia, Gdynia 2007, p. 23.
[23] Z. Kopacz, W. Morgaś, J. Urbański, *Wybrane zagadnienia międzynarodowego systemu bezpieczeństwa morskiego i bezpieczeństwa nawigacji…*, op. cit., p. 85.

VTS systems, starting from simple systems which only provide information (equipped with sensory modules only), to systems which are applied to control vessel traffic (telematics systems equipped with analytical and effector modules). A system of vessel traffic surveillance is composed of the following elements:[24]

- radars which define the operational range of a particular VTS system,
- a VHF radio station for communication with vessels,
- a VTS Management Centre (VTS Centre),
- objects, that is namely vessels approaching the VTS operational zone and vessels arriving and leaving the port.

The main elements of the architecture and the flow of information in the VTS systems are presented in Figure 3-9.

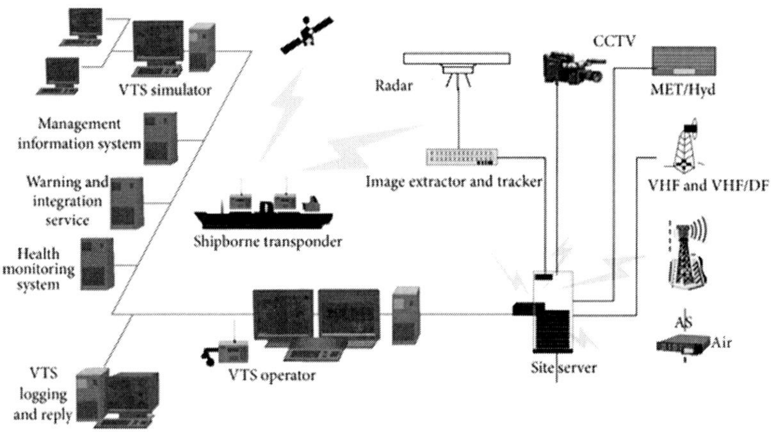

Figure 3-9. The architecture and the flow of information in a VTS system

Source: adapted from B. Lee, N. Park, *Performance Improvement Based Authentication Protocol for Intervessel Traffic Service Data Exchange Format Protocol Based on U-Navigation System in WoT Environment*, Journal of Applied Mathematics, Volume 2014 (2014), Article ID 734768, p. 7.

---

[24] R.K. Miler, *Systemy zobrazowania żeglugi i aplikacje informatyczne transportu morskiego* [in:] J. Dąbrowski (ed.), *Wyzwania i problemy transportu morskiego*, Fundacja Rozwoju UG, Gdańsk 2011, p. 215

Basic tasks of a VTS system include:[25]

- streamlining traffic of particular vessels and improving throughput and capacity of the traffic flows in restricted and limited water areas,
- decreasing congestion and lowering the probability of nautical risks, such as collisions and stranding into obstacles,
- lowering threats of maritime environment pollution generated by vessels,
- improving economic efficiency of port operations.

Figure 3-10 presents an example of navigation traffic in the approach fairways to the port of Odessa.

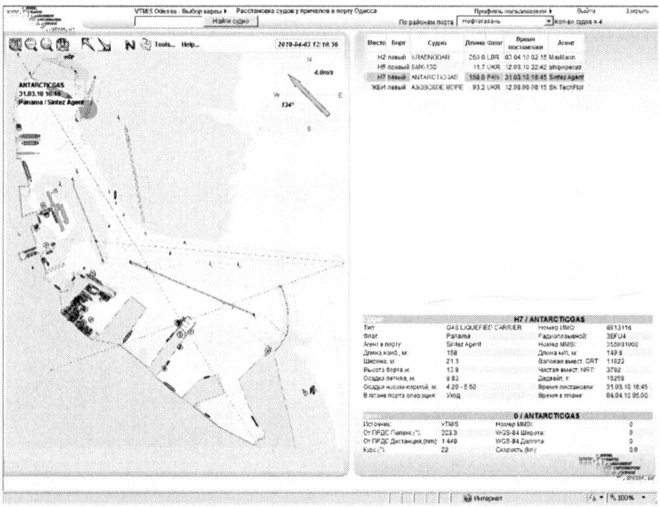

Figure 3-10. The VTS system of the port of Odessa displayed on the operator's console

Source: adapted from materials provided by the port of Odessa http://www.vtmis.odessa.ua/Demo/03_port1.aspx [accessed on 23rd May 2016]. [c]

Information acquired from the radars is transmitted to the VTS Centre via the radio or an optical fiber cable. Then the data are processed in the central processors and displayed on the VTS operators' stations.[26]

[25] Z. Kopacz, W. Morgaś, J. Urbański, *Europejski system monitoringu ruchu statków i informacji…*, op. cit., p. 46

[26] R.K. Miler, *Systemy zobrazowania żeglugi i aplikacje informatyczne transportu morskiego…*, op. cit., pp. 216–217.

The reason for the implementation of the LRIT (Long Range Identification and Tracking) system was an initiative to provide anti-terrorist protection after the attacks in the USA in 2001. The initiative was presented during the IMO conference, in December 2002, and it was promoted by the USA. The USA is ultimately going to track each movable object which moves in its territorial waters or in their nearest vicinity. As a result of numerous discussions and agreements, the basic parameters of the LRIT system were decided. The range of the system for vessels of the flag state and for the port state is unlimited, whereas the range of the system for coastal countries and for the requirements of SAR is approximately 1000 nautical miles.[27]

In the LRIT system, vessels do not send any information themselves (contrary to the AIS system), but they respond to the question calls. It has been established that information delivered in the LRIT system must include at least:[28]

- international call sign of the vessel (identification number),
- the position of the vessel,
- the time of the position.

Figure 3-11 presents the architecture and the flow of information in the LRIT system.

The most important components of the LRIT system are devices applied for:[29]

- basic data transmission,
- providing communication services of the system (on-board communication station, including satellite communication),
- providing an access to all the system services (on-board computer station),
- devices of other LRIT centres (including peripheral devices).

In accordance with the V/19.1 Chapter of the SOLAS Convention, the LRIT system is installed on:[30]

- passenger vessels and high speed passenger crafts (HSC),
- cargo vessels and high speed crafts (HSC) of 300 GT and more,

---

[27] http://www.lrit.de/pol.php [accessed on 17th July 2015].

[28] R. Miler, *Use of Vessel Monitoring Systems in the Logistics Chain of Supply within the EU Seas* [in:] J. Jaworski, A. Mytlewski (eds), *Funkcjonowanie Systemów Logistycznych*, Prace Naukowe WSB w Gdańsku, vol. 2 / 2009, p. 123.

[29] M. Kościelski, R. Miler, M. Zieliński, *European Maritime Safety Agency (EMSA) in the way to enhance safety at EU seas*, Zeszyty Naukowe AMW no. 2, Gdynia 2008, p. 56.

[30] http://www.lrit.de/pol.php [accessed on 17th July 2015].

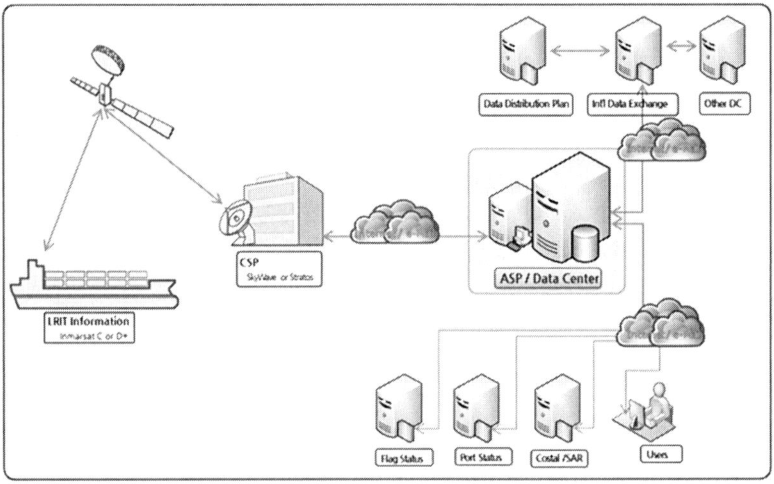

Figure 3-11. The architecture and the flow of information in the LRIT system

Source: adapted from http://www.modootel.co.kr/en/LRIT/ContentIndex?contentID=122&groupID=228 [accessed on 28th May 2018].

- movable crafts of deep sea mining (oil platforms and rigs used for drilling and exploration),
- vessels constructed after 31st December 2008.

Figure 3-12 presents the interface of the LRIT system.

The national centres of data reception are authorised to receive signals and to track vessels which:[31]
- are not further than 1000 nautical miles from their coast,
- are heading towards the ports of their country,
- are under their state flag and operate on the waters of the World Ocean.

The Ship Reporting System (SRS) is a system which defines reporting procedures for vessels. The procedures define who and when a vessel should report to, what should be included in such reports and how to format such reports. The current regulations of the SAR 79 Convention impose an obligation on

[31] R.K. Miler, *Systemy zobrazowania żeglugi i aplikacje informatyczne transportu morskiego…*, op. cit., pp. 222–223.

Figure 3-12. The interface of the LRIT system

Source: adapted from http://www.fulcrum-maritime.com/lritsystems.nsf [accessed on 24th April 2015]. [c]

its signatory parties to establish vessel reporting systems in their territorial waters in order to improve the efficiency of SAR operations.[32]

Other factors which have contributed to the development of the SRSs include a necessity of improvement in protection of maritime environment which is threatened by vessels transporting oil and hazardous cargo, and a tendency towards the unification of procedures applied in all maritime services.

In 1983, the International Maritime Organisation (IMO) issued its first resolution concerning vessel reporting systems (Resolution A.531(13)). The document defines a standard format of reports submitted by vessels. It also precisely attributes the particular content of the report to the particular alphabet letters, forming a report system (reporting system).

Apart from the content of reports, the standards of vessel reporting system established by the IMO also define the type of reports, the time and their addressees. The system includes the following types of reports:[33]

[32] Z. Kopacz, W. Morgaś, J. Urbański, *Europejski system monitoringu ruchu statków i informacji...*, op. cit., p. 50.

[33] Ibidem, p. 51.

- sailing plan;
- position report,
- deviation report,
- final report,
- hazardous cargo report,
- harmful substances report,
- pollution report,
- other reports.

Maritime Assistance Services (MAS) is an auxiliary system applied to support operations at sea during collisions and failures. Its tasks are stated in the IMO Resolution of 2005 (Resolution A.950(23)). The aim of the system is:[34]
- to receive reports, recommendations and notifications resulting from the requirements established by the IMO,
- to monitor the situation of a vessel,
- to perform as a contact point between the captain of a vessel and the particular coastal country.

It has been assumed that the operation of the MAS system is generated by the following situations:[35]
- the vessel has taken part in a maritime accident which should be reported, however the vessel has maintained its ability to continue its voyage;
- the vessel needs assistance, but it is not in danger;
- the vessel is in danger, but the people on board have been already rescued.

The MIS (Management Information System) is the system that has capabilities allowing it to process information passed on by the VTS intelligently, or fed directly into operator's database. The efficiency of controlling and monitoring particular activities depends on the type and the treatment of the information that has been generated. Information collected by the VTS system must be processed, detailed and made available in order to improve and to facilitate the management of port operations.

The SafeSeaNet (SSN) is a computer system of navigational data exchange in the European Union (administered by the EMSA), and it comes as a basic component of the VTMIS system. It is not only a monitoring system, but

[34] Z. Kopacz, W. Morgaś, J. Urbański, *Europejski system monitoringu ruchu statków i informacji...*, op. cit., p. 51.
[35] R.K. Miler, *Systemy zobrazowania żeglugi i aplikacje informatyczne transportu morskiego...*, op. cit., pp. 222–223.

it also comes as a system applied to acquire information from databases, database systems and Internet resources. It is organised on the basis of the Article 14 of the Directive 2002/59/EU on the establishment of the VTMIS.[36]

Basic tasks performed by the SafeSeaNet bring numerous operational benefits which are directly related to maritime transport safety, including:[37]

- minimisation of the response time required by relevant maritime services in the particular EU member countries to react to maritime accidents and pollution of maritime environment;
- early identification of vessels which may come as a threat (prevention aspects);
- development of databases and statistics referring to vessel traffic and cargo transported in the particular water areas (establishing a pattern of behaviour and its analysis with regard to anomalies).

Collection and exchange of information in the SafeSeaNet includes arrivals and departures of the following vessels:[38]

- vessels over 300 GT,
- fishing, tourist and cargo vessels of the length over 45 meters,
- all vessels transporting hazardous cargo.

Apart from maritime institutions and services, also other state institutions, administrative bodies and commercial entities are authorised to use the national SafeSeaNet system. They are namely:[39]

- Border/Coast Guard,
- Customs Service,
- ship-owners,
- shipbrokers,
- inspectors of sanitary services, protection of natural environment, fishing and others,
- authorised state services, etc.

The display of vessel traffic is acquired through the module of the SafeSeaNet, the Traffic Information Relay and Exchange System (STIRES), which uses the GIS platform. The STIRES interface (with additional information from the LRIT system) is presented in Figure 3-13.

---

[36] http://emsa.europa.eu/operations/safeseanet.html [accessed on 17th July 2015].

[37] Ibidem.

[38] R. Miler, *Use of Vessel Monitoring Systems in the Logistics Chain of Supply within the EU Seas…*, op. cit., p. 46.

[39] Ibidem, p. 47.

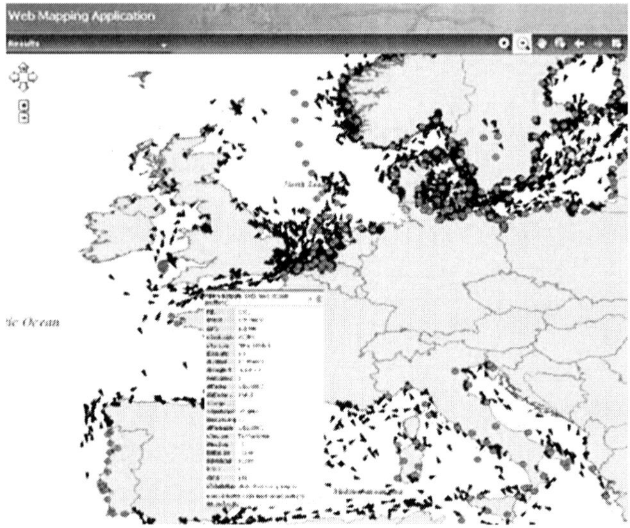

Figure 3-13. Navigation traffic displayed in the STIRES application

Source: adapted from http://www.lhg.is/media/vaktstod_siglinga/SafeSeaNet_January_2009.pdf [accessed on 20th February 2010]. [c]

The CleanSeaNet (CleanSN) comes as an extension of the range of information that can be acquired by the EMSA and a development of capabilities required to collect databases (in this case, information from the field of ecological threats – oil spills). The system operates using satellite images which practically cover all the EU water areas.[40] The satellite images are analysed with the use of advanced IT technologies in order to find any potential spills on the sea surface. If any pollution is identified in a particular water area, including territorial waters of the EU countries, an alert report is sent to the relevant services of the country in question. A complete, precise and analysed image is available in the system after about 30 minutes since the moment of the satellite passing above the polluted area. Figure 3-14 presents the operational principle of the CSN with regard to time regimes.

The images acquired from the system undergo further analysis and additional informational processing. System operators supplement the received satellite images with additional meteorological data (e.g. the speed

[40] J. Szcześniak, A. Weintrit, *Europejskie systemy kontroli i śledzenia ruchu statków – geneza, zasady funkcjonowania…*, op. cit., p. 89.

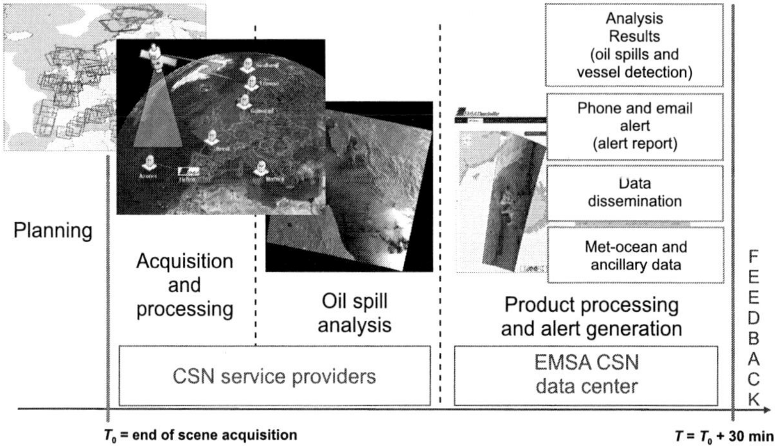

T_0 – sequence finishing time of a satellite image (provided in 1400 km/400 km segments); T – the system response time to the identified anomaly (T_0 + 30 min)

Figure 3-14. The concept presenting operation of the CleanSeaNet system

Source: adapted from J. de Sousa, *EMSA's Integrated Maritime Environment – a tool for improved Maritime Domain Awareness*, EMSA Publication, Lisbon 2012, p. 15. [c]

and direction of wind, the sea state), oceanographic data (e.g. the speed and direction of sea currents and tides) and data acquired from navigation monitoring (e.g. positions of vessels from the AIS system). Such activities are performed to provide a comprehensive summary of all the acquired information, which may result in:

- identification of oil spills on the water surface, stated with high probability,
- precise identification of the size, character and level of pollution with the direction of its movement,
- identification of the vessel responsible for the spill.

All the data acquired during the CleanSeaNet system operation are stored in the EMSA database, where they can be used for further analysis and statistics.

The CleanSeaNet system comes as an excellent example of an evolution of telematics systems in maritime transport, which is aimed at their implementation in analytical and decision-making processes. It has required integration with complementary systems, and it has also provided a possibility of applying the acquired information for commercial purposes.

3.2.2. Semi-integrated, complementary and commercial systems in maritime transport security and safety monitoring

The systems presented so far are mostly elements related to acquisition of information from the field of maritime navigation at the sensory level. An additional valuable attribute of these applications is a possibility of displaying a particular situation on the screens of operators' stations and transmitting the information in one unified standard (most often an AIS stream) to other users of the system. In accordance with the concept presented above, some VTMIS systems have a highly developed structure (analytical, decision-making and effector modules incorporated), and therefore, they meet the criteria of usefulness not only in the process of monitoring but also in the process of making decisions related to safety and optimisation of logistical processes in maritime transport. Some examples of such partially integrated systems which fit into the implemented doctrines of maritime safety and navigation protection are:

- at the national level – Shipping Monitoring Data Exchange (SMDE) – type systems, e.g. SWIBŻ PL,
- at the regional – the European level – EMSA – IMDatE,
- at the sub-regional level, as experimental tools of the MSA concept developed by NATO – BRITE.

The systems of the SMDE-class are implemented in the particular coastal countries within the conceptual framework of national/domestic systems of maritime safety (NSMS) (an example of a national system is SWIBZ-PL; see Figure 3-15).

The system and the software are based on an information flow provided by the AIS which generates the display of seagoing vessel traffic within the VTS, with particular consideration of vessels which move along the approach fairways of the main sea ports. The SMDE-class system should provide an access to very detailed and specific information; it should allow the interested parties to use the SafeSeaNet / CleanSeaNet tabs and a dedicated sub-system of maritime administration (PSC/FSC control of arrivals and checking officer and seafarer certificates).

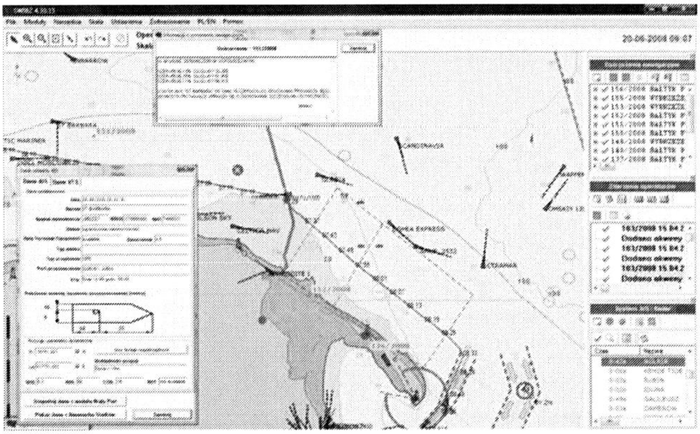

Figure 3-15. An example of the SMDE-class system at the national level – the interface of the SWIBŻ-PL application

Source: adapted from materials provided by the Maritime Office in Gdynia, Poland at http://www.umgdy.gov.pl [accessed on 14th February 2016]. [c]

As an example of an integrated analytical and decisional system, the SMDE-class system performs the following functions:[41]

a. an auxiliary system supporting operations in the areas of maritime jurisdiction of a particular country in the following situations:
   - crisis and anti-terrorist situations under the ISPS Code in cooperation with the Central Contact Point, Regional Contact Point, Port Security officers and operational services of law enforcement ministries,
   - alarm and emergency situations in cooperation with local Crisis Management Centres,
   - rescue operations carried out to save human life and property at sea and fight against spills,
   - monitoring of vessel traffic and surveillance of safe exploration, exploitation and research in the Polish territorial water areas,
   - providing Maritime Assistance Service (MAS) to vessels which need assistance or shelter,
   - central application for the integrated NSMS system at the national level;
b. a system of collecting and transmitting a wide range of data acquired from vessel traffic monitoring and surveillance systems and systems of registering and transmitting information to domestic and foreign co-users;

[41] http://www.umgdy.gov.pl/ [accessed on 14th February 2016].

c. a system of traffic registering and vessel traffic surveillance in national territorial water areas, including registration of vessels coming in/out small ports and harbours (Port/Harbour Authority Office);
d. a national component of the SafeSeaNet system – an electronic system applied to exchange information about vessel traffic monitoring and to exchange data with the European Union (EMSA – European Maritime Safety Agency);
e. supporting the organisation of port and maritime traffic in favour of enterprises related to maritime economy (Departments of Shipping Services, Pilotage Offices and Sea Port Authorities in particular).

The SMDE-class system also provides a possibility of acquiring data from the field of nautical safety in its broad sense, e.g. navigation warnings, hydro- and meteorological conditions, etc. (Figure 3-16).

Figure 3-16. Specific functions performed by the SMDE-class system – the interface of the SWIBŻ-PL application in the tab of nautical information exchange

Source: adapted from materials provided by the Maritime Office in Gdynia, Poland at http://www.umgdy.gov.pl [accessed on 14th February 2016]. [c]

Working over the integration of systems for monitoring safety of maritime transport and marine environment in the EU water areas includes tasks referring to Integrated Maritime Data Environment (IMDatE) which comes as a platform and an IT tool for data exchange (discussed in detail along with the concept and strategy of the CISE in Chapter 3.4). Figure 3-17 presents the basic architecture of an integrated system applied for acquiring, processing and exchanging data related to maritime safety in the EU water areas.

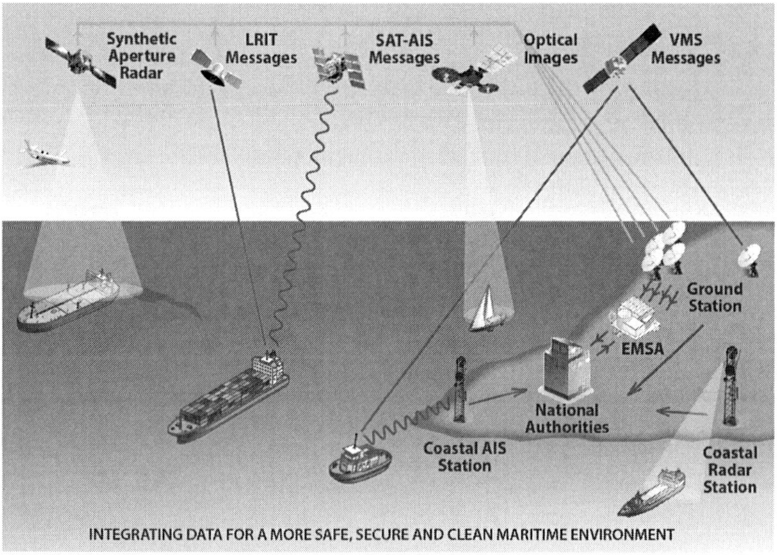

Figure 3-17. The basic architecture of IMDatE

Source: adapted from materials provided by EMSA at https://portal.emsa.europa.eu/web/imdate [accessed on 24th January 2015].

During the implementation of the MSA (NATO) concept which assumes development of a system for the exchange of navigational information, an experimental tool has been developed: the BRITE (Baseline for Rapid Iterative Transformational Experimentation) application.[42] It is a comprehensive solution based on telematics technology and self-learning algorithms, which is:

- developed by ACT (Allied Command Transformation),
- free to all NATO nations,
- equipped with Version and baseline configuration management,
- Wiki-based ticket tracking, product roadmap, and read-only source code access,
- based on and compliant with TIDE (Technology for Information, Decision, and Execution superiority) specifications,

[42] R. Demczuk, M. Kościelski, R. Miler, *Miejsce, rola i znaczenie Centrum Operacji Morskich w tworzeniu Krajowego Systemu Bezpieczeństwa Morskiego oraz w procesie wdrażania natowskiej koncepcji MSA*, Zeszyty Naukowe AMW, 172 K/1 2008, p. 54.

- established on Service Oriented Architecture (SOA),
- based on Different Data Sources.

As a systemic tool, the BRITE (at present, it is the BRITE 2.0) is capable of not only displaying navigational data (monitoring) but also of providing data analysis (identification of anomalies and regularities). In this way, it becomes an element of the early warning system referring to the situation at sea. The system uses the AIS signal, and it provides a display of commercial vessels above 300 GT which are equipped with an AIS transponder (Figure 3-18).

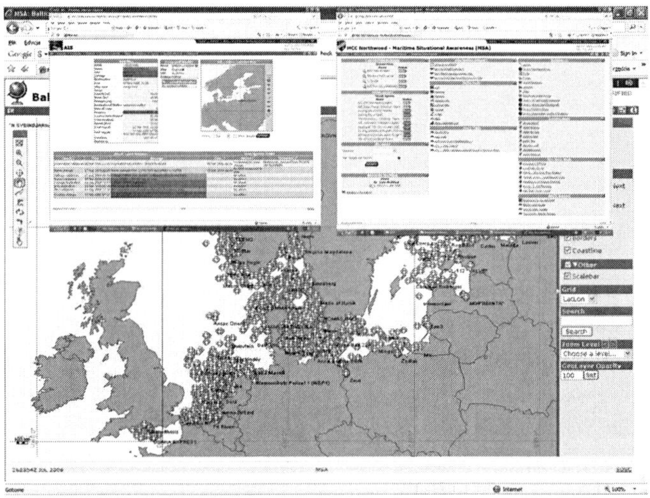

Figure 3-18. The interface of the BRITE application

Source: adapted from materials provided by the Centre of Naval Cooperation and Guidance for Shipping (NCAGS) at Maritime Operation Centre (MOC), Gdynia.

As soon as an operator identifies an anomaly during the monitoring of specified parameters, the BRITE is capable of automatic identification of any deviations by comparing the current data with the archival data acquired, for example, from Lloyd's database. Detected anomalies are displayed in the form of screen alerts visible on the operator's control, with the use of the colour code in which red indicates deviations.[43] Practical application of the system for automatic indication of anomalies is presented in Figure 3-19.

[43] R.K. Miler, *Systemy zobrazowania żeglugi i aplikacje informatyczne transportu morskiego...*, op. cit., p. 212.

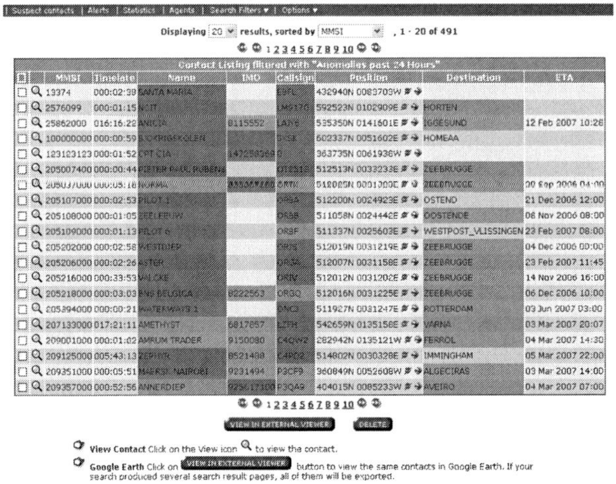

Figure 3-19. The interface of the BRITE system for indication of anomalies (positions in red)

Source: adapted from materials provided by NATO *Shipping Centre* (NSC), HQ MC Northwood, UK. [c]

It is a highly practical system which provides automatic display of screen alerts and preliminary evaluation of the situation. Further data processing involves checking an anomaly, its acceptance or rejection, and it requires a 24-hour watch of a professional operator.[44] The next steps which must be undertaken by the operator are presented in Figure 3-20.

The BRITE has also a non-public, classified layer containing information about vessels which are objects of special interest (contact of interest – COI).[45] However, its key advantage is the fact that there are intelligent algorithms incorporated into the system, which are applied for data evaluation, data comparison with reliable data or a specified required status (referred to as smart agents), along with the automation of decision-making processes. The capability of identifying deviations from the assumed principles of navigational behaviour patterns (such as deviations from the planned route, any delays in the estimated time of arrival – ETA, speed which is unusual for a particular type of vessels, two or more vessels standing next to each other

44 Ibidem, p. 212.

45 M. Kościelski, R. Miler, *Naval Co-operation and Guidance for Shipping (NCAGS)*, "Baltic Transport Journal", Gdynia no. 3 /2009, p. 12.

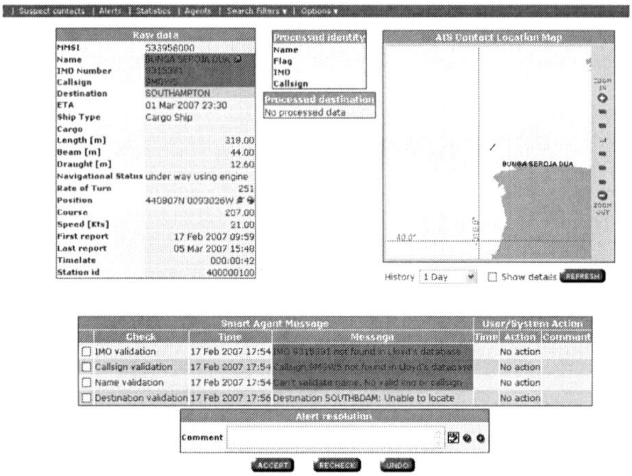

Figure 3-20. The interface of the BRITE system in the function of an individual alert (the options for further procedures to be followed by an operator are: *accept, recheck, undo*)

Source: adapted from materials provided by NATO *Shipping Centre* (NSC), HQ MC Northwood, UK.

having stopped their engines, entering restricted zones closed for navigation, crossing the reporting lines without sending out a relevant report and many other cases) allows the system to provide a quicker response, and it gives more time to check the reason for a particular anomaly and to start relevant preventing actions (Figure 3-21).[46]

Understanding the above-mentioned principles results in the implementation of the BRITE system analytical functions which improve the capability of its immediate response and almost automatic reaction to identified anomalies. It mainly refers to the capability of the system to initiate defensive action in order to prevent any negative phenomena in the field of maritime transport safety.[47] The conceptual solutions of the BRITE system find their reflection in analytical functions present in other telematics systems, including the IMDatE.

The above-mentioned partially integrated analytical and decisional systems applying navigation monitoring are not the only systems which are currently

[46] R. Miler, *Bezpieczeństwo transportu morskiego…*, op. cit., p. 218.
[47] Ibidem, p. 220

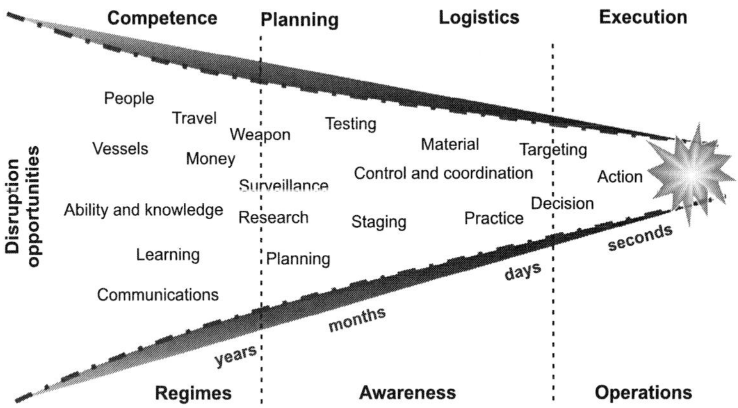

Figure 3-21. The time axis referring to the formation of threats to maritime safety in the aspect of terrorist attacks

Source: adapted from materials provided by NATO *Shipping Centre* (NSC), HQ MC Northwood, UK.

used for the monitoring of maritime transport safety. There is a group of systems (which are complementary to the monitoring systems described above) which include the following (for example):

- international systems:
  - GMDSS (Global Maritime Distress and Safety System),
  - MSSIS (Maritime Safety and Security Information System),
  - NAMESIS (Naval Merchant Shipping Information System),
  - VMS (Vessel Monitoring System);
- national systems:
  - Coast Guard systems,
  - Maritime Administration systems.

The GMDSS (Global Maritime Distress and Safety System)[48] is a global system of safety and alert at sea, the operation of which is based on a set of approved safety procedures. It requires vessels to be equipped with radio and

[48] Implemented on 1st February 1992, at present the system performs all the functions related to alarming in dangerous situations, locating places of disasters, providing communication required for coordination at places of rescue operations, providing general communication and spreading information related to navigation safety. The minimal radio equipment defined in the SOLAS Convention makes sending a distress signal independent from only one type of radio service. Depending on the water area, vessels can send such signals using at least two types of devices operated on the VHF, HF or MF bands,

satellite communication devices that can ensure safe navigation. The system provides fast and efficient alarming during accidents at sea; it transmits information important for vessel safety and provides communication during search and rescue operations. In accordance with the GMDSS assumptions, each vessel, regardless of its position at sea, must have all the radio communication functions which are indispensable for its own safety and safety of any other vessels moving in the same water area. The system also provides radio communication for security purposes. It allows its users to spread information referring to maritime safety/security, such as navigational or meteorological warnings, weather forecasts and other urgent information.

The GMDSS system is a set of communication sub-systems which use a wide spectrum of frequencies. The components of the system are:[49]

- the satellite communication sub-system INMARSAT,
- the EGC (Enhanced Group Call) and the DSC (Digital Selective Call) sub-systems,
- the COSPAS-SARSAT sub-system,[50]
- the Search and Rescue Transponder (SART) radar transponder,
- the EPIRB (Emergency Position Indicating Radio Beacon) satellite emergency beacon,
- the NBDP (Narrow Band Direct Printing) – the narrow band telex system with direct printing,
- the NAVTEX (NAVigational TEXt).[51]

The components of the GMDSS system and its basic architecture reflecting its functional concept are presented in Figure 3-22.

---

tele-printer radiotelegraphy with automatic reception with the use of digital selective call (DSC) or the INMARSAT satellite system – the Author's note.

[49] http://www.gmdss.republika.pl/index.html [accessed on 17th July 2015].

[50] The COSPAS-SARSAT international programme agreement was signed by Canada, France, the former Soviet Union and the United States on 1st July 1988. It entered into force on 30th August 1988 (however the implementation of the programme was started in 1979). As a result of the combination of the COSPAS satellite system of vessel search (Russian: Cosmicheskaya Sistyema Poiska Avariynykh Sudov), developed by the former Soviet Union, and the SARSAT (Search And Rescue Satelitte-Aided Tracking) rescue system, a new satellite system was developed and dedicated to tracking emergency beacons transmitting signals on 121,5 MHz and 406 MHz frequencies. The system enables all the search and rescue organisations to operate in any emergency situation on land, in air and at sea – the Author's note.

[51] It is an automatic system applied to transmit meteorological and navigational warnings, weather forecast and any other urgent information related to navigation safety (MSI – Maritime Safety Information) to vessels – the Author's note.

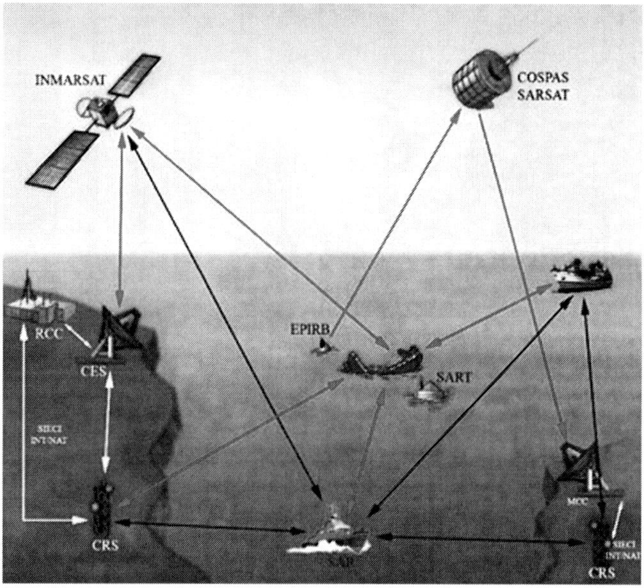

RCC – Rescue Coordination Center; CES – Coast Earth Station of the INMARSAT; CRS – Coast Radio Station using MF/HF/VHF; MCC – Mission Control Center of the COSPAS-SARSAT system; SAR – Search And Rescue

Figure 3-22. The concept of the GMDSS system operation

Source: adapted from J. Czajkowski, *System GMDSS*, Wyd. Skryba Sp. z o.o., Gdańsk 2002, p. 5.

Summing up, as a modern communication system of the global range, the GMDSS, performs all the functions which it has been designed for. The most important of them are:[52]

- alarming – fast and efficient notification about a danger threatening a vessel at sea;
- coordinating connection – indispensable for the coordination of all the operations undertaken by sea and air units which participate in search and rescue actions;
- providing connection at the place of an accident – provided mainly by radiotelephony with the use of specially dedicated VHF channels;
- identifying the place of an accident – through the tracking of signals sent from emergency beacons;

[52] http://www.gmdss.republika.pl/index.html [accessed on 16th March 2015].

- transmitting information related to navigation safety – implemented with the use of the NAVTEX sub-system, the EGC satellite system;
- providing general connection and connection between two vessels (bridge to bridge) – reliable communication between vessels and on-shore networks providing the exchange of information with pilotage and towage services, correspondence referring to exchanging charts, reporting repairs, etc.

The Maritime Safety and Security Information System (MSSIS) has been developed by the U.S. Department of Transportation's Volpe Systems Center. The basic source of data is the AIS signal. The countries (most often the NATO members) which feed the system with data flows from their own AIS signal receivers also obtain data flows in return from other users of the system. The AIS signal is collected from both stationary and mobile receivers. The acquired data are neither analysed nor stored – they are displayed as the pure AIS signal. In this way, a display of vessel traffic all over the world is provided (compare Figure 3-23). The system can be accessed through the Internet connections protected with passwords.[53]

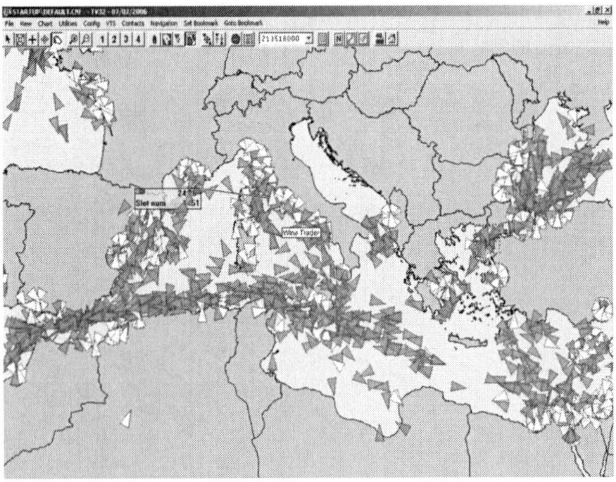

Figure 3-23. The interface of the MSSIS application

Source: adapted from materials provided by the Centre of Naval Cooperation and Guidance for Shipping (NCAGS) at Maritime Operation Centre (MOC), Gdynia Poland. [c]

[53] http://www.fisa.cl/marketing/EXPONAVAL_2008/PDF%20Conferencistas/GMDA/dia_4/2nd%20WHMDAWS%20CDR%20Dave%20Wirth.pdf [accessed on 17th July 2015].

Most European countries are contributors and beneficiaries of that system. The AIS data flow from the national system feeds the MSSIS through the VOLPE server. The flow of the signal is presented in Figure 3-24.

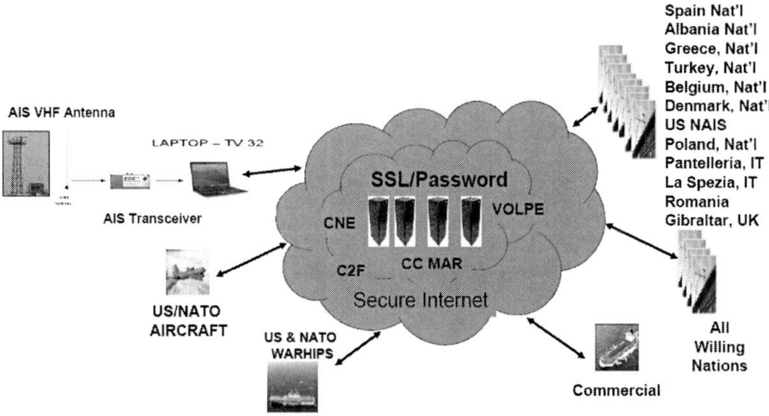

Figure 3-24. The flow of the signal in the MSSIS system

Source: adapted from https://mssis.volpe.dot.gov/Main/home/ [accessed on 20th February 2014].

Having received an access to the MSSIS system, users immediately get a chance to improve potential capabilities of navigation monitoring in a particular country, its agencies and maritime operators. The images are displayed with the use of the TV 32 software (Figure 3-25).

The TV 32 application is operated in the Windows environment, and because of its advantages, it can be used as:[54]

- a project platform for early warning about accidents at sea (Maritime Domain Awareness),
- an auxiliary system for pilotage (Pilot Navigation),
- a system for vessel traffic management (Vessel Traffic Management),
- a management system applied in a situation of oil spills and pollution (Oil Spill Modelling),
- an auxiliary system for after-collision procedures (Accident Investigation),
- a monitoring system of navigational mark positioning (Buoy Positioning),
- a system for recording and storing nautical information (Data Logging),

[54] R.K. Miler, *Systemy zobrazowania żeglugi i aplikacje informatyczne transportu morskiego…*, op. cit., p. 214.

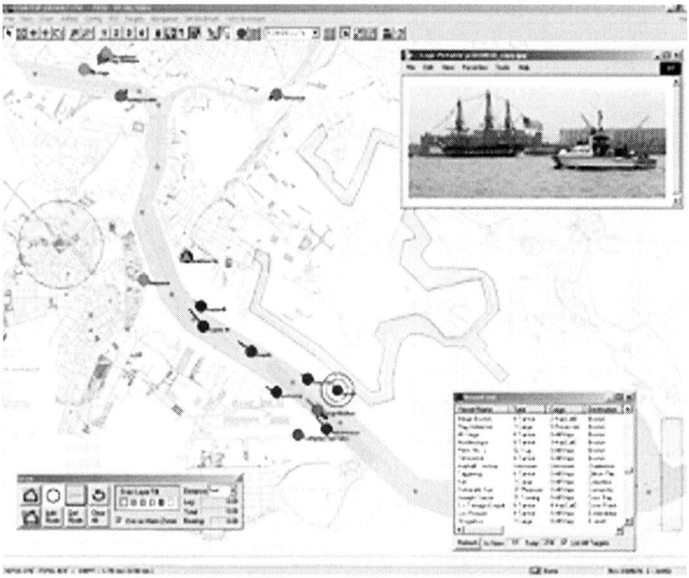

Figure 3-25. An image of vessel traffic in a selected area, as displayed in the TV 32 application for the MSSIS system

Source: adapted from https://mssis.volpe.dot.gov/Main/home/ [accessed on 23rd August 2016]. [c]

- a system for translating and exchanging data (Data Translation/Exchange),
- an auxiliary system for security of port facilities (Port/Harbour Security).

Another international complementary system, which is operated in the domain of maritime transport safety, is the NAMESIS (Naval Merchant Shipping Information System).[55] It has been developed for the requirements of the Royal Navy, and at present, it is used for military cooperation and guidance for navigation (NCAGS – Naval Cooperation and Guidance for Shipping)[56] by the Navy forces of other NATO countries. The display of vessel traffic is based on the mathematical reckoning of the movement parameters of a particular vessel (its course and speed) which are entered into the system by its operator (Figure 3-26).

[55] See more on: http://www.rncom.mod.uk/template.aspx?pageid=764&terms=namesis [accessed on 17th July 2015].

[56] M. Kościelski, R. Miler, *Naval Co-operation and Guidance for Shipping (NCAGS)*, op. cit., p. 16.

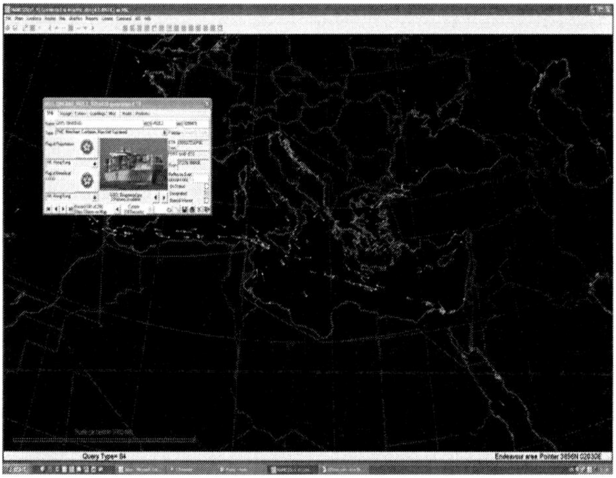

Figure 3-26. The interface of the NAMESIS application

Source: adapted from materials provided by the NSC (NATO Shipping Center), Northwood, Great Britain and the Centre of Naval Co-operation and Guidance for Shipping at Maritime Operation Centre COM-DKM, Gdynia Poland. [c]

The system offers advanced analytical capabilities which can be used to process the entered data through the use of dead reckoning[57] and "estimate movement" display, along the planned route (estimated position). However, operation of the system requires many operators, especially when it is necessary to enter large amounts of new data. Due to the manual handling of the system, it works in a situation when no other systems are accessible (e.g. when the AIS or LRIT signals are switched off or jammed).

The VMS (Vessel Monitoring System)[58] comes as another representative example of international complementary systems. It is dedicated to the monitoring of fishing vessel movement, mostly to prevent illegal, unreported and unregulated fishing (IUU fishing).[59] This system is compulsory for the EU

[57] In navigation, dead reckoning is the process of calculating one's current position by using a previously determined position, and advancing that position based upon known or estimated speed over the elapsed time and course – the Author's note.

[58] http://www.commercial-fishing.org/electronics/vms-reimbursement-program-news-a174.html [accessed on 13th September 2017].

[59] IUU fishing diminishes the amount of fish, destroys marine habitats, disturbs competition, and it is detrimental to the social interest. Still, because of the imposed fishing limits, IUU fishing becomes more and more common – the Author's note.

vessels of the length over 15 m (since 1st January 2012 – for vessels of the length over 12 m) and for vessels under non-EU flags which operate in European fisheries. Hence, vessels of the same size, operating on the EC water areas under a non-EU flag, must have a serviceable satellite tracking device installed on board.[60]

A signal carrying information about a vessel, its position, course and speed, with data referring to the fishing, is sent by satellite to the national Fisheries Monitoring Centres. Thanks to that fact, the range of the system is practically unlimited (Figure 3-27).[61]

The EU countries and highly developed countries in the world (including the USA) have been working intensively on preventing unfair operators from earning profits on IUU fishing.[62]

At present, control over fishing does not involve only checking the activities performed by fishermen at sea, as it turns out to be disproportionally expensive, and it does not provide any guarantee that legal regulations are respected. Inspections are carried out at each stage, starting from fishing vessels and ending with fishmongers. Fishing vessels cannot even leave their ports without a valid fishing permit.

Although inspections are still carried out at sea, the main control operations refer to:

- ports where fish is unloaded or handled,
- fish transport,
- fish processing factories,
- markets where fish is sold.

[60] R.K. Miler, *Systemy zobrazowania żeglugi i aplikacje informatyczne transportu morskiego…*, op. cit., p. 216.

[61] R. Miler, *Bezpieczeństwo transportu morskiego…*, op. cit., p. 249.

[62] Based on the information provided by regional organisations for fishing management, a list of vessels involved in IUU fishing is regularly published. The EU ordinance on IUU fishing defines a possibility of placing countries which tolerate illegal fishing on the black list. The EU operators who are involved in illegal fishing in any place in the world and under any flag must face high financial penalties which are proportional to economic value of such fishing, and they are deprived of all the profits earned in this way. At each stage of the production chain, for each batch of fish, relevant information should be submitted to prove that fishing has been performed legally – the Author's note.

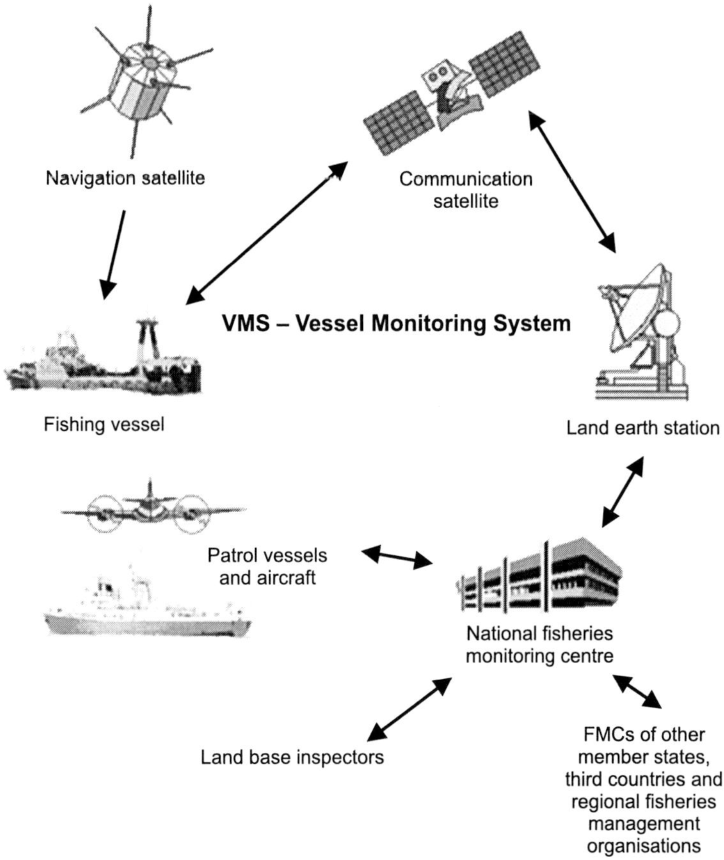

Figure 3-27. A flow of signals in the VMS system

Source: adapted from http://ec.europa.eu/fisheries/cfp/control_enforcement/vms_en.htm [accessed on 3rd March 2015].

The fishing control system broadly applies modern technologies in order to guarantee efficient monitoring and control over the fishing fleet. Such technologies are intended to facilitate an access to actual data related to fishing and to provide multi-aspect verification of information coming from various sources. The ERS (Electronic Reporting System) is one of the solutions applied in this field. The ERS is used for recording fishing data (fishing catch, unloading, sale) and passing such data to the authorities responsible for fishing in the member countries. The system has its counterparts in highly developed countries. It replaces paper logbooks and sales documents; hence, it is often referred to as an e-logbook. Figure 3-28 presents the functions of the system.

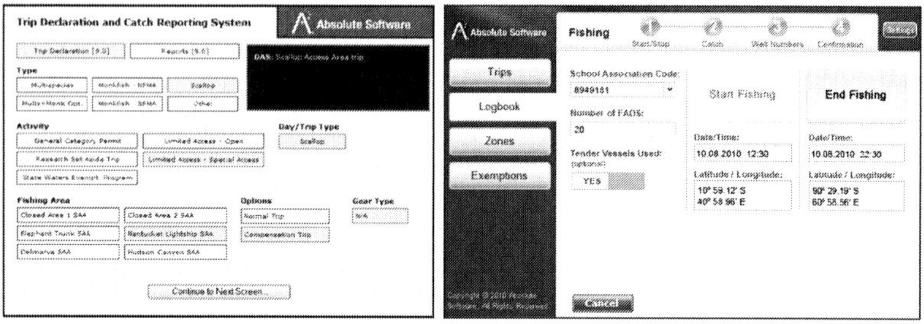

Figure 3-28. The Electronic Reporting System (an e-logbook tab)

Source: adapted from materials provided by NOAA's National Marine Fisheries Service ("NOAA Fisheries") at http://www.nmfs.noaa.gov [accessed on 11th February 2015].

The EU and highly developed countries foster the common use of the Vessel Detection System (VDS). Using satellites (providing satellite images of sea areas), the system can track and identify fishing vessels at sea and at ports.[63] The VDS interface (the operator's main screen) is presented in Figure 3-29.

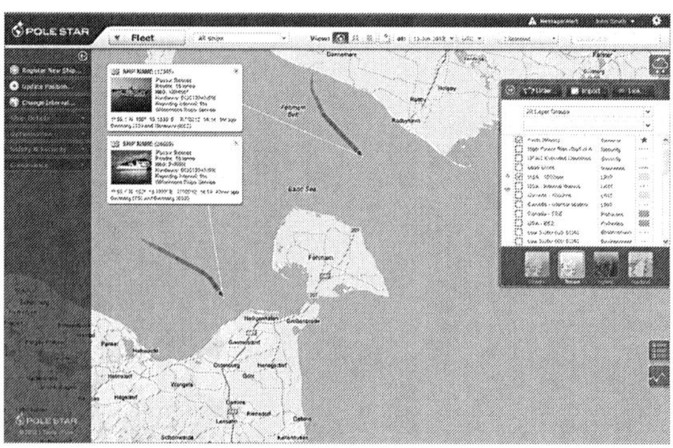

Figure 3-29. The VDS operator's screen

Source: adapted from http://fisheries.polestarglobal.com/vessel-monitoring [accessed on 22nd February 2015]. [c]

[63] In accordance with the EU regulations (the EU Ordinance 1224/2009) the bodies responsible for fishing monitoring should be provided with proper technical equipment that would allow them to use the VDS – the Author's note.

The efficient use of modern technologies in the field of monitoring and surveillance of fishing is implemented by the Fishing Monitoring Centre (FMC). The tasks of the Fishing Monitoring Centre involve in particular:[64]

- entering data on fishing into the Electronic Reporting System (ERS), based on information received directly from fishing vessels (e.g. in the vCatch system – Figure 3-30);

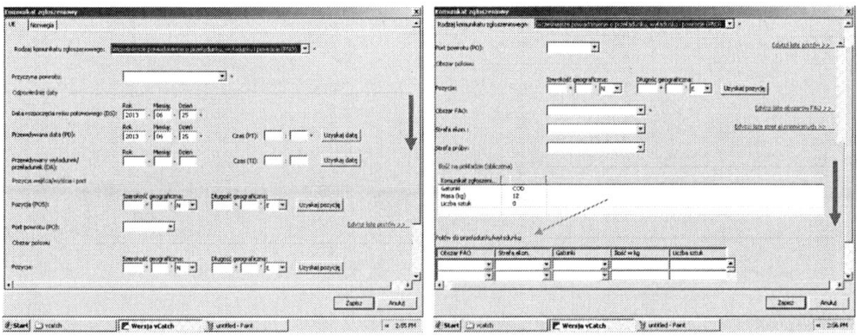

Figure 3-30. The interface of the ERS vCatch system

Source: adapted from http://www.cmr.gov.pl/files/cmr/ogloszenia/pno.pdf [accessed on 22nd February 2015].

- satellite monitoring of fishing vessels (VMS):
  - fishing vessels under their own state flag, regardless of the water areas or ports where they stay,
  - fishing vessels of the EU member countries and fishing vessels of the third countries, operating in the territorial water areas or water areas under the jurisdiction of a particular country (Figure 3-31);

- recording and storing VMS and ERS data received from fishing vessels under their own state flag or registered at their national ship register, and data received from fishing vessels under the flag of another member state (MS) in a digital form for three years;
- transmitting data on fishing vessels under their own state flag, operating in the water areas of a particular country to the FMC, VMS and ERS data in particular;
- cooperation with inspection services of other EU member countries, the European Commission and the European Fisheries Control Agency (EFCA);

[64] http://www.cmr.gov.pl [accessed on 24th March 2015].

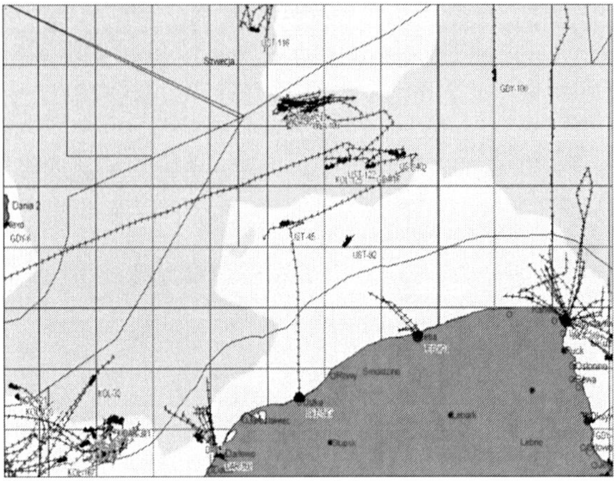

Figure 3-31. The interface of the VMS application

Source: adapted from materials provided by the Fishing Monitoring Centre in Gdynia, Poland. [c]

- administering, managing and maintaining the continuous operation of the following IT systems: ERS, VMS, SMS notifications, SIRM, Sales Note, vCatch; entering those data into the EU Fisheries Data Exchange System (FIDES III).

The example of the VMS indicates that numerous countries implement their national strategies for providing maritime safety (safety, security and eco), developing diversified tools for the monitoring and control over the processes which take place in transport and, more broadly, in the maritime domain. Such systems are implemented in the Border Guard and Maritime Administration in order to achieve their intended aims.

Hence, the institution which – by definition – must be continuously able to generate images of the situation in the domain of maritime safety (and maritime transport safety) at the national level is the Border Guard. Maritime Border Guard units are equipped with the Coastal Monitoring Solution (CMS)-class systems.[65] The Transas CMS is one of such systems. Based on the specific Transas Vessel Traffic Monitoring software, the system is capable of providing the Coast Guard/Board Control with real time data on any activities occurring

[65] http://www.transas.com/products/vtms/coastal-surveillance-systems/CoastalMonitoring [accessed on 27th July 2017].

in coastal zones. A multiple sensor (AIS and Radar) tracking instantly detects even the smallest targets, supporting decision-making processes. It helps to prevent illegal activity in a costal zone and to protect critical on-shore infrastructure from intruders. The main stakeholders of this systems are the Coast Guard, the Navy, Port Authorities and Coastal Administration. The Transas CMS is equipped with all the necessary sensors and functionalities to provide:[66]

- accurate detection, tracking and identification of small and high-speed targets even in the worst weather conditions (darkness, fog, heavy rain),
- use of the additional Thermal Vision sensor (ideally suited for low-light situations),
- full control over the entire coastal area and nearby territories,
- real-time Vessel Traffic Monitoring,
- detection of piracy, smuggling and illegal activities,
- configurable line (border) crossing,
- automatic alarm generation and display,
- target classification (types of ships),
- SAR aircraft tracking,
- data provision for SAR operations,
- synchronous recording and playback of all data and tracks for incident analysis,
- images and movement captured for documental evidence.

Another specific type of (national) systems includes systems applied by Maritime Administration. Data for such systems are provided by offices, ship-owners, ship-operators, agents and captains, and they are addressed to the Minister responsible for maritime economy and (most often: Maritime Administration) institutions the Minister has under surveillance and control. Communication of the system with external users (agents, ships) is provided by the Internet, so it can take place from any place in the country or in the world.[67] Figure 3-32[68] presents a representative system dedicated for the requirements of Maritime Administration, the PHICS[69] system.

---

[66] Ibidem.

[67] R. Miler, *Bezpieczeństwo transportu morskiego...*, op. cit., p. 246.

[68] More information available at http://www.ums.gov.pl/modules.php?name=News&file=article&sid=366 [accessed on 27th July 2017].

[69] PHICS Polish Harbours Information and Control System has been operated since 2004. It is an electronic system applied to exchange documents related to surveillance and control over maritime transport provided by the Polish Maritime Administration. It has been developed based on a twinning agreement entered by the Polish and British governments.

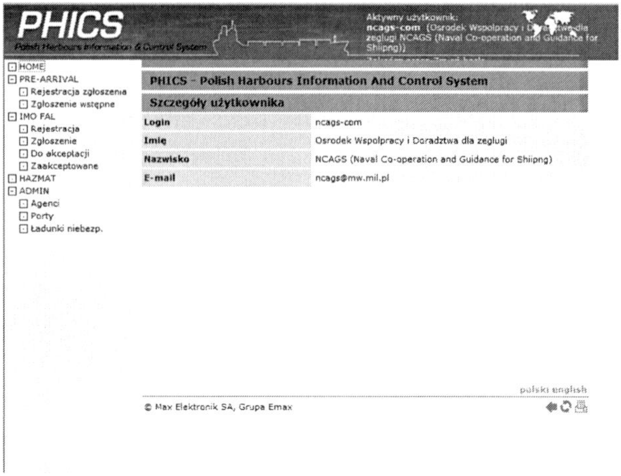

Figure 3-32. The interface of the PHICS application

Source: adapted from materials provided by Centre of Naval Co-operation and Guidance for Shipping (NCAGS) at Maritime Operation Centre (MOC), Gdynia.

The system consists of five substantively separate elements which refer to:[70]

- control over vessel arrivals (IMO FAL),
- passenger registration (PAS-REG),
- monitoring seagoing vessel traffic in terms of hazardous cargo transport (HAZMAT),
- control over foreign vessels provided at the country (PSC),
- documents proving qualifications, certificates required for navigation and seamen's examination (STCW).

The system is closely related to another operational application which supports work of Maritime Administration at a more integrated level: the THETIS system. It is an informational system applied to support a new inspection regime, the Port State Control[71] (PSC) which has been implemented by the EU Directive

The development and implementation of the PHICS system consist in the adjustment of standards applied in the operation of the Polish services of Maritime Administration to the standards applied in the EU countries – the Author's note.

[70] Materials provided by the Maritime Office in Gdynia http://www.umgdy.gov.pl [accessed on 17th June 2014].

[71] The Directive 2009/16/EC of the European Parliament and of the Council of 23 April 2009 on Port State Control imposes a necessity of establishing such an operational system; the EMSA was responsible for the project management, funding and development of the

on the Port State Control. The THETIS is dedicated not only to the EU member countries but also to a larger group of interested countries which have ratified the Paris Memorandum of Understanding on Port State Control (Paris MoU).[72]

In order to facilitate scheduled PSC inspections, the THETIS is connected to the SafeSeaNet. The SafeSeaNet provides information about vessels, their estimated time of arrival at ports of the member countries (72- and 24-hour notifications). The THETIS indicates vessels on which the inspection should take place first, and it allows its users to record the outcomes of the inspection. Based on the THETIS, the reports on the inspections carried out on particular vessels become accessible for all the PCS services of the member countries and the countries of the Paris Memorandum.[73]

The THETIS also uses numerous databases related in various ways to maritime safety, including those provided by classification associations, the EU systems and national systems of the particular member countries. Collected in such a way, the data allow inspectors to form a full picture of a vessel.

Another group of advanced systems for monitoring maritime transport safety (with particular consideration of maritime navigation and port-terminal operations) are commercial systems developed on the basis of the AIS signal. Commercial systems are usually developed after the implementation of central systemic regulations and after the implementation of new technologies in national monitoring systems (non-commercial by definition) of maritime transport safety (the AIS has become such a system – the Author's note). The representative examples of such systems include the following applications:[74]

- AIS Live,
- Marine Traffic,

---

system. The development work started in November 2008, in close cooperation with the member countries and the European Commission. The work on system development, including the connection of the THETIS with the SafeSeaNet and its deliverance, ended at the end of November 2010. The THETIS was recognised as fully functional and ready to take on a new inspection system on 1st January 2011. It was preceded by comprehensive training of PSC inspections with regard to the new regime – the Author's note.

[72] These are Canada, Croatia, Iceland, Norway and Russia. See more in: L. Szcześniak, A. Weintrit, *Europejskie systemy kontroli i śledzenia ruchu statków – geneza, zasady funkcjonowania oraz perspektywy rozwoju*, op. cit., pp. 86–88.

[73] L. Szcześniak, A. Weintrit, *Europejskie systemy kontroli i śledzenia ruchu statków – geneza, zasady funkcjonowania oraz perspektywy rozwoju…* op. cit., p. 87.

[74] R.K. Miler, *Systemy zobrazowania żeglugi i aplikacje informatyczne transportu morskiego…*, op. cit., p. 221

- Vessel Tracker,
- exactAIS™ (based on the satellite AIS-S signal),
- Gatehouse AIS Display System.

The AIS Live is the first commercial service which uses its own receivers of the AIS system in order to provide full monitoring of navigation in the water areas which are crucial in terms of economic interests.[75] The interface of the application which is accessible through the Internet is presented in Figure 3-33.

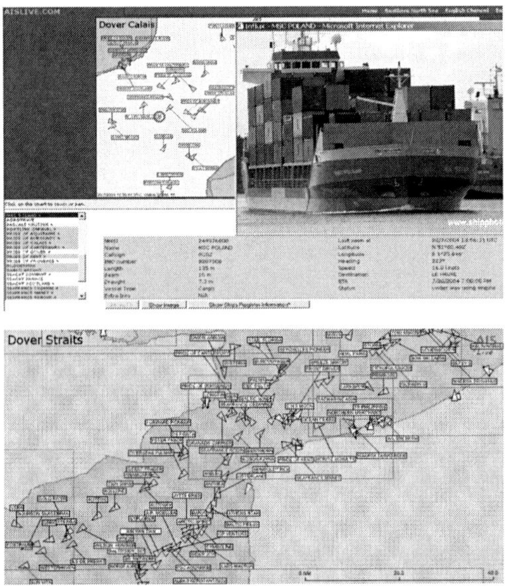

Figure 3-33. The interface of the AIS Live application, Dover Straits traffic from AIS Live

Source: adapted from materials provided by the AIS Live at http://www.aislive.com [accessed on 3rd December 2008]. [c]

The AIS Live is available against a subscription fee, it requires a login and a user's password; when these elements are submitted correctly, the system provides authenticated navigational information and vessel traffic monitoring from the water areas of Europe, North America, the Caribbean Sea, the Mediterranean Sea and regions of the Far East. The key functions of the application include:

[75] http://www.aislive.com [accessed on 28th March 2015].

- full vessel monitoring along all the communication routes,
- display and vessel monitoring until the moment when vessels reach their ports and berthing places,
- the AIS flow in real time (in accordance with the AIS standard),
- e-mail notifications about events taking place in the previously defined zoncs,
- direct access to the database of HIS Fairplay,
- stable chart tiles (Tile Based Chart).

The website www.MarineTraffic.com allows users to enter another application for navigation monitoring, based on the AIS signal. The application offers data which refer to the current position and the register data of vessels, their estimated route, destination ports, estimated time of arrival (ETA). The current photos of vessels are also available with some data on their history. The interface of the application is presented in Figure 3-34.

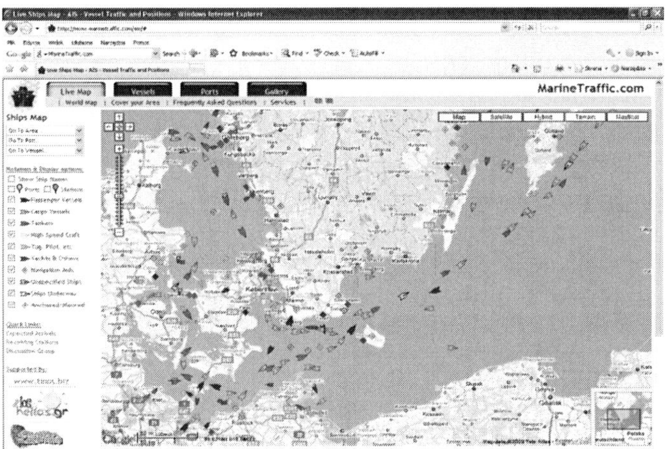

Figure 3-34. The interface of the Marine Traffic application

Source: adapted from materials provided by Marine Traffic at http://www.marinetraffic.com/ais/ [accessed on 16th February 2014]. [c]

The access to the basic system of vessel traffic display is free of charge and registering is not required. However, after user's registration some options become available which considerably extend the functionality of the system. They include:[76]

- My Fleet – a possibility to define the user's own group of vessels that will undergo the process of close monitoring;
- My Notifications – a possibility to receive e-mail notifications with information about the previously defined vessels reaching their port of destination, the current position of a selected vessel (ultimately also SMS notifications) or estimated time of arrival at a particular port (ETA information);
- My Points – a possibility to define points or areas of special interest and to receivie notifications about each arrival of the vessel to a particular point or area.

Another commercial application rooted in the Internet and based on the AIS signal is the VesselTracker. It provides a full range of navigational information which can be applied by logistics operators related to land-sea supply chains.[77] The VesselTracker was developed in 2006, and at present it has over 210 thousand registered users and over 1 million hits per month. Information from the AIS is completed with additional information from the radar systems. The basic function of that software (navigation display) does not require user registration, but after registration users are provided with an access to some additional functions of the application, such as:[78]

- vessel tracking in real time with the data referring to vessel movement (the course, speed, ETA, NPOC (next port of call);
- access to technical specifications and history of a vessel;
- focus on the previously defined areas;
- possibility of applying Google Earth chart tiles in order to display the details of a port and berthing places, in accordance with the positions in the AIS Live system (Figure 3-35);
- history of port calls with the precise time of stay;
- a notification system providing information about previously defined events through e-mail, SMS or phone (voice message);
- management of a previously defined fleet of vessels;

[76] R.K. Miler, *Systemy zobrazowania żeglugi i aplikacje informatyczne transportu morskiego...*, op. cit., p. 217.
[77] http://www.vesseltracker.com/en/Home.html [accessed on 16th February 2014].
[78] R.K. Miler, *Systemy zobrazowania żeglugi i aplikacje informatyczne transportu morskiego...*, op. cit., p. 218.

- a mobile application for phones (smartphones and tablets) equipped with an intelligent operational system e.g. Android;
- the XML system which provides information for the internal systems of the user without the necessity of changing and buying some expensive software.

Figure 3-35. The interface of the VesselTracker for the application based on the Google Earth tiles

Source: adapted from materials provided by VesselTracker at http://www.vesseltracker.com/en/Home.html [accessed on 23rd May 2016].

Another commercial system is the ExactAIS™. It is the first fully global commercial system based on the micro-satellite technology. The main idea of the system is application of the AIS display transmitted by satellites – not by on-shore stations of the horizontal AIS system which has been applied so far.[79] The system fosters the initiative of the MSA, and it is applied by numerous government institutions and selected economic institutions to provide them with an access to managerial information in the field of navigation safety. The full access to information is granted after user registration via the server of the central monitoring and the data processing system administered by the Exact Earth consortium and the GateHouse company. Figure 3-36 presents the display of the exactAIS™ system (in its global version).

---

[79] http://www.exactearth.com [accessed on 23rd May 2016].

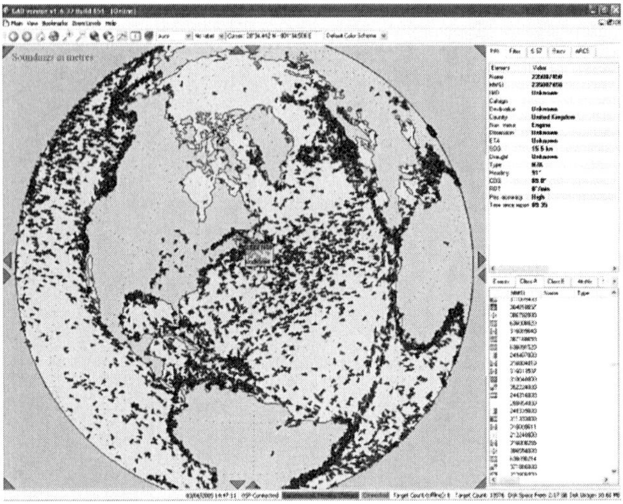

Figure 3-36. The interface of the exactAIS™ system

Source: adapted from materials provided by Exact Earth at http://www.exactearth.com [accessed on 23rd May 2016].

The ExactEarth and its product, exactAIS™, meet all the additional requirements set for modern solutions in the field of navigation monitoring systems described in the previous systems. Furthermore, the following features are provided:[80]

- full customisation of the service upon the user's request (all the modules of selection, anomaly detection in previously defined areas, etc. are accessible);
- full systemic support for users in terms of software and hardware, offered online 24/7;
- full compatibility with national systems functioning within the MSA initiative.

The GateHouse is another company which commercially deals with navigation monitoring systems. It cooperates with the exactEarth in terms of using the AIS satellite signal to provide a display of vessel traffic (Figure 3-37). The GateHouse offers a wide range of additional functions of the system which

---

[80] R.K. Miler, *Systemy zobrazowania żeglugi i aplikacje informatyczne transportu morskiego...*, op. cit., p. 222

reach beyond the field of tracking and vessel monitoring.[81] In cooperation with the IALA (International Association of Lighthouse Authorities), the GateHouse has developed a tool for evaluation of risk and nautical threats related to congestion – IWRAP MK II.

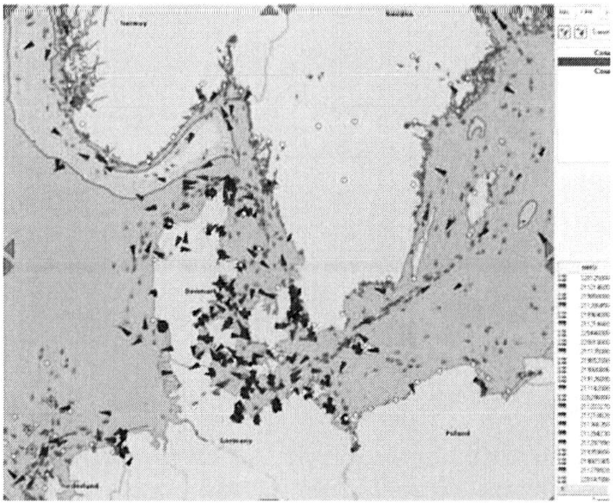

Figure 3-37. The interface of the GateHouse system

Source: adapted from materials provided by GateHouse at http://www.gatehouse.dk/Maritime-Surveillance-479.aspx [accessed on 17th March 2014].

The GateHouse offers a dedicated system of display which does not result in any system overloads and prevents information entropy. The access to the application is limited (registered users only) and customers may use the following tools:[82]

- defined filtering of images after the individual criteria (filters) have been set, e.g. vessels moving at speed above 10 knots, tankers, vessels under a particular flag, vessels of a particular draft, vessels carrying a particular type of cargo (according to the IMO categories) or heading to a particular port, etc.;
- possibility of sending text messages to all vessels in a specified area or to vessels grouped in one fleet (Figure 3-38);

[81] http://www.gatehouse.dk/Maritime-Surveillance-479.aspx [accessed on 17th March 2014].
[82] R.K. Miler, *Systemy zobrazowania żeglugi i aplikacje informatyczne transportu morskiego...*, op. cit., p. 223.

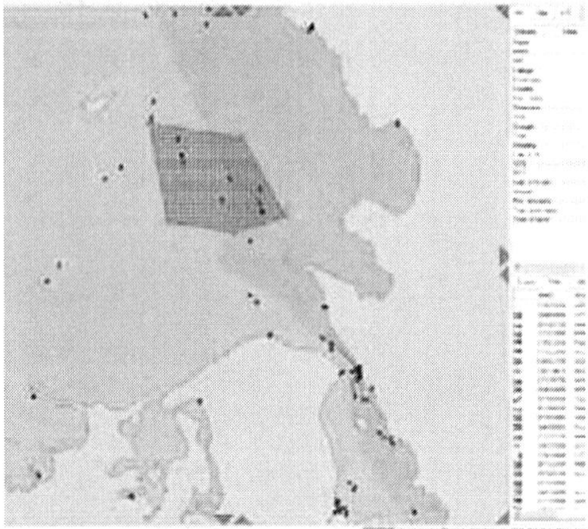

Figure 3-38. The interface of the GateHouse system shown in the tab of the text message handling system (sending SMS, txt, fax, e-mail to all the vessels in the defined water area)

Source: adapted from materials provided by GateHouse at http://www.gatehouse.dk/Maritime-Surveillance-479.aspx [accessed 17th March 2014].

- a histogram of crossing a defined line, with consideration of the precise intersection point of that line and the course of a vessel, which is applied to monitor traffic volume and potential congestion (Figure 3-39).

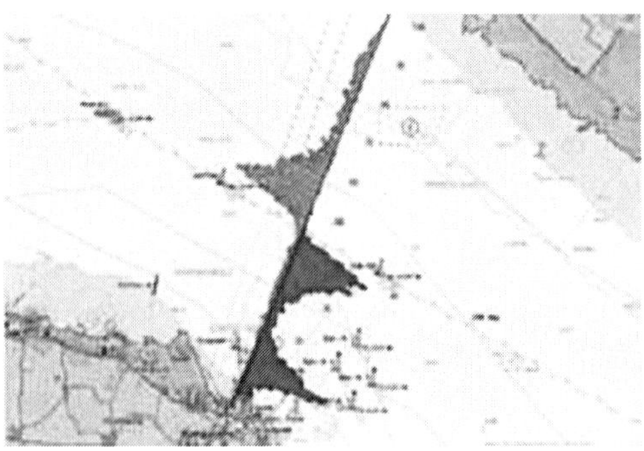

Figure 3-39. The interface of the GateHouse system in the histogram tab

Source: adapted from materials provided by GateHouse at http://www.gatehouse.dk/Maritime-Surveillance-479.aspx [accessed 17th March 2014].

In order to provide a complete analysis of systems applied in the process of the monitoring and maintenance of the assumed safety level in maritime transport, it is necessary to refer to one more group of telematics systems used for communication and transmission of information between vessels and the shore. Frequently, such systems come as the sub-systems of other more developed systems, such as the GMDSS, for example:

- INMARSAT (the system of satellite communication),
- NAVTEX (the NAVigational TEXt Messages).

The system developed by the British INMARSAT company provides satellite connection between vessels, maritime operators, ship-owners and other entities of maritime economy.[83] It is one of the oldest and the most reliable connection systems, based on geo-stationary satellites, which offers a full range of modern connection services in the shore-vessel and vessel-vessel relation (Figure 3-40), regardless of atmospheric conditions and the current position of a vessel. As the only satellite system, it meets the GMDSS criteria. Undoubtedly, the key factors of that system include its global range, reliability, universality, safety of data transmission, according the VPN standard (coding and decoding).

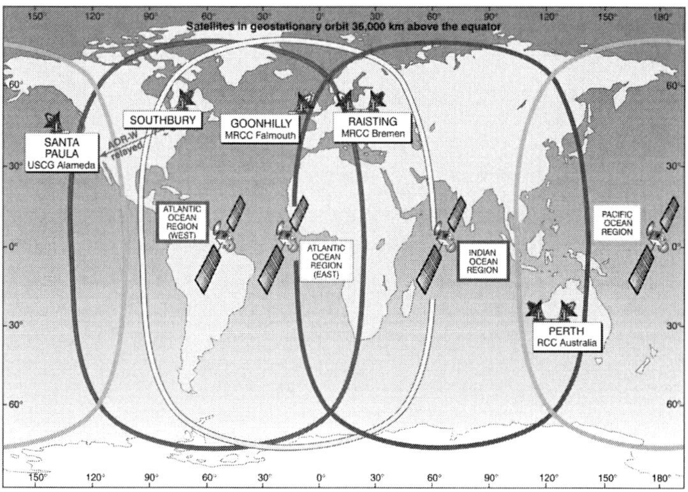

Figure 3-40. The range of satellite connection provided by the INMARSAT system

Source: adapted from materials provided by INMARSAT at http://www.inmarsat.com [accessed on 4th April 2013]. [c]

[83] http://www.inmarsat.com [accessed on 4th April 2013].

The INMARSAT offers a number of standard solutions which are adjusted to the requirements of the customers, including:[84]

- Fleet Broadband – parallel voice connection and packet data transfer up to 432 kbps;
- Fleet 77 – packet: voice, fax, 64/128 kbps ISDN, packet data, GMDSS;
- Fleet 55 – packet: voice, spot beam fax, 64 kbps ISDN, packet data;
- Fleet 33 – packet: voice, spot beam fax, 9.6 kbps circuit-switched data, packet data;
- FleetPhone – low cost voice connection and data transfer up to 2.4 kbps.

Additionally, the solutions currently handled by the INMARSAT include:[85]

- e-mail and webmail service, SMS, other formats of short text messages,
- the update of digital maritime charts and weather forecast in real time,
- transmission of the GMDSS signal,
- access to the broadband Internet and the Internet,
- safe communication (coding and decoding),
- transfer of large files,
- a communication system for the crew,
- a telemetric system and a system of remote operation of the drive (telemetry of engines),
- video-conferences.

The NAVTEX ("NAVigational TEXt Messages") is also a part of the integrated GMDSS system. It provides telexing information, which is important for navigation safety, from on-shore stations dedicated for such transmission to vessels which are equipped with relevant receivers. The system is operated on the 518 (main) and 490 kHz (alternative) frequencies, and depending on the conditions, its range is approximately 400 nautical miles from the transmission station.

Each station has its own letter code (one letter). Messages are attributed to the following categories:[86]

- distress call,
- navigational warning,

---

[84] R.K. Miler, *Systemy zobrazowania żeglugi i aplikacje informatyczne transportu morskiego…*, op. cit., p. 224.

[85] Information has been obtained on the basis of a description of the possibilities offered by the system at http://www.inmarsat.com [accessed on 27th July 2016] – the Author's note.

[86] R.K. Miler, *Systemy zobrazowania żeglugi i aplikacje informatyczne transportu morskiego…*, op. cit., p. 224.

- weather forecast,
- operation of navigation systems etc.

Additionally, each category has its own letter code. The receiver installed on a vessel should be set to receive messages from a particular station (which transmits information in the water area in which the vessel operates or enters), and it is possible to define which information is not to be received (e.g. referring to navigation systems which are not installed on that particular vessel) – as presented in Figure 3-41.

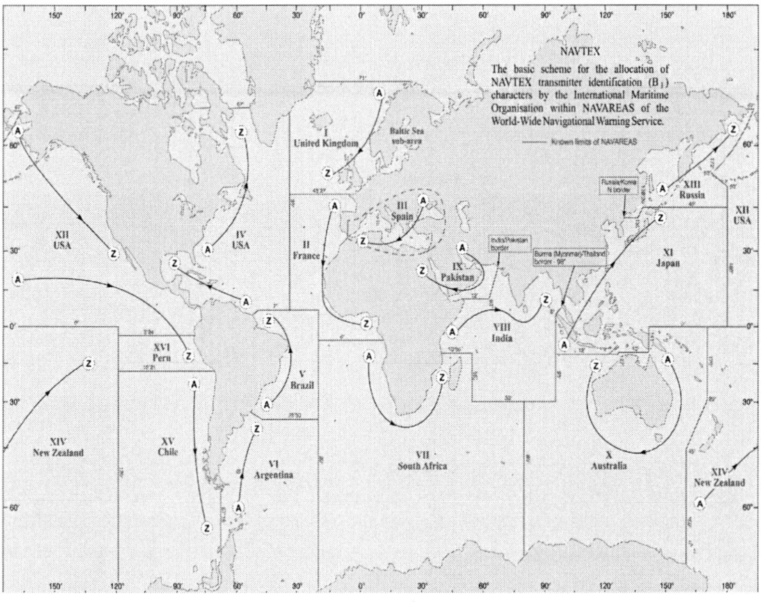

Figure 3-41. The range/regions of the Navtex system

Source: adapted from materials provided by Navtex at http://www.gmdss.com.au/navtex.htm [accessed on 16th June 2012].

The receiver on a vessel automatically receives all the information, it eliminates errors as far as possible, it recognises the letter code to define whether the message comes from a proper station and whether it belongs to a proper category, and then it prints out the message.[87]

[87] http://www.gmdss.com.au/navtex.htm [accessed on 16th June 2012].

Despite the experimental integrating approach (IMDatE), the concept implemented so far assumes that all the monitoring systems have been developed independently, and there is no obligation imposed on these systems to cooperate in order to provide a multilayered and integrated display of the situation in the domain of maritime transport. The architecture of such a system based on the co-existence of the systems which are independently and individually developed is presented in Figure 3-42.

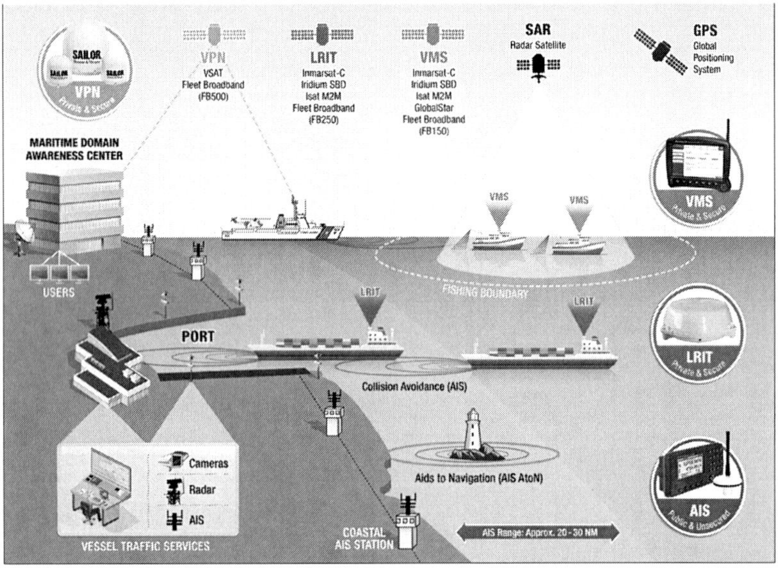

Figure 3-42. The architecture of independent monitoring systems of maritime transport safety

Source: adapted from http://sevenseas-marine.com/services/ [accessed on 5th February 2015].

Still, maritime transport is not only navigation – it also involves port and terminal operations. In order to face the growing scale of difficulties related to operations of the fastest loading and unloading (port/terminal handling of a vessel), it is necessary to support that process with intelligent telematics systems (monitoring and management of processes attributed mostly to the field of logistics) of maritime transport.

3.3. Monitoring and tracking systems of the main processes in maritime transport

3.3.1. TOS (Terminal Operation Systems) and ship-owner/transport operator systems in maritime transport (including monitoring emission)

Depending on the type and purpose of a port, each of its terminals is visited by several or several dozen ships daily. Hence, the processes of loading and unloading must be handled with exceptional precision. They must provide for the full scope of process comprehensiveness, that is namely a synchronised sequence of logistical activities at the point where maritime and land transport operations meet. The key aspect of timely and safe operations performed at port terminals is a possibility of precise *a priori* planning such processes, with consideration of real time information acquired from the systems of navigation monitoring discussed above, with the use of integrated TOS solutions (Terminal Operation System).[88] The TOS is an integrated and fully automated software system designed to manage container terminals,[89] to control delivery, storage, container processing and unloading operations at the container terminal, and to manage container terminal documentation in real time. The efficiency of the system is characterised by:

- transparent management and control of all operations,
- reduction of operating costs,
- faster processing of containers,
- optimisation of equipment usage,
- increased capacity of a container terminal,
- prevention of container loss,
- prevention of container theft,
- increased performance of the staff,
- faster processing of vessels, trains and trucks.

---

[88] There is a larger group of TOS systems (depending on the software producer), for example: SOTek, Navis SPARCS N4, nGen (developed by Hongkong International Terminals, the flagship terminal of Hutchison Port Holdings), CATOS Computer Automated Terminal Operating System applied at the port of Busan (Korea) and other places – the Author's note.

[89] According to Drewry Maritime Research: Global container fleet = 32.9 million TEU in 2012, dry container 92%, refrigerator 7%, tank 1% – the Author's note.

As intelligent software developed for efficient container terminal management, the TOS is dedicated to its main users, such as:

- marine container terminals,
- inland container terminals (dry ports),
- rail cargo terminals,
- ro-ro terminals and ports,
- break-bulk ports and terminals,
- multi-purpose facilities.

The main functions (modules) of the TOS are presented in Figure 3-43.

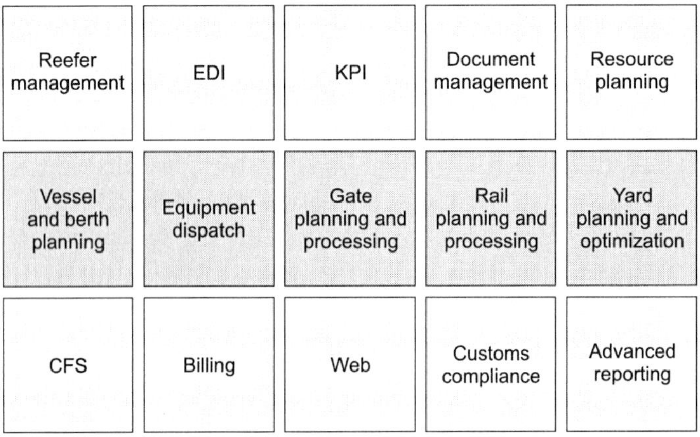

Figure 3-43. The main functions (modules) of the TOS

Source: adapted from http://www-old.solvo.ru/en/products/tos/ [accessed on 24th March 2018].

Another example of the TOS-class software is the SPINNAKER system (developed by Tide Works of the USA) which is able to provide IT support to the container handling process and, at the same time, also:[90]

- to provide time optimisation through the graphic presentation of a scenario referring to loading and unloading operations, with regard to the capabilities of hoisting facilities and gantry cranes;
- to provide proper sequence of container stacking and arrangement, with regard to the handling operations to be performed at the next port of call (NPOC);

[90] http://www.tideworks.com/products/spinnaker/ [accessed on 29th January 2016].

- to provide proper consideration of vessel stability and its metacenter (Figure 3-44).

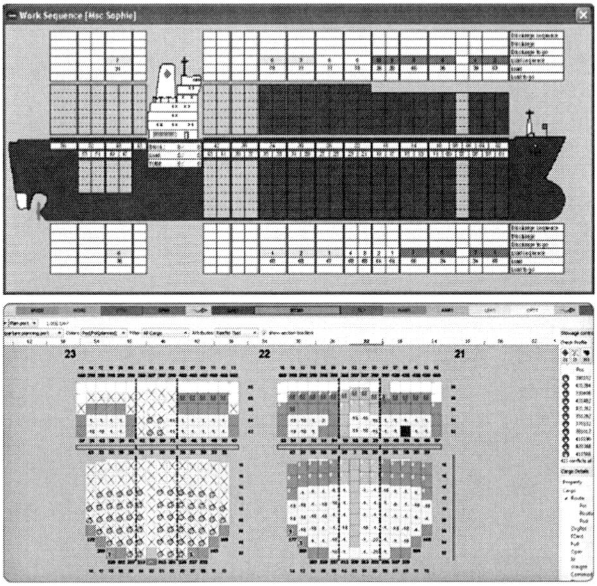

Figure 3-44. The SPINNAKER application window in the vessel planning function; the stowage plan

Source: adapted from materials provided by Tideworks at http://www.tideworks.com/products/spinnaker/ [accessed on 29th January 2016]. [c]

The system offers a number of operational benefits, and it considerably limits safety risks (including physical risks and security risks) during the performed operations through:[91]

- achieving high precision of operation planning with maintaining the required flexibility,
- shortening the stay at berth to its indispensable minimum,
- optimising the use of the own cargo handling facilities,
- shortening time required for operation planning and elimination of potential errors generated by the human factor,
- current monitoring of previously defined threats.

[91] R.K. Miler, *Systemy zobrazowania żeglugi i aplikacje informatyczne transportu morskiego…*, op. cit., p. 216

The Spinnaker Planning Management System® (the TOS management function) provides a possibility of integrating and harmonising basic information which affects the optimisation of the use of terminal space and facilities in one system.[92] Such information includes:

- the parameters of vessels and their cargo,
- occupancy and parameters of wharfs, piers and quays,
- the use of cargo handling facilities and storage places,
- availability and timing which refer to the use of land transport means (railway, trucks, container semi-trailers with maneuvering tractors, etc. – Figure 3-45).

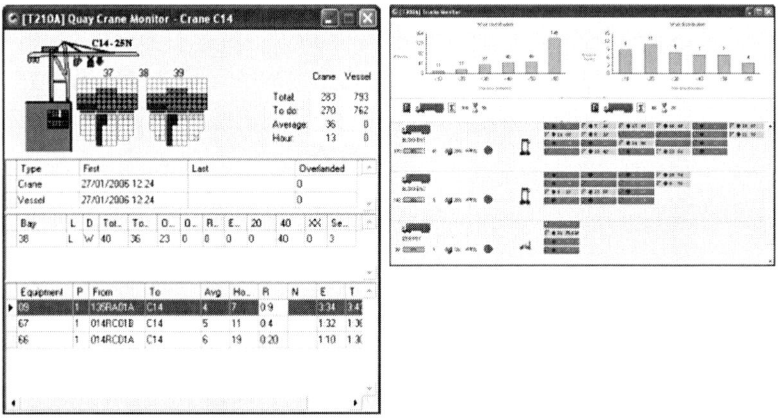

Figure 3-45. The windows of the Spinnaker Planning Management System® application

Source: adapted from materials provided by Tideworks at http://www.tideworks.com/products/spinnaker/ [accessed on 29th January 2013]. [c]

The Spinnaker Planning Management System® offers the following systemic tools:[93]

- Graphical Planning Tools – provides intuitive management of terminal space and container flows in the vessel – wharf – means of land transport relation;

[92] http://www.tideworks.com/products/spinnaker/ [accessed on 29th January 2013].

[93] R.K. Miler, *Systemy zobrazowania żeglugi i aplikacje informatyczne transportu morskiego...*, op. cit., p. 218.

- Vessel Berthing – provides the optimal use of wharf space, with regard to their cargo handling capability and vessel size;
- Yard Navigator – provides the maximal use of storage yards and operational space of a terminal; it allows operators to immediately locate the stored containers and to minimise the processes required to move containers in the terminal area;
- Vessel, Yard &Rail Editor – provides a possibility of defining parameters and detailed characteristics of vessels, resources of a container terminal and capacities of railway lines in order to optimise their mutual relations;
- EDI (Electronic Data Interchange) – allows entities which are involved in operations of a logistical chain to exchange information and documents in the EDI standard;
- Advanced Reporting – it streamlines the flow of documents in customs and control operations;
- Administrator-Controlled User Rights Management and Authentication Capabilities – allows the interested parties to manage and administer the system with a possibility to define procedures and access levels for various users of the system.

Another group of important IT systems which support maritime transport at the level of auxiliary processes (as viewed from the perspective of the ship-owner) includes applications supporting processes at the level of fleet management.

The AMOS application (Asset Management Operating System) developed by the SpecTec is an adaptation of a classic solution involving an integrated system of a logistical management system of the ERP-class (Enterprise Resources Planning) dedicated to ship-owners and maritime operators.[94] The system refers to the specificity of the industry and a number of particular requirements (e.g. classification requirements). It also provides a possibility of planning sailing routes based on the way points and data exchange among the system users through the AMOS mailing system and the Shipdex™ protocol.[95]

---

[94] http://www.spectec.net/catalogue.asp?t=maritime_businesssuite&lang=en [accessed on 28th August 2014].

[95] Shipdex™ is a set of business principles developed in order to provide standardisation of processes and procedures in the environment of maritime economy. It provides electronic information exchange based on the XML standards. See more in: A. Nerć-Pełka, P. Pełka, *Informatyzacja procesów zaopatrzenia na statkach morskich* [in:] M. Chaberek, A. Jezierski

The AMOS application is also equipped with the module for stock management based on the barcode system and other useful applications characteristic for the ERP-class systems, e.g. a module for planning and operational control (Figure 3-46).

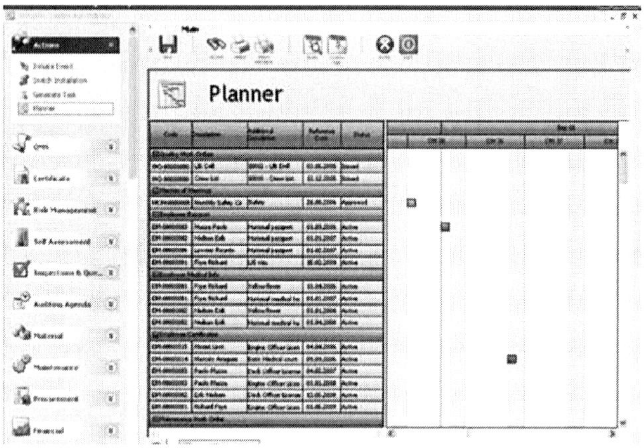

Figure 3-46. The AMOS application window presenting the Planner function

Source: adapted from materials provided by SpecTec at http://www.spectec.net/downloads/newsletters/SPEC_NEWS_MAY-Finalwithgraphchangejune2.pdf [accessed on 28th August 2014].

The AMOS software provides the use of basic modules, such as:[96]

- supply and material purchase,
- maintenance of technical efficiency and material stock and spare parts management,
- providing high quality of processes and safety,
- human resource management,
- monitoring the condition of devices, management of operational regimes and technical life in the AMOS Mimic CBM (condition-based monitoring) module,
- self-evaluation, evaluation and constant improvement of the managed system.

(ed.), *Informatyczne narzędzia procesów logistycznych,* CeDeWu Wydawnictwa Fachowe, Warszawa 2010, pp. 61–68 – the Author's note.

[96] http://www.spectec.net/catalogue.asp?t=maritime_mimic&lang=en [accessed on 28th August 2014].

The above-mentioned CBM (condition-based monitoring) modules come as another group of ship-owner sub-systems which are applied to optimise transport processes. They also considerably improve maritime transport safety (through monitoring and control over the risk related to drive system failures which might stop the vessel and as a result – might generate safety threats). The CBM systems use the combination of on-line monitoring (applying the above-mentioned connection systems) and apply sensory indicators presenting the condition of the key elements of the vessel systems. The technical condition of the main drive is monitored (mechanically through the analysis of vibrations), including temperature and oil pressure, system efficiency and many other factors of current exploitation (including energy efficiency and fuel consumption). All the components of the CBM system are presented in Figure 3-47.

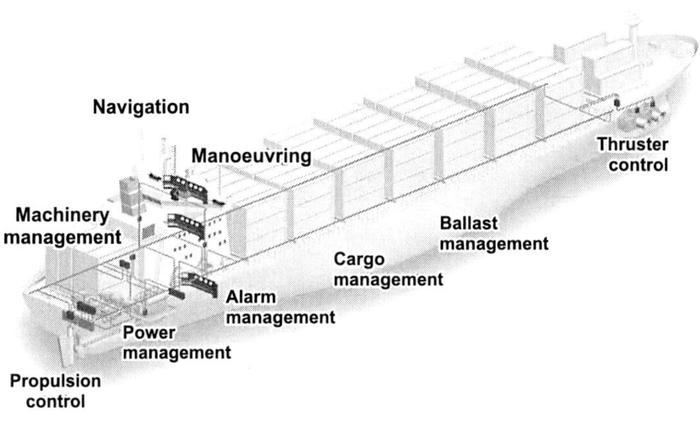

Figure 3-47. The components of the CBM system

Source: adapted from http://www.shippipedia.com/ship-automation-control-system/ [accessed on 24th March 2013].

Information acquired from the sensory sources on the board of the vessel is transmitted in real time (24/7) to the on-shore CBM system management centres in order to allow intelligent software to provide a current analysis and evaluation of the parameter status and make all the necessary decisions. The system provides a possibility of parameter archiving and providing an analysis of historical comparison (e.g. to a particular period or to a particular trip, etc.). The principle states that providing daily reports from the operational centre to the vessel is necessary along with providing immediate alerts whenever any significant anomalies and failures, which might affect further operation of the main vessel systems, are detected. The most often monitored systems are the vessel drive systems. It is especially important nowadays, during the implementation of unmanned engine rooms. The structure of such a system is presented in Figure 3-48.

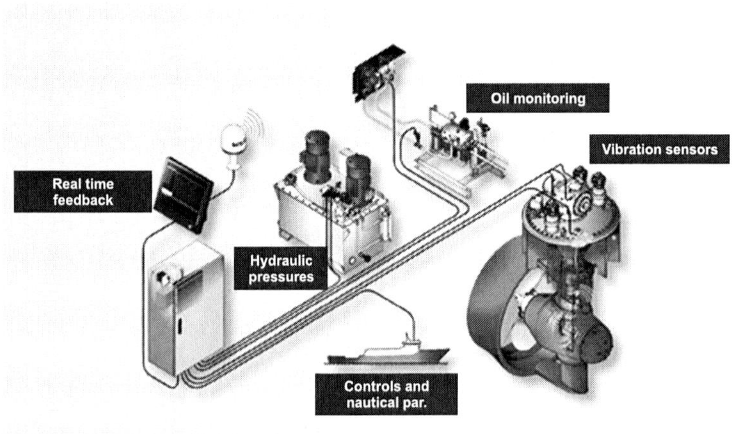

Figure 3-48. The conceptual scheme of the PCMS sub-system (Propulsion Condition Monitoring Service)

Source: adapted from M. Santala, *Condition monitoring provides key to ship automation*, Wärtsilä Corporation e-publication 2012, http://www.wartsila.com/en/condition-monitoring-provides-key-to-ship-automation [accessed on 24th March 2013].

The above-mentioned sub-systems are an important step towards the development of systemic and structural premises for the implementation of fully automated solutions which will allow interested entities to operate unmanned/autonomic seagoing vessels in the UMS (Unattended Machinery Space) regime in the near future, as it has been described in the previous chapters.

The last category of telematics systems in maritime transport (or more precisely: in maritime navigation) is closely related to the category of ecological safety monitoring.[97] Viewed in the context of anthropopressure factors

[97] R. Miler, A. Bujak, *Implementacja systemów monitoringu zużycia paliw i efektywności energetycznej statku jako pochodna dyrektyw proekologicznych UE w transporcie morskim*, "Logistyka" no. 2/2014, p. 24.

of transport, at present problems of natural environment protection have become extremely significant. International maritime transport accounts for 3% of GHG emission in the world and for 4% of emission in the EU, and this sector is the only mode of transport on which no obligatory reduction has been imposed so far in this respect. Despite this fact, since 2015, in accordance with the Directive of the European Commission on sulphur, vessels operating in the basin of the Baltic Sea, the North Sea and the English Channel have been obliged to use fuel with a drastically lowered sulphur content (0.1%). Additionally, a necessity has been gradually imposed on ship-owners and ship operators to monitor and to report on verified amounts of CO_2. The question refers to the volume of that gas emitted by vessels which head to the EU ports, from and between such ports. Ship-owners are also obliged to provide some other specific information, such as data indispensable for the calculation of vessel energy efficiency under the Ship Energy Efficiency Management Plan – the SEEMP. The guidelines of that plan are applied to lower GHG emission due to lower fuel consumption. Such an approach has provided favourable conditions for conceptual and implementation work on new on-board systems for monitoring fuel consumption and energy efficiency.

In order to implement the Directives of the European Commission on the limitation of GHG emission, some special systems for fuel consumption monitoring (in relation with the drive and energy production: main engine, auxiliary engines, boilers) are installed on majority of vessels. The data obtained from such devices are transmitted to the bridge and to the engine room of a vessel.[98] In modern telematics systems, such data can be also accessible in real time to relevant ship-owner's on-shore units. In this way, also the ship-owner is able to decide about the most favourable option with regard to the limited fuel consumption and GHG emission.[99] Figure 3-49

---

[98] http://www.polsteam.com/mniej-gazow-cieplarnianych-ze-statkow-pzm [accessed on 23rd December 2014].

[99] See more at: http://morzaioceany.pl/zegluga-morska/pzm/1070-mniej-gazów-cieplarnianych-ze-statków-pżm.html, Jurdziński M., *Innowacje technologiczne na statkach morskich w celu redukcji zużycia energii i emisji* CO_2, Zeszyty Naukowe Akademii Morskiej no. 77, Gdynia 2012; BAS-WASGiPU/WAPEiM-1909/ The opinion on the communication of the Commission to the European Parliament, the European Council, the European Economic and Social Committee and the European Committee of Regions: *Włączenie emisji z transportu morskiego do unijnej polityki w zakresie ograniczenia emisji gazów cieplarnianych* (COM(2013) 479 *final*) – non-published materials provided by the Bureau of

presents an example of a system applied for the management of fuel consumption, installed on a vessel and connected with a system for GHG (CO_2, NO_x) emission control.

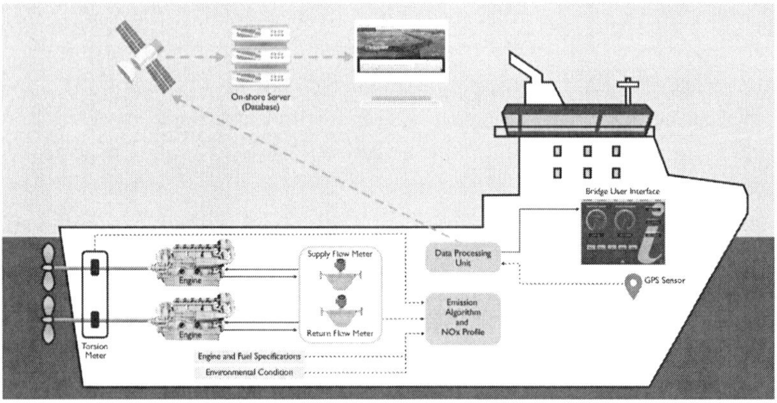

Figure 3-49. Fuel management system upgraded with emission monitoring technology

Source: adapted from https://safety4sea.com/fuel-management-system-upgraded-with-emissions-monitoring-tech/ [accessed on 24th March 2018].

One of the possible solutions that can streamline the processes of fuel consumption monitoring is the APISS Economizer system. The system provides a possibility of measuring light and heavy fuel consumption in real time. It automatically archives the data for further analysis in any spreadsheet format.[100] Additionally, due to the measurements from the receivers installed on a vessel, it is possible to record a number of other parameters which are required for the calculation of energy consumption. Such data can be sent to the ship-owner via the Internet. The system displays the following measurement and calculated data:[101]

Research and Analysis at the Polish Parliament; BSR Innoship, Project Seminar *Limitation of the emission of toxic substances generated during fuel combustion on vessels operating in the area of the Baltic Sea and ports*, non-published materials of the Polish Register of Ships, Gdańsk 2012.

[100] http://www.apiss.pl/produkty/kalkulator-zuycia-paliwa-dla-statkw/ [accessed on 27th July 2017].

[101] R. Miler, A. Bujak, *Implementacja systemów monitoringu zużycia paliw i efektywności energetycznej statku jako pochodna dyrektyw proekologicznych UE w transporcie morskim*, "Logistyka" no. 2/2014, p. 25.

- fuel consumption per hour or per nautical mile,
- global consumption,
- VU indicators of fuel consumption, the course of the vessel, wind direction,
- indicators of operating vessel devices, such as boilers, aggregates, alternators, engines,
- GHG emission indicators.

Another important field in which monitoring systems of maritime transport safety are applied includes ports, terminals and port facilities where such systems are implemented, being responsible for the implementation of the ISPS code recommendations. Such tasks can be performed by the Geoinformatics Port Security System (GPSS) which uses the GIS technology to provide security to operational activities related to monitoring and protection of ports from the sea-side and the shore-side, in the restricted access areas (RAA). The main element of the system is a map module which consists of high quality digital 2D/3D charts (compatible with the ECDIS). The second module is a system of sensors in the clothes worn by security officers. The sensors apply technologies which allow them to track the users of the system with the measurement of some selected life parameters in time and place of the operation (active biometric documents – worker identification cards, WIC). An indispensable module of the system is the connection module which ultimately integrates the sensors, navigation, monitoring and communication systems. The whole system is completed with a module of the user's interface which diversifies functions of the particular stations. The last element is an administration module which is applied to control and to manage other modules, to define the information flow between them and to provide the access rights for the users.[102]

The common and global application of the above-mentioned systems has considerably contributed to the achievement of the desirable safety (and indirectly security) level in maritime navigation and to the improvement of economic efficiency of maritime transport. Integrated and synchronised in three-module arrangement (sensory, analytical-decisional and effector) VTMIS-class systems have defined new standards and new quality of maritime safety. They have introduced an automatic and standardised

---

[102] A. Stateczny, W. Kazimierski, N. Wawrzyniak, *Analiza funkcjonalności geoinformatycznego systemu ochrony portu…*, op. cit., p. 402.

method of procedure whenever some anomalies are identified. In this way, more possibilities of prevention and counteraction against negative phenomena in the field of maritime navigation safety have been provided. Commercial systems which support logistical processes taking place in maritime navigation and sea ports have streamlined and optimised cargo handling operations. Security systems at ports and port facilities have considerably lowered the risk of asymmetric threats. As a result, measurable outcomes have been observed in the evaluation of economic efficiency, competitiveness and reliability of maritime transport, especially since the above-mentioned systems started to acquire information from other commercial systems applied in land-sea logistical chains – monitoring systems of container loading units.

3.3.2. Monitoring and tracking systems of the container loading units in maritime transport

Nowadays, containerisation has become a domain of global processes related to transport and logistics. Considering its role in global trade, maritime transport, which is broadly represented by container transport, requires some special monitoring systems. Tracking containers and monitoring each identified anomaly in real time (e.g. opening container doors, damaging container walls, changes of temperature, etc.), theft detection, reporting the movement status (a place in the logistical chain) are only some selected possibilities offered by modern systems of container monitoring, which come as the most important tools for management of logistical processes in maritime transport.[103]

Considering the global scale, international trade is implemented with the use of approximately 16 million containers per day, regardless of the season, weather, means of transport or a stage of a logistical chain. Viewed from the economic perspective, each mistake made in the proper handling of a container (with regard to its scheduled "path") brings about direct consequences, such as increased operational and transactional costs, elongated time of the operation, increased risk of further errors. There are also indirect consequences involved when a container with its cargo is not delivered on time – most often these are problems with maintaining the continuity of processes, limited commercial offers, decreased com-

[103] R. Miler, *Bezpieczeństwo transportu morskiego…*, op. cit., p. 225.

mercial reliability, damaged reputation on the market or a lower level of competitiveness.[104]

Unfortunately, while millions of containers are translocated during the global processes of maritime transport, a problem of "a needle in a haystack" can be observed too often. Too few containers undergo inspections and too few transport and forwarding companies use advanced systems of container monitoring, which would not slow down logistical processes (as it happens during time-consuming inspections) but which would allow the interested parties to speed up processes of container handling. Modern systems must meet the requirements of container monitoring in real time and the requirements related to the immediate access to tracking/monitoring data provided to all the stakeholders of the system – starting from the entities of maritime transport, maritime operators and forwarding agents and ending with relevant state control agencies of the countries involved in the process.[105] Therefore, integrated systemic solutions applied in this field must be equipped with:[106]

- optimisation tools – improving the efficiency of logistical processes and lowering their costs;
- technological tools – providing a possibility of monitoring one container during the whole logistical process;
- sensory tools – providing the monitoring and signaling of any deviations from numerous parameters defined for the process (e.g. the route, technological regimes of transportation, such as temperature, humidity, light exposure);
- integrating tools – providing a possibility of using current technologies for positioning and automatic identification (GPS, RFID – Radio Frequency Identification).

Modern systems of container monitoring are often referred to as Electronic Container Tracking Service (ECTS),[107] and they use LEO (Low Earth Orbital)

---

[104] http://www.keytonetech.com/solutions/Cargo-Monitoring.html [accessed on 30th January 2015].

[105] R. Miler, *Electronic Container Tracking System as a Cost-Effective Tool in Intermodal and Maritime Transport Management,* "Economic Alternatives", Issue 1(2015), Sofia 2015, p. 42.

[106] http://www.keytonetech.com/solutions/Cargo-Monitoring.html [accessed on 30th January 2015].

[107] At present, the ECTS comes as an integral part of the ITS-class (Intelligent Transportation Systems) systems which provide informational support to logistical processes of supply chain management (SCM) – the Author's note.

satellite systems to provide positioning under the GPS. Information can be transmitted with the use of a proper interface to the computer systems applied by shipping companies and operators of maritime transport (even to the systems installed on mobile devices).[108] Figure 3-50 presents a scheme of the ECTS (Electronic Container Tracking Service) process.

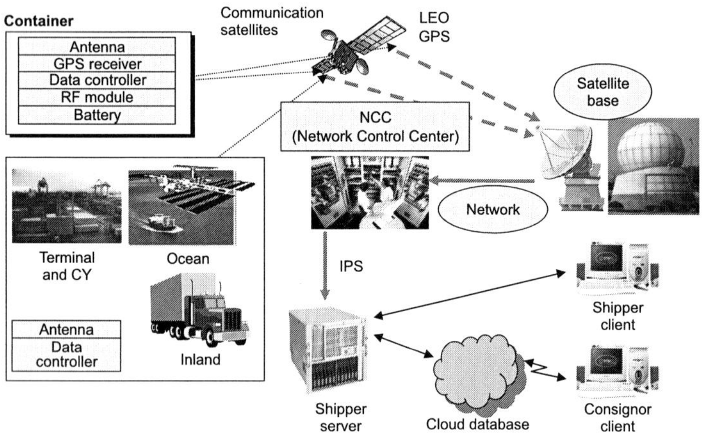

Figure 3-50. A functional scheme of the Electronic Container Tracking Service system

Source: adapted from S-B. Ahn, *Container Tracking And Tracing System To Enhance Global Visibility*, Proceedings of the Eastern Asia Society for Transportation Studies, 2005 Vol. 5, p. 1722.

Electronic protection of the integrity of container doors sealing and systems of container tracking are more and more often based on the RFID technology. The main task of the ECTS-class systems is to provide safe container transport during the whole logistical supply chain. The administrator must have an access to full information in real time – starting from the process of container consolidation (e.g. consolidation of cargo for the FCL shipping) at a logistical centre in the sender's country, transportation processes (with consideration of the specificity of maritime transport – vessel monitoring at all the water areas of the World Ocean) and ending with the processes of temporary storage, consolidation at port terminal and logistical centres in the addressee's country.[109] Summing up, it should be stated that all the logis-

[108] Cf.: S-B. Ahn, *Container Tracking And Tracing System To Enhance Global Visibility…*, op. cit., pp. 1719–1727.

[109] R. Miler, *Bezpieczeństwo transportu morskiego…*, op. cit., p. 226.

tical operations involving containers during the transportation and logistics processes must be monitored under the door-to-door delivery service.[110] One of the most integrated and advanced ECTS-class systems applied in maritime transport is the AVANTE system.[111]

The system includes several key functions, the most important of which are the following:[112]

- monitoring the integrity of a loading unit, based on multi-sensory system of the container surveillance (each of its six sides) which practically eliminates any possibility of raising false alarms;
- monitoring activities of employees responsible for transport and safety of a container, based on the transportation worker identification card (TWIC), which allows the interested entities to monitor not only the position of a container in real time, but also the location of the employees responsible for its safety; additionally, while facing an emergency, an employee can raise a discrete alarm, using the "panic" function attributed to their card;
- monitoring all the means of transport, including seagoing vessels, which involves the constant vessel monitoring with the use of GPS, GPRS, SATCOM systems, assuming that information about each detected anomaly that affects safety and security of the container and the vessel is transmitted to the monitoring centre in real time;
- efficient system of data display on the devices specified by the user (including mobile devices, such as smartphones);
- a unique and doubled monitoring system applied for HVG (High Value Goods) containers.

The ECTS AVANTE system is equipped with sensors which are able to detect practically any kind of anomalies which can be defined for a container in maritime transport, such as:[113]

---

[110] S. Mahlknecht, S.A. Madani, *Architecture of Low Power Wireless Sensor Networks for Container Tracking and Monitoring Applications*, 5th IEEE International Conference on Industrial Informatics (Volume 1), Vienna 2007, pp. 353–358.

[111] See more in: http://www.avantetech.com/products/shipping/, a competitive system to the American AVANTE *Intermodal Real-time Container Tracking Solution* is the TRITON *Real-time Container Tracking System* developed by STARCOM Systems of Great Britain. It is characterised by very similar parameters. – the Author's note.

[112] http://www.avantetech.com/products/shipping/ [accessed on 23rd September 2014].

[113] See more at: http://www.avantetech.com/uploads/pdf, and R. Miler, *Electronic Container Tracking System as a Cost-Effective Tool in Intermodal and Maritime Transport Management…*, op. cit., p. 44.

- anomalies related to the temperature, humidity and pressure specified for a particular container;
- any types of shocks after hitting, rapid translocation, falling, etc.;
- sound anomalies (e.g. sounds characteristic for sawing, oxy-fuel cutting, etc.);
- anomalies involving increased brightness inside the container (e.g. after unauthorised door opening).

A general principle of the system operation is presented in Figure 3-51.

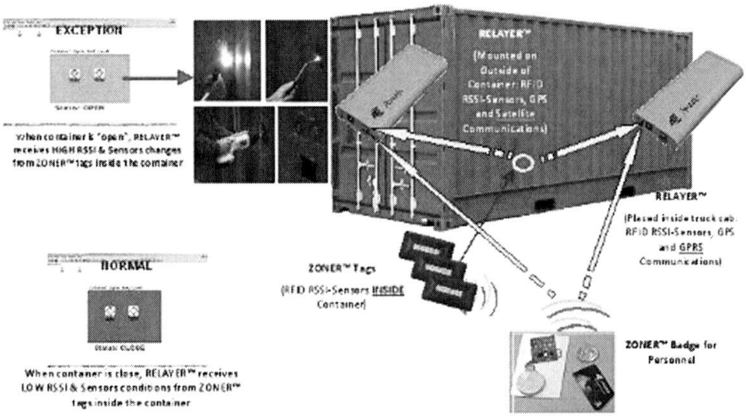

Figure 3-51. A conceptual scheme of the sensor operation in the ECTS AVANTE system

Source: adapted from materials provided by Avante at http://www.avantetech.com/products/shipping/ [accessed on 12th November 2013].

All anomalies, such as unauthorised door opening, are detected by proper sensors (which are most often operated in the RFID technology) and immediately transmitted to relevant institutions (which have been previously specified). The system operates on the EXCEPTION principle – deviations from the pre-defined standards and transport regimes (NORMAL) detected by a properly configured sub-system (RELAYER, ZONER).[114] Figure 3-52 presents the operational principle of the system.

[114] Cf.: R. Miler, *Bezpieczeństwo transportu morskiego…*, op. cit., p. 228; R. Miler, *Electronic Container Tracking System as a Cost-Effective Tool in Intermodal and Maritime Transport Management…*, op. cit., p. 44.

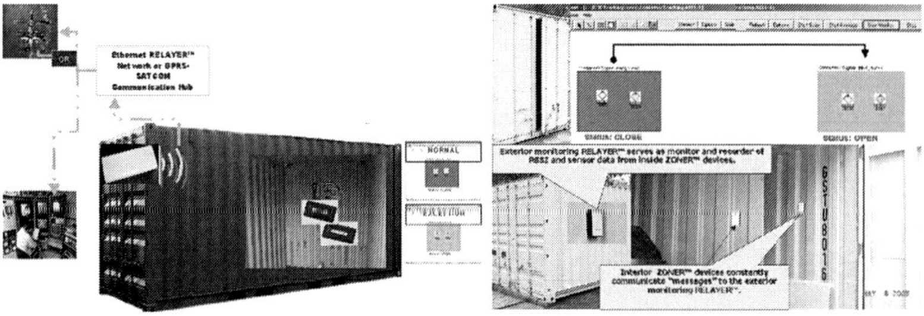

Figure 3-52. The operational principle of the ECTS AVANTE system (alarm initiated by specified anomalies)

Source: adapted from materials provided by Avante at http://www.avantetech.com/products/shipping/ [accessed on 12th November 2013].

As it has been already mentioned, the ECTS AVANTE system is equipped with sensors which detect previously specified anomalies and also with a communication sub-system which transmits information about a threat to the specified address or allows the interested entities to remotely access the monitoring display at a dedicated website (accessed with a correct login and a password). Figure 3-53 presents the notification methods applied to report temperature anomalies.

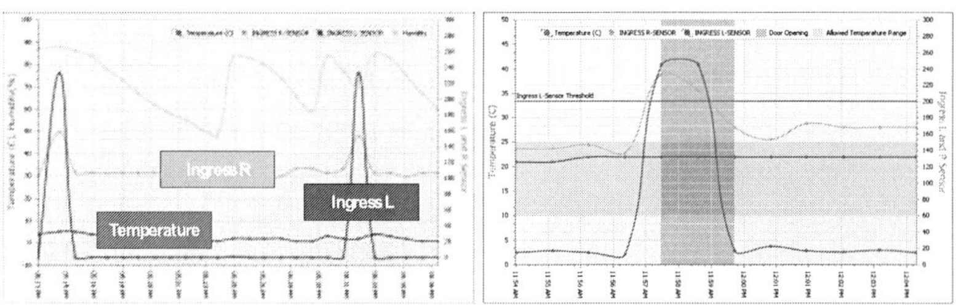

Figure 3-53. The interface of the ECTS AVANTE system applied to display temperature monitoring inside a container

Source: adapted from materials provided by Avante at http://www.avantetech.com/uploads/pdf [accessed 12th November 2013].

Figure 3-54 presents an example of an interface providing information about translocation, position and status of a container in the logistical process of maritime transport.

Figure 3-54. The AVANTE system of container location in maritime transport

Source: adapted from materials provided by Avante at http://www.avantetech.com/products/shipping/ [accessed on 12th November 2013]. [c]

The system described above comes as a representative example of the potential and technical capabilities of electronic systems applied for container monitoring in maritime transport. Apart from some obvious functions referring to the improvement in safety of transportation process, such systems allow their users to receive additional economic benefits, such as:[115]

- direct benefits for private companies which provide:
  - increased efficiency of logistic processes,
  - increased reliability and competitive position,
  - decreased expenditures required for other safety and security systems;
- direct benefits for the public sector which provide:
  - increased efficiency of control processes implemented by government institutions,
  - increased maritime safety of the state,
  - increased safety level of logistic processes implemented by the public sector,
  - decreased external costs of maritime transport, including ecological costs,
  - decreased congestion in maritime transport and costs incurred in relation to it;
- indirect benefits for the sector of maritime transport which provide:

[115] See: http://www.avantetech.com/products/shipping/, R. Miler, *Electronic Container Tracking System as a Cost-Effective Tool in Intermodal and Maritime Transport Management…*, op. cit., p. 44.

- scale benefits resulting from decreased unit costs and an increased volume of containers handled in maritime transport,
- increased potential and capacities of maritime transport hubs,
- positive effects of supporting transport-related and port-related operations in maritime economy.

The examples of telematics systems in maritime transport which have been discussed so far come as an excellent illustration of efforts made by particular countries which intend to achieve an acceptable level of risk and to provide safety to all the processes of maritime transport. Unfortunately, such efforts are largely determined by the specificity of the particular countries (the level of wealth, the level of maritime transport importance identified for the achievement of economic targets, the level of accessible techniques and technologies of monitoring: sensors, analytical systems, effectors, the level of determination to provide safety), and they are mostly made independently by the particular countries. As a result, there is a lot of diversification in the current systems, the data streams are not compatible, there are difficulties with regard to integration, there is the lack of access to national classified data and data protected by trade secrets of maritime entities (commercial sensitive data) and multiplication of financial expenditures (duplication of costs related to the development and maintenance of separate systems). Therefore, since 2009 a concept has been developed in the EU on the integration of current national monitoring systems of maritime transport safety (most often systems for navigation monitoring). It is referred to as a platform for integrated maritime surveillance (which is also rooted in the MSA strategy) under its working name: the Common Information Sharing Environment (CISE), along with an experimental tool: the Integrated Maritime Data Environment (IMDatE).

3.4. Integration of telematics systems in maritime transport

3.4.1. The concept of CISE (Common Information Sharing Environment)

Maritime risk and threats exist regardless of any national or administrative borders. Progressing in the water areas of the EU, the idea related to the openness of the sea is an additional risk factor. The optimal risk management is based on the access to information. The improvement in the exchange of information between the bodies of maritime surveillance becomes one of the key strategic aims of the European Union under the integrated maritime policy, and also an important component of the European strategy for maritime safety.[116] Reinforcement of maritime safety at the EU seas and oceans comes as a fundamental part of the programme involving activities that foster the blue growth, which contributes to the economic growth and to formation of new jobs, with simultaneous respect towards the principles of ecology and decreased antropopressure in maritime economy and transport.[117] A doctrine referred to during the implementation of such strategic aims is the doctrine of integrated maritime surveillance,[118] the CISE (Common Information Sharing Environment). In the strategy assuming an increase in the potential of the EU maritime transport safety, the main aim is to obtain a full picture of the situation at sea involving activities which are significant for maritime safety and security, protection of navigation, border checks, marine environment and its pollution, fishing control, exercising the law in its general scope, for defence capabilities, and for all economic interests of the European Union.[119] The exchange of information should take

---

[116] Joined communication of the Commission and EEAS of 6th March 2014, JOIN (2014)9 final.

[117] Communication of the Commission to the European Parliament and the European Council, *Improvement of situational orientation due to improved cooperation of maritime surveillance bodies: next steps under the common mechanism of information exchange in the EU maritime areas;* COM(2014) 451 final, Brussels 2014, p. 2.

[118] Communication of the Commission to the European Parliament and the European Council, on *The draft operational schedule for the establishment of the common mechanism for information exchange required for the surveillance of the EU maritime areas*; COM (2010) 584 ultimate version, Brussels 2010, pp. 2–6.

[119] The main aim is to provide maritime surveillance information collected by one maritime body which is indispensable for operational activities performed by other entities. Such information should be mutually accessible and should be used multiple times – it should not be collected and prepared several times or collected and stored only for one-off reference. Application of the EU instruments of data protection will be necessary whenever such data identify a natural person or make it possible to identify a natural person. Due to the transition to the application of data for numerous purposes and the achievement of interoperability of current systems dedicated to maritime surveillance, the collection of data will be less time-consuming and resource-consuming, and in the optimal case, the bodies will be always provided with the best available information about the situation at sea. Indirectly, the solution should limit multiplication of activities in the field of data collection resulting from suboptimal cooperation of maritime administration bodies and operators of maritime transport. It may also affect purchase, maintenance and location of the surveillance means such as radars, communication systems, vessels, helicopters,

place according to the demand for information and the legal obligation of providing access to the required information. This fact, first of all, forces the EU countries to establish and to follow relevant legislation.[120]

As the above-mentioned system of information exchange is intended to function in the pan-European context, all the groups of users will participate in it on the same principles. Their sectoral aims and limitations must be taken into consideration, and the development of a common mechanism of information exchange should provide each sectoral group of users with an added value.[121] It is estimated that the added value from integration of maritime surveillance, understood in this way, shall consist in the fact that current images of the situation and activities at sea carried out by sectoral groups of users[122] from the EU member countries and the EEA countries will be more detailed due to some additional inter-sectoral and cross-border data referring to broadly understood maritime transport safety. In this way, they will considerably contribute to the limitation of safety and security incidents in maritime transport.

During the work on the strategy, there have been six stages defined, which come as indispensable activities in the development of the CISE mechanism:[123]

---

planes and satellites. Improved information exchange may help to avoid doubled purchase of means, doubled monitoring of the same water area or collection of the same information for several times and performance of the same missions at sea – the Author's note.

120 Owners of the information should compare the requirement of information exchange (especially in the case of a direct threat) to a threat which results from the lack of such exchange. The improved picture of the situation at sea will increase the efficiency of state member bodies and profitability of activities they have been carrying out, also in economic categories, because considering the context of growing maritime risk which must be faced by those bodies and limited operational and financial means, optimised information exchange may prove to be more efficient and rational in terms of incurred costs – the Author's note.

121 Since the communication on the CISE in 2010, some progress has been made towards the implementation of the action plan on the CISE. The member countries have closely cooperated under the projects such as "BlueMassMed", "MARSUNO" and "Cooperation", which have been, in fact, the pilot software of a new project. See more in www.bluemassmed.net, www.marsuno.eu, http://www.coopp.eu – the Author's note.

122 A detailed list of all the groups of users is referred to as a list of stakeholders in the suggested model of the integrated monitoring system of maritime safety – the Author's note.

123 The Communication of the Commission to the European Parliament and the European Council, on *The draft operational schedule for the establishment of the common mechanism for information exchange required for the surveillance of the EU maritime areas*; COM (2010) 584 ultimate version, Brussels 2010, pp. 2–6.

- Stage 1 – defining all the users – considering a diversified structure of the form and administrative legitimacy of the stakeholders identified in the EU member countries and the EEA countries, it is necessary to focus on the functions based on the already existing groups of users rather than on the types of national bodies.
- Stage 2 – mapping data sets and providing an analysis of divergence observed in the purposes of data exchange; it is implemented in order to provide the added value to the CISE mechanism through a map of the existing data exchange which takes place at the EU level and the national level (data stream inventory) and an analysis of divergence, performed in order to define sectoral demand for data, which are not supplied currently.
- Stage 3 – common levels of data classification provided in order to solve a possible problem related to the classification of the same type of data in a different way by the sectoral groups of users. Without changing the national levels of data clasiffication and in order to provide data exchange under the CISE mechanism, Stage 3 will allow the interested parties to define which national levels of data classification correspond to each other, and in this way, to establish a common base for data exchange with the use of the CISE mechanism.
- Stage 4 – developing the support framework for the CISE mechanism: this is going to be a technical support framework for the CISE mechanism, which will allow the interested parties to define interfaces between the current and intended sectoral systems in order to provide inter-sectoral data exchange. The respresentatives of various sectoral groups of users should be responsible for the development of such a solution, based on the available results of the 7PR and pilot projects (e.g. MARSUNO, BluemassMed, EUROSUR pilot projects which are related to communication networks, GMES, PT MARSUR, pilot projects based on the SafeSeaNet).
- Stage 5 – establishing the access rights is related to the definition of the users belonging to various sectoral groups, who will be granted the rights to have the inter-sectoral access to various data sets. It exclusively refers to the data which should be exchanged under the CISE mechanism in the EU and the EEA.[124]
- Stage 6 – following legal regulations must be provided in order to ensure a clear legal framework for data exchange that will define at least: the char-

[124] The access to environmental data should be open, in accordance with the Convention of Aarhus – the Author's note.

acter of data, the possibilities and the rights of data suppliers and recipients in the field of data exchange, the aims (and methods) of data exchange. It will also include indispensable protection necessary to guarantee confidentiality and safety of (some) data and protection of personal data in all applicable cases. It is also necessary to define difficulties in data exchange which exist in the EU legislation and to develop proper solutions to such problems.

Considering a significant number of potential users of the CISE mechanism, the variety of legal frameworks and potentially exchanged information, it is very unlikely that one technical solution will be suitable to all the information exchange that will take place via the CISE mechanism. Considering the architecture of the CISE mechanism, the required solution should be developed as […] *cost-efficient, decentralised combination of various information layers which would increase the efficiency of maritime surveillance systems through the complementation of information gaps observed in Europe and through the avoidance of data multiplication at the same time.*[125]

So far, the project analysis has indicated that the maritime CISE mechanism has some clear advantages, such as:[126]

- improving knowledge and situational orientation at sea – both these elements can definitely improve prevention, readiness and response to incidents in the field of maritime safety related to cross-border and organised crime (illegal trade, illegal fishing, piracy, armed aggression, terrorism), protection of the sea and illegal discharges into the sea or accidental pollution. A possible outcome can be limitation of such risks and threats averagely by 30% (an excellent example can be the exchange of information on illegal immigrants entering the Schengen zone via the Mediterranean Sea, between civil and military bodies, which improves the efficiency of crisis management);
- considerable limitation of efforts related to data collection – the interested parties (the EU countries) indicate that there is high demand for additional data exchange, because over 40% of data collected in the EU

---

[125] Communication of the Commission to the European Parliament and the European Council, *Improvement of situational orientation due to improved cooperation of maritime surveillance bodies: next steps under the common mechanism of information exchange in the EU maritime areas;* COM(2014) 451 final, Brussels 2014, p. 2.

[126] *Improvement of situational orientation due to improved cooperation of maritime surveillance bodies: next steps under the common mechanism of information exchange in the EU maritime areas*, op. cit., p. 4.

(referring to vessels unwilling to cooperate) are gathered by several bodies at the same time;

- limitation of administrative costs and operational activities in the field of maritime surveillance. The calculation provided by experts from the member countries indicates that any potential cost savings resulting from the improved information exchange would contribute to the advantage for European economy that would reach the level of approximately EUR 400 million annually and would bring direct savings to public entities at the amount of at least EUR 40 million annually. Relevant investment costs would reach the level of EUR 10 million annually during the first decade.

In order to support the efficient process of decision making, based on the comprehensive information acquired with the use of the CISE mechanism, the concept of the Integrated Maritime Data Environment, IMDatE, has been developed. It comes as a telematics platform and an IT tool for data exchange.

3.4.2. The IMDatE instrument (Integrated Maritime Data Environment)

Despite the fact that the CISE and the IMDatE are at an early stage of their implementation, and at present they still come as an experimental solution based on the pilot partial projects, it is possible to notice the logic of the whole concept.[127] Organised information flows and development of the added value gained on the integration of data streams are presented in Figure 3-55.

The IMDatE integrates data streams and analytical capabilities of four applications which are operated in the field of maritime transport safety in the EU water areas, that is namely the following systems:[128]

- SafeSeaNet (SSN),
- EU LRIT DC (LRIT),[129]

[127] R.K. Miler, A. Bujak, *The concept of the Integrated Maritime Data Environment (IMDatE) as a framework for European integrated and comprehensive shipping monitoring data exchange system*, Springer, Berlin 2015, p. 123.

[128] M. Mylly, *EMSA's Integrated Maritime Environment – A tool for Maritime Awareness...*, pp. 16–19; J. Szcześniak, A. Weintrit, *Europejskie systemy kontroli i śledzenia ruchu statków – geneza, zasady funkcjonowania oraz perspektywy rozwoju*, Zeszyty Naukowe Akademii Morskiej w Gdyni, no. 77, Gdynia 2012, pp. 89–90.

[129] See: J. Januszewski, *Satellite Navigation Systems in the Transport, Today and in the Future*, Archives of Transport, Vol. 22, No. 2, 175–187, 2010.

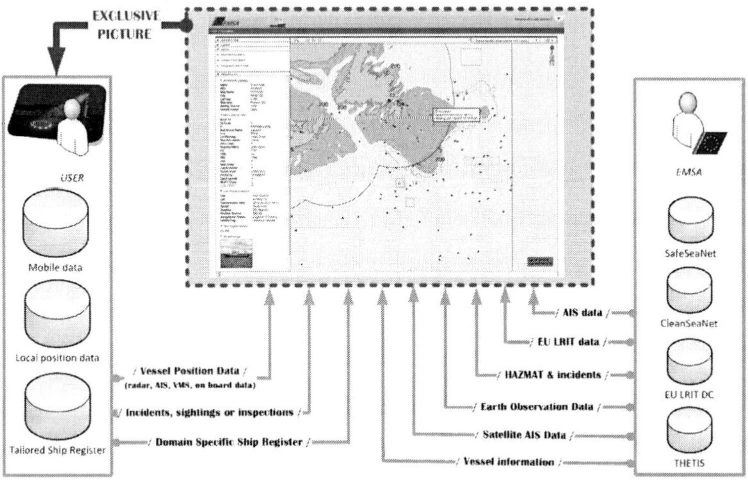

Figure 3-55. The concept of an integrated image of maritime safety in the EU, under the CISE IMDatE mechanism

Source: adapted from M. Mylly, *EMSA's Integrated Maritime Environment – A tool for Maritime Awareness*, EMSA Publication, Lisbon 2014, pp. 16–19.

- CleanSeaNet DC (CleanSN),
- THETIS.

The IMDatE has been developed with the assumptions referring to the open character of its architecture, based on the principles of the Service Oriented Architecture (SOA) and Integrated Product Development (IPD). The operational result of the integrated information acquired from the cooperating systems can be an opportunity of developing informational services in the following fields:[130]

- Integrated Ship Profile Service – integrated database referring to all the available information on a particular vessel;
- Area Centric Service – integrated database which contains nautical and oceanographic data; it consists of numerous layers, and it is available for a selected water area;

[130] Cf.: R.K. Miler, A. Bujak, *The concept of the Integrated Maritime Data Environment (IMDatE) as a framework for European integrated and comprehensive shipping monitoring data exchange system…*, p. 34 and http://emsa.europa.eu/operations/maritime-monitoring/86-maritime-monitoring/1520-integrated-maritime-data-environment-imdate.html [accessed on 5th February 2015].

- Maritime Surveillance Service – integrated database which contains data about vessel traffic in the selected water area with an access to the history of vessel traffic and any incidents which took place in the past;
- EU Common Maritime Space Monitoring Service – integrated database of the CMS (Common Maritime Space) European satellite system which uses satellite images for the monitoring of vessels and accidents at the EU water areas.

The IMDatE concept, its components and the role of so called services in achieving operational outcomes of the system are presented in Figure 3-56.

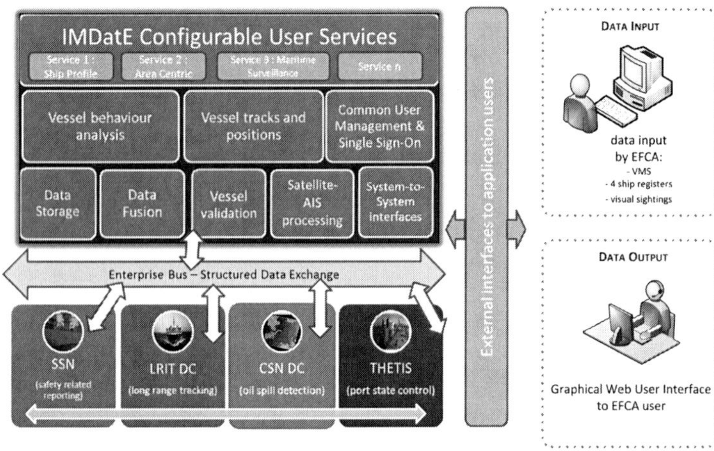

Figure 3-56. The IMDatE components

Source: adapted from M. Mylly, *EMSA's Integrated Maritime Environment – A tool for Maritime Awareness*, EMSA Publication, Lisbon 2014, pp. 16–19.

The IMDatE concept will ultimately lead to a generation change in the approach towards problems related to maritime transport safety: from the current non-integrated systems, through their integration and development of a monitoring centre, to end up with a selective access to integrated data streams and their analysis provided to relevant entities in the EU member countries which are responsible for maritime safety.[131] Understood in such a way, the idea of integration is presented in Figure 3-57.

[131] See more in: *Maritime Surveillance In Practice – Using Integrated Maritime Services*, EMSA Publication, Lisbon 2014, p. 7.

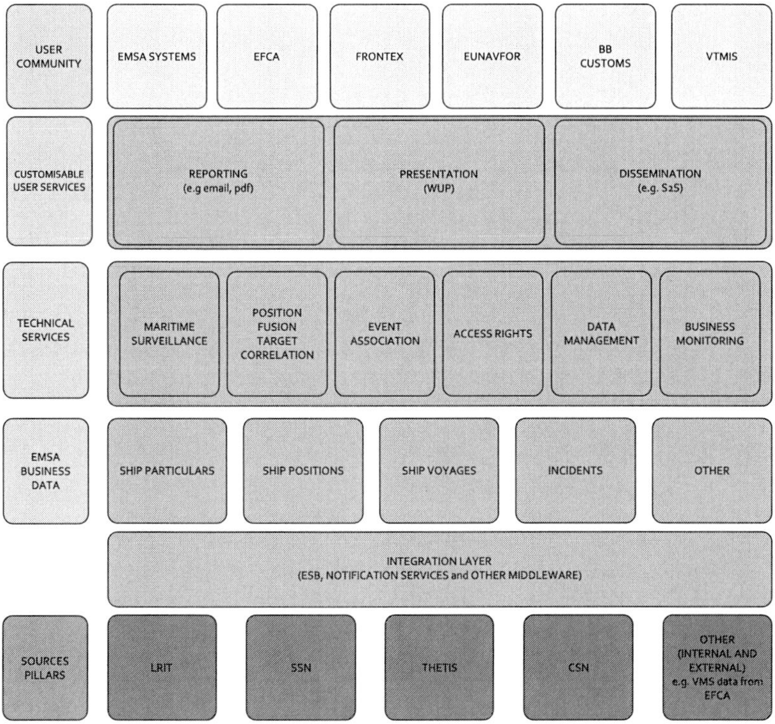

Figure 3-57. The idea of system integration under the IMDatE

Source: adapted from J. de Sousa, *EMSA's Integrated Maritime Environment – a tool for improved Maritime Domain Awareness*, EMSA Publication, Lisbon 2012, p. 14.

Configured in this way, the platform which provides integration, visualisation, analysis and processing of information gained in the field of maritime transport safety from numerous independent sources allows the interested parties to provide the added value in the form of the following functions:[132]

- a graphical time slider,
- a timeline.

The graphical time slider provides visualisation of vessel traffic in its historical layer by indicating a particular vessel and – using the time slider – tracking (animating) its movements for some inspection purposes, for example. The interface of this function is presented in Figure 3-58.

[132] *Maritime Surveillance In Practice – Using Integrated Maritime Services…*, op. cit., p. 7.

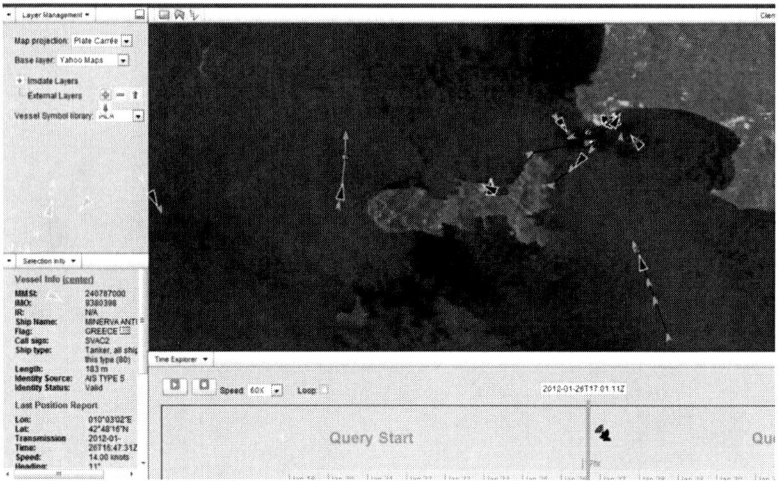

Figure 3-58. The interface of the time slider graphic function

Source: adapted from materials provided by EMSA at *Maritime Surveillance In Practice – Using Integrated Maritime Services*, EMSA Publication, Lisbon 2014, p. 8. [c]

Another function of the IMDatE is the timeline which provides the re-focusing of the map on an element indicated in the window of the timeline. It considerably speeds up the situational analysis and allows the interested parties to make optimal decisions. The interface of the timeline function is presented in Figure 3-59.

The new functions identified above and a possibility of integrating available information (with regard to the fact that it is an experimental stage and only a small part of the ultimate information stream has been undergoing the process of integration and analysis) allow the interested parties to notice the first results of such an approach. The best developed tools which come as a result of the assumed concept are the following:

- analysis of oil spills (Clean SN – the procedures described in more detail in Chapter 3.2.2),
- analysis of vessel traffic in the areas threatened by pirate attacks – Anti-piracy (MarSurv – white picture),
- analysis of fishing vessel traffic with regard to illegal fishing.

The IMDatE function which provides detection and analysis of oil spills comes as a development and further improvement in the current functions of the

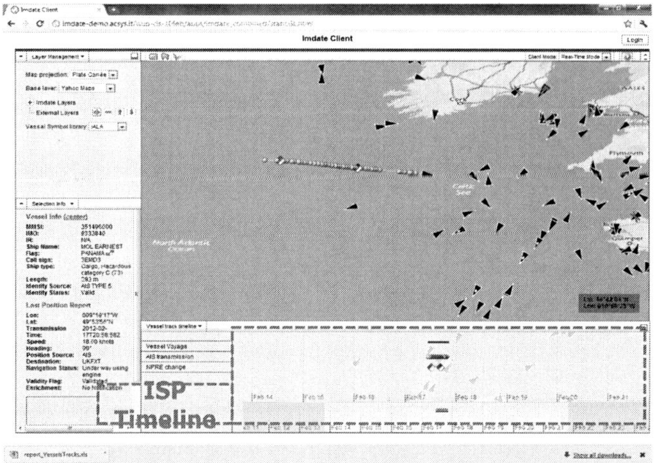

Figure 3-59. The interface of the timeline function

Source: adapted from materials provided by EMSA at *Maritime Surveillance In Practice – Using Integrated Maritime Services*, EMSA Publication, Lisbon 2014, p. 9. [c]

Clean SN system. The sequence of analytical processes starts with detection of a spill (red alert) (Figure 3-60), then it refers to the current traffic and historical movement of vessels in the area of the spill (Figure 3-61) and ends up with a clear indication of the vessel responsible for the spill (Figure 3-62).

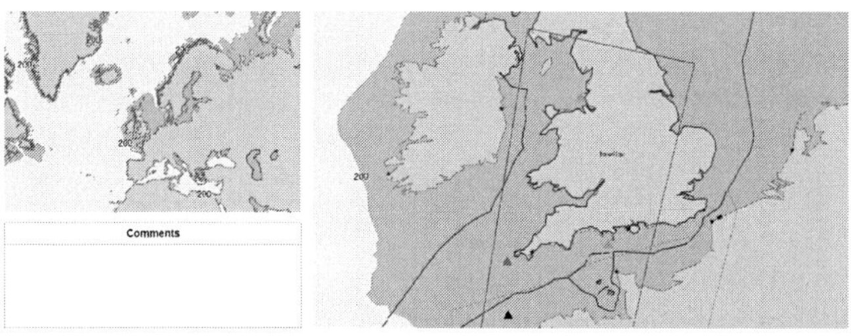

List of possible spills

Spill # on map	Spill identifier	Centre Position		Area	Length	Width	Alert	Oil Spill Warning Issued	Possible Source	
		Latitude	Longitude	(km²)	(km)	(km)			Detected	Identified
1	OS_19294_2	49° 59' 37" N	006° 00' 12" W	20.79	36.4554	8.8846	Red	N/A	Yes	No
2	OS_19294_3	53° 38' 47" N	003° 14' 03" W	7.34	4.1814	2.6221	Green	N/A	Yes	No
3	OS_19294_4	50° 22' 58" N	002° 14' 32" W	1.72	5.396	0.7452	Green	N/A	Yes	No

Note: Possible spills outside alert area are presented on map as ▲ - Additional spills may also have been reported outside the map - Please consult GIS Viewer

Figure 3-60. Detection of an oil spill in the SSN tab of the IMDatE system

Source: adapted from M. Mylly, *EMSA's Integrated Maritime Environment – A tool for Maritime Awareness*, EMSA Publication, Lisbon 2014, p. 28. [c]

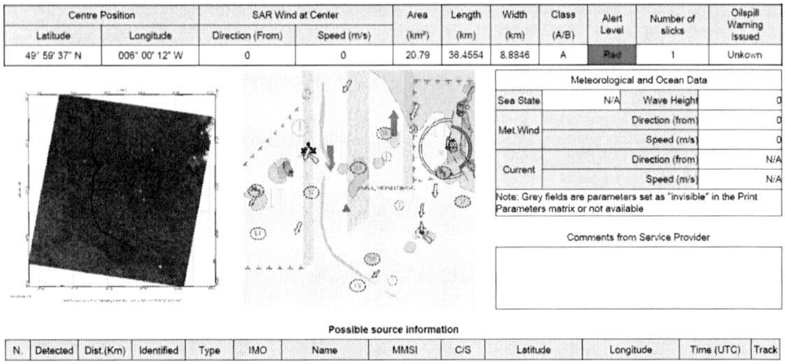

Figure 3-61. Integration of the information layer referring to an oil spill with the layer of movement and other data in the SSN tab of the IMDatE system

Source: adapted from M. Mylly, *EMSA's Integrated Maritime Environment – A tool for Maritime Awareness*, EMSA Publication, Lisbon 2014, p. 29. [c]

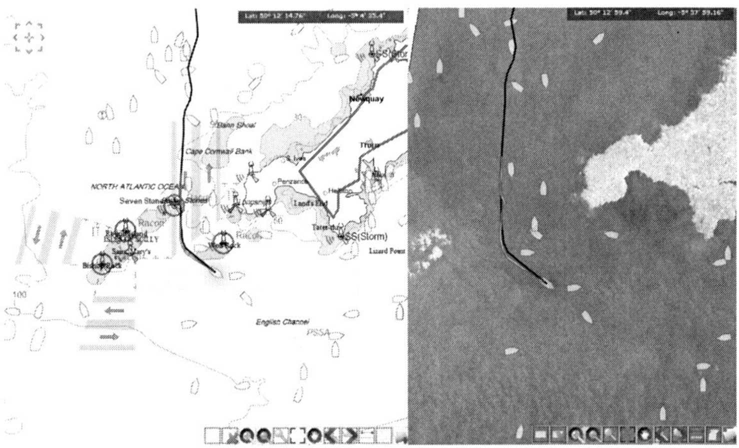

Figure 3-62. The analysis of the information layer referring to an oil spill with the layer of movement and indication of the vessel responsible for the spill in the SSN tab of the IMDatE system

Source: adapted from M. Mylly, *EMSA's Integrated Maritime Environment – A tool for Maritime Awareness*, EMSA Publication, Lisbon 2014, p. 30. [c]

After further development of this application, the users will ultimately receive a full report integrated with other data (including nautical, hydrographic, meteorological and historical information in the vessel traffic layer). It will refer to an oil spill, not only indicating the vessel responsible for the spill but also facilitating its removal and prevention of ecological loss. It will also lower the direct costs of the operation (action) as well as the costs included in the Total Social Cost (TSC) category (external and ecological costs).

The analysis of vessel traffic in the areas threatened by pirate attacks (MarSurv – white picture) is another innovative application of the experimental IMDatE system. The display of commercial vessel traffic in the risk area (referred to as the white picture) is obtained through the integration of a display acquired from many sources, including satellite ones (LRIT, AIS-S) and the combination of that display with a situation recognised by the EUNAVFOR forces, referring to the position and movement of suspicious vessels in relation to the movement of other commercial vessels operating in the area, in order to estimate the potential risk. Figure 3-63 presents the interface of this function.

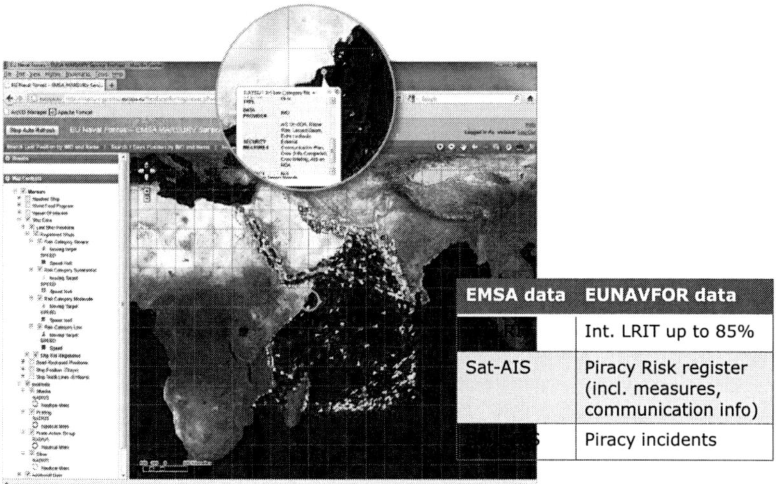

Figure 3-63. The anti-piracy function interface in the IMDatE system

Source: adapted from M. Mylly, *EMSA's Integrated Maritime Environment – A tool for Maritime Awareness*, EMSA Publication, Lisbon 2014, p. 26. [c]

Further development of the above-mentioned application, including the function of providing satellite images in the Optical Satellite Image (OSI) technology, would considerably improve the efficiency of this tool with regard to risk estimation in a dynamic operational situation in the areas threatened with piracy. The improvement in analytical capabilities of the system and the addition of the movement display of fast motor boats would decrease the level of risk, and it would improve the efficiency of the whole system in prevention of piracy incidents.

The analysis of fishing vessel traffic with regard to illegal fishing is the last innovative function of the IMDatE. In accordance with the MarSurv-3 standards, information about fishing vessel movements is acquired and displayed. The Illegal Fishing function is dedicated for efficient identification of fishing vessels which operate in fishery areas without any relevant permits, catch limits or which operate in restricted areas using unauthorised types of fishing nets or fishing methods. Figure 3-64 presents the interface of that application.

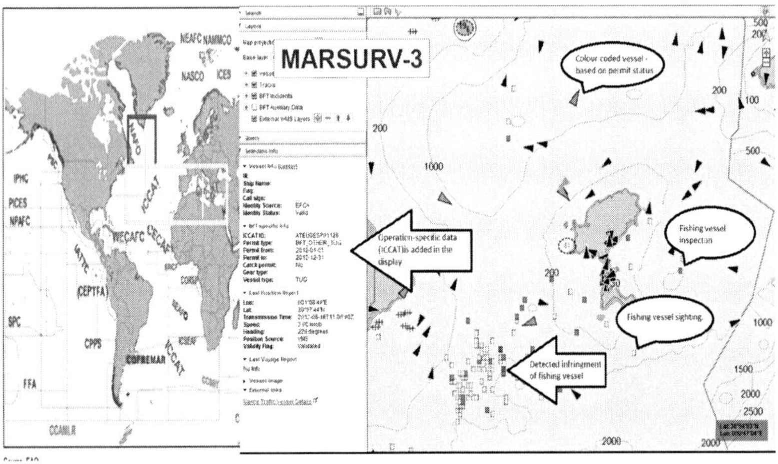

Figure 3-64. The interface of Illegal Fishing function in the IMDatE system

Source: adapted from M. Mylly, *EMSA's Integrated Maritime Environment – A tool for Maritime Awareness*, EMSA Publication, Lisbon 2014, p. 24. [c]

The above-mentioned display should be integrated with the IMAP (Integrated Maritime Picture), as it would definitely improve the ability to provide faster control and prevention. This, in turn, should considerably contribute to a potential decrease in TSC costs and pressure exerted on marine environment. It would also allow the interested parties to increase the efficiency of legal fishing (the fishery areas would not be overfished).

To sum up, the IMDatE is not intended to replace the existing systems. It develops new databases of the monitoring systems applied mainly for the surveillance of fishing vessels (Vessel Monitoring Systems VMS Data) and satellite data of the AIS. It has been assumed that the IMDatE platform will be operated by a special Data Processing Centre for the AIS satellite system, and it will facilitate the processing and distribution of data acquired from all new sources which can be developed in the future.[133]

After the complete implementation of the system, its new functions will allow the interested entities to achieve all operational benefits related to the implementation of the IMDatE which have been previously identified. They will involve, first of all, better visualisation of the situation at sea, full access to data streams provided by various systems and data warehouses after one-time signing in, automated monitoring of vessels, monitoring of the situation in the selected areas or the condition of marine environment and full verification of data based on numerous sources providing nautical information.

The identification of the IMDatE functions ends our considerations in this chapter. Having indicated the tasks and morphological elements of the systems applied to monitor maritime transport safety, the chapter presents a taxonomic division of such systems, based on the criteria which have been identified and assumed originally. Some particular attention has been paid to the VTMIS-class systems (Vessel Traffic Monitoring and Information System) with their basic sub-systems, such as VTS, AIS, SRS, MAS, and DGPS, INMARSAT, GMDSS auxiliary systems. Taking the subsequent criteria of the division of the systems applied for monitoring of maritime transport safety into consideration has allowed the Author to identify the groups of systems which are partially integrated, such as the systems at the national level of Shipping Monitoring Data Exchange (SMDE), e.g. SWIBŻ PL, IMDatE (EMSA – UE), BRITE (applied as an experimental tool of the MSA concept developed by NATO) and complementary systems, such as international systems, e.g.: GMDSS (Global Maritime Distress and Safety System), MSSIS (Maritime Safety and Security Information System), NAMESIS (Naval Merchant Shipping Information System), VMS (Vessel Monitoring System), and national systems (applied by the Border Guard and Maritime Administration).

---

[133] Cf.: R.K. Miler, A. Bujak, *The concept of the Integrated Maritime Data Environment (IMDatE) as a framework for European integrated and comprehensive shipping monitoring data exchange system…*, op. cit., p. 36.

Commonly applied standards have contributed to full commercialisation of some IT solutions; hence, commercial systems used for the monitoring of maritime transport safety have been also identified (they are represented by such applications as AIS Live, Marine Traffic, Vessel Tracker, exactAIS™ based on the AIS-S satellite signal, Gatehouse AIS Display System). Another group of systems identified with the use of the assumed criteria are terminal and ship-owner systems, including the TOS (Terminal Operation System), applications supporting processes at the level of fleet management (e.g. AMOS – Asset Management Operating System), CBM (condition-based monitoring) applications used for the monitoring of drive parameters, fuel consumption and GHG emission, and Geoinformatics Port Security System (GPSS) which implements the ISPS recommendations. The last identified category includes systems for monitoring container loading units in maritime transport: Electronic Container Tracking Service (ECTS).

A similar development of information sources and integration of the current telematics systems can be expected in the field of inland waterway transport.

CHAPTER 4

TELEMATICS SYSTEMS IN INTELLIGENT INLAND WATERWAY TRANSPORT

4.1. Classification of telematics and vessel monitoring systems in inland waterway transport

4.1.1. Tasks of telematics and vessel monitoring systems in inland waterway transport and their morphological elements (including the river terminal Improved Port Ship Interface – IPSI)

Over the last decade, one of the key (strategical) lines of the EU transport policy and of all the developed countries in the world has been a decrease in a detrimental impact (anthropopressure) of transport on natural environment. It involves implementation of procedures referring to the internalisation of transport costs (with regard to the factor related to the impact exerted on natural environment, such as, for example, emission and noise). It is related to the (legal, financial, organisational and conceptual) support provided to all environmentally friendly modes of transport and transportation technologies, including inland waterway transport which is considered to be the friendliest to natural environment (it is ecological, it involves low emission and low anthropopressure indicators along with low external costs).

In order to achieve the strategic aim which has been defined in such a way, it is necessary to provide a SWOT analysis for inland waterway transport in Europe, with special regard to the identification and counteraction against outcomes resulting from the weak points of that mode of transport (low speed, the risk of low water levels, problems related to the safety of navigation and cargo, problems related to inter- and multimodality). As a result, it will allow interested entities to face the current challenges and to use inland waterway navigation in a broader way. Figure 4-1 presents the SWOT matrix for inland waterway transport.

Strengths	Weaknesses
High level of safety Reliable services Lowest external costs Relatively large network of waterways	Limited accessibility Relatively slow speed (transit time) High intermodal costs, low priority in ports and limited cooperation Minimum sailing distance Limit in efficiency scale High fixed costs
Opportunities	**Threats**
Improved accessibility in big ports Economies of scale Increased quality of logistics supply chain management Opening new markets Fiscal and administrative stimulations	High competitiveness level of other transport modes (road) Strict environmental regulations Insufficient amount of qualified personnel Port congestion Natural constrains (high dependence on water level)

Figure 4-1. The SWOT matrix for the IWT

Source: adapted from R. Gort, *Design of an autonomous loading & unloading inland barge. A concept for container transport on the Albert canal*, Delft/Antwerp, October 2009, p. 9.

Considering the above-mentioned facts, it should be noticed that in order to achieve the assumed results and to reinforce the role and significance of inland waterway transport in the European logistics and transport system, new concepts of services and specific technical solutions have been generated. They pertain to the field of the support provided to operational processes and (operational, efficiency-oriented, safety and risk and ecological) management in inland waterway transport. So far, the lines of innovative modifications have mainly referred to the operational activities, including:[1]

- solutions applied in the field of inland vessel construction (e.g.: the INBAT-type vessels or vessels constructed in the River Hooper system – river pallet carriers);
- solutions applied in the field of inland vessel drive (e.g.: river vessels/pusher tugs equipped with the solar, hybrid, electric or LNG drive);

[1] M. Kaup, *Znaczenie systemów telematycznych w funkcjonowaniu transportu wodnego śródlądowego na przykładzie Systemu Informacji Rzecznej (RIS)*, "Logistyka" nr 6/2014, p. 5279–81.

- solutions applied in the field of the linear and nodal infrastructure construction (e.g.: locks, vessel lifts – as for example the Falkirk Wheel);
- automation of cargo handling processes (under the integrated logistical process) at inland ports (e.g.: the Improved Port Ship Interface project – IPSI).

Developed in the EU, the 5th Framework Programme involves the implementation of the INBAT (Innovative Barge Trains for Effective Transport on Shallow Waters) research project.[2] The aim of the project is to develop a new generation of barge pushed trains for shallow waters. The main water region for such barge trains is the Oder River and the Elbe River (Poland and Germany). The following outcomes have been achieved with regard to vessels which have been operating in that area since 2005:[3]

- an increase in carrying capacity by 31% for the Oder and by 26% for the Elbe,
- a decrease in exploitation costs by 32%,
- an improvement in general efficiency by 25.4% for the Elbe and by 20.5% for the Oder,
- a reduction of power per unit of transportation work from 27 kWh per 1000 ton-km to 20 kWh per 1000 ton-km.

Figure 4-2 presents a graphical concept of a new barge pushed train of the INBAT type.

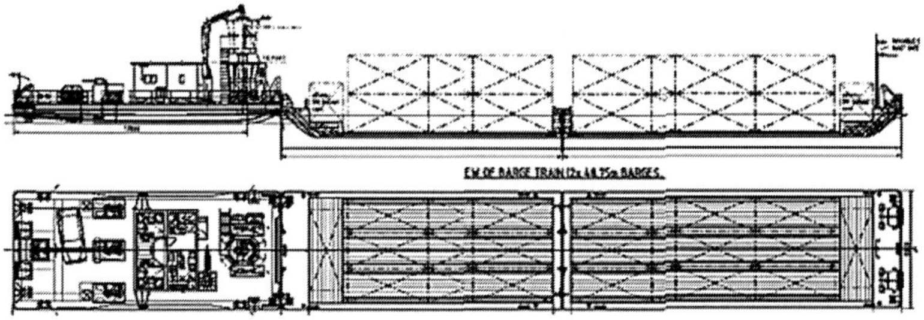

Figure 4-2. The INBAT barge train

Source: adapted from https://www.zegluga.wroclaw.pl/news.php [accessed on 16th September 2016].

---

[2] It allows operators to use barge trains up to 118 meters long, and two containers layers – Author's note.

[3] https://www.zegluga.wroclaw.pl/news.php?readmore=58 [accessed on 2nd August 2018].

Considering the annual scale, one INBAT barge train can carry approximately 28700 tonnes of cargo along the Oder River (so far, it has been only 16640 tonnes). The difference results from:[4]

- higher average carrying capacity of an INBAT barge train (1025 to 798),
- higher number of sailings per year (28 to 20),
- the fact that an INBAT barge train can be operated at the draft which is only T = 0.8 m.

Such results have been achieved through a decreased weight of barges and increased efficiency of transport at a very small barge draft (below 1.1 m). Barges are adjusted to transportation of bulk cargo and containers. The obtained results indicate some significant opportunities for an increase in transport efficiency in bad navigational conditions. The analysis indicates that inland waterway transport with the use of the INBAT system, at a draft of only 0.9 m, can generate profit for ship-owners.[5] Unfortunately, the costs which must be incurred during the construction of new INBAT-type barges are relatively high, and this fact comes as a considerable barrier to their common use.[6]

In Europe, the development of logistical centres at sea-river and river ports has become a challenge and, at the same time, an opportunity for inland water navigation. Due to this fact, the incorporation of inland navigation into the processes of providing supplies to cities with the use of small river vessels (such as the River Hopper[7]-type river pallet carries, Figure 4-3) has become a new application field which allows interested entities to avoid congestion and to decrease safety risk on roads.

Being a significant factor in human settlement, waterways usually run through towns, and therefore, they allow interested parties to provide such locations with efficient service from logistical distribution centres or sea ports (supplies, waste collection, broadly understood urban logistics). Both of the above-mentioned solutions require further implementation

---

[4] https://www.zegluga.wroclaw.pl/news.php?readmore=58 [accessed on 2nd August 2018].

[5] J. Kulczyk, T. Lisiewicz, T. Nowakowski, *Flota nowej generacji na odrzańskiej drodze wodnej*, Prace Naukowe Politechniki Warszawskiej, Z. 82 Transport (2012), pp. 55–67.

[6] Ibidem, pp. 65–67.

[7] In 2002 a Dutch company, River Hopper, launched a pallet carrier which was able to transport 680 pallets. This small vessel (about 63 m long and 7.2 m wide) was able to navigate along waterways of low parameters (IV class) (Cf.: *Freight transport embarking on a new course. Inland Waterway Transport. A transport solution that works*. European Communities. Brussels 2003) – the Author's note.

Figure 4-3. A concept of a River Hopper pallet carrier

Source: adapted from materials provided by Bureau Voorlichting Binnevaart at www.intrasea.org (Inland Transport on Sea Routes – Final Report, Dreamforce Infomedia AB in collaboration with the Swedish Maritime Administration, Sweden 2014), p. 14. [c]

work, however the potential of such innovations in their operational field cannot be underestimated.

Another aspect, or a factor, which determines not only the development of inland waterway transport but also telematics systems applied to improve management of operations involved in this mode of transportation, is an aspect related to ecological drive systems implemented on barges and pusher tugs. Similarly to the maritime transport, a lot of effort is made towards the implementation of LNG, hybrid (diesel fuel + electrical drive), electrical or even solar drive systems. It is extremely important in passenger inland navigation in which the legislator requires double drive systems with regard to safety reasons. Constructional solutions applied in the linear and nodal infrastructure (e.g.: locks or modern vessel lifts, such as the Falkirk Wheel, described in the previous chapter) come as the indicators defining the lines in the development of inland waterway transport, which strongly determine opportunities and a necessity of applying telematics systems to provide support for the processes of the optimal use of such infrastructure.

However, automation of cargo handling processes (under the integrated logistical process) at inland sea-river ports (e.g. the Improved Port Ship

Interface project – IPSI) comes as the largest challenge of modern operational processes in inland navigation. Considering their immanent features and the strong and weak points identified in the SWOT analysis, port operations are much more time-consuming, and they are related to a higher number of operations (involved in inter- and multimodal transport).

In maritime transport, the transfer of one container from an oceangoing vessel to a truck (and vice versa) requires two moves. The first one is from the container vessel to the stack area. In fact, the container is lifted from the vessel to the quayside, from where it is transported to the stack. The company which operates the terminal generally defines this as one move. The second move is translocation of the container by a straddler carrier from the stack to the truck (in this situation, it is assumed that the truck is the final transportation mode).[8] Figure 4-4 presents a typical two-move scheme of container movement (ship-to-shore/stack and yard/stack-to truck).

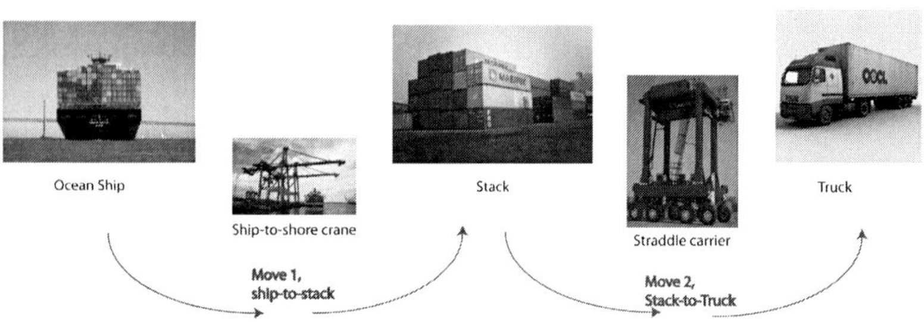

Figure 4-4. Two-move container handling in the maritime-land logistics chain of supply overview

Source: adapted from R. Gort, *Design of an autonomous loading & unloading inland barge. A concept for container transport on the Albert canal*, Delft/Antwerp, October 2009, p. 10.

Considering transportation by an IWT barge, the logistical system becomes more complicated. Figure 4-5 shows a four-move container handling cycle.

Inland navigation is hardly ever the final transport mode. Transporting containers by barge results in extra intermodal moves which are often very expensive and time-consuming. Hence, solutions aimed at the elimination of that negative aspect of river port operations have been lately have become highly sought-after.

[8] R. Gort, *Design of an autonomous loading & unloading inland barge. A concept for container transport on the Albert canal…*, op. cit., p. 11.

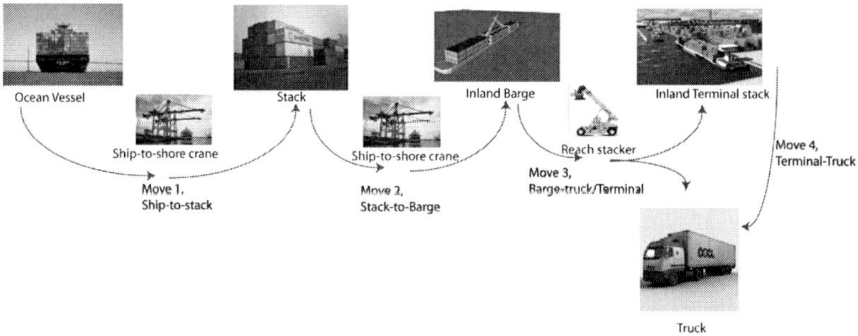

Figure 4-5. Four-move container handling in the maritime-land logistics chain of supply overview (with use of an inland water transport barge)

Source: adapted from R. Gort, *Design of an autonomous loading & unloading inland barge. A concept for container transport on the Albert canal...*, op. cit., p. 11.

River and sea-river ports must be characterised by higher and higher efficiency and by lower costs in the pursuit of an efficient combination of all the transportation modes which meet in their area. Therefore, the IPSI project has been developed. The main purpose of the IPSI project is the implementation of modern concepts related to the development of inter-modal terminals, vessels used for short sea shipping (SSS) and river barges used for related inland navigation.[9]

Considering intermodal terminals, the project mainly emphasizes the development of modern cargo handling systems. Such systems should provide efficient cargo handling operations which involve the most common loading units in the world, such as ISO standard containers, container semi-trailers, swap-bodies and containers placed on special cassettes or pallets. Cargo handling operations at a terminal should be performed with the use of automatic guided vehicles – AGVs which use cassettes specially designed for the IPSI systems. The concept of AGV cassettes is presented in Figure 4-6.

In accordance with the IPSI project (depending on hydrological conditions), a terminal should be equipped with one of two specially designed ramp systems:[10]

[9] M. Kaup, M. Chmielewska-Przybysz, *Analiza i ocena technologii wykorzystywanych do obsługi kontenerów w portach morskich*, Logistyka 5/2012, p. 501.
[10] Ibidem, p. 502.

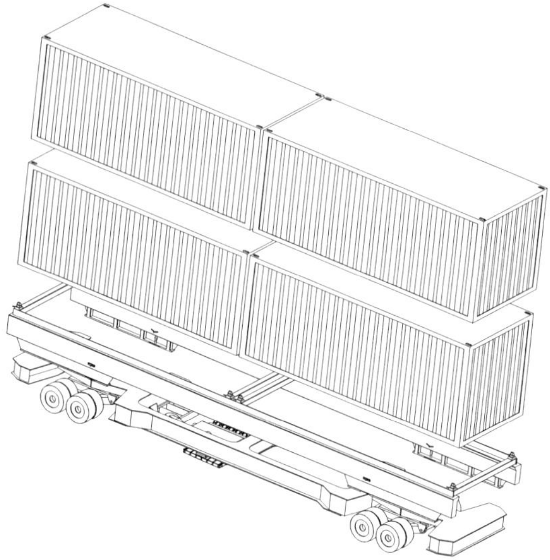

Figure 4-6. The concept of transporting a larger number of containers on an AGV (AGV + cassette + containers)

Source: adapted from *Improved Port/ship Interface* (IPSI), Deliverable D 6005 IPSI WP6200, Final Report for Publication.

- a permanent two-level ramp installed at the wharf for cargo handling during the time of small differences in the water level/tides (up to 2 m),
- a movable two-level ramp installed at the wharf for cargo handling during the time of large differences in the water level/tides (up to 4.3 m and potentially even more).

The IPSI project involves a development of two special types of vessels:[11]

- IPSI vessels[12] for short sea shipping (SSS) (Figure 4-7),
- IPSI barges[13] for inland water transport, navigating along the rivers.

---

[11] Improved Port/ship Interface (IPSI), Deliverable D 6005 IPSI WP6200, Final Report for Publication, p. 36.

[12] An IPSI vessel is characterised by a single-hull equipped with stern ro-ro ramps installed at its full beam. It can have two or three decks of simple loading lines and internal ramps for faster and more efficient translocation of the cargo. It can be also equipped with some facilities for automatic cargo stowing – the Author's note.

[13] An IPSI barge is equipped with a bow ro-ro ramp. It can have one or two decks of simple loading lines and some facilities for efficient cargo stowing – the Author's note.

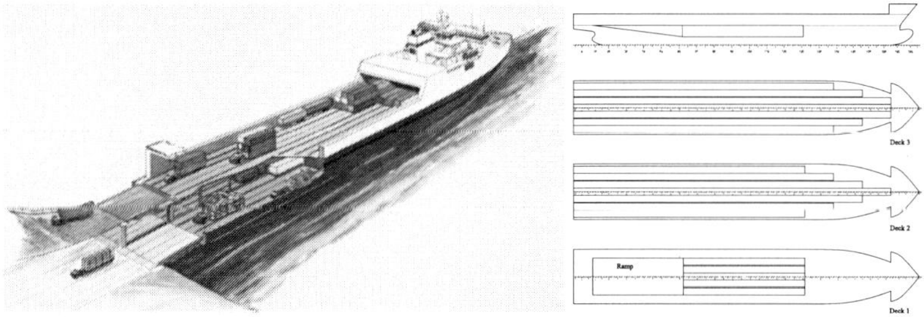

Figure 4-7. An IPSI vessel – general characteristics

Source: left: adapted from M. Kaup, M. Chmielewska-Przybysz, *Analiza i ocena technologii wykorzystywanych do obsługi kontenerów w portach morskich*, Logistyka 5/2012, p. 502; right: adapted from Improved Port/ship Interface (IPSI), Deliverable D 6005 IPSI WP6200, Final Report for Publication.

Despite the necessity of adjusting terminals and vessels/barges to the assumptions of the IPSI project (which is obviously related to considerable financial expenditure), it should be considered that the suggested solutions for terminals and IPSI vessels are relatively flexible. In fact, IPSI terminals can also handle standard ferries and ro-ro vessels (including such types of vessels as the Enisys vessels[14] and the Interbarge vessels[15]). Also, IPSI vessels can enter any port equipped with infrastructure and facilities for standard ro-ro cargo handling.[16]

Thus, the IPSI terminal is a multimodal ro-ro solution, with (AGV) operations of which enable the standard terminal to handle 2400 TEU per day or

[14] The Enisys ship – a high speed and slim body ship design with low fuel consumption and efficient cargo handling equipment. The concept has excellent sea-keeping and maneuvering capabilities due to the "azimuth" technology and bow thrusters. The ship handles containers, trailers, swap bodies, cassettes, with automatic lashing. The vessel provides high reliability and flexibility, and it is environmentally friendly due to low emission – the Author's note.

[15] The Interbarge – a barge design with the efficient ro-ro technology. The barge construction has reduced steel weight through the use of curved steel plates providing better cargo carrying capacity and economy for transporters compared to conventional barges of the same size. The Interbarge new construction concept provides a possibility to build ship hulls, and especially barge hulls, easier and faster with less use of steel, fewer and easier operations during production, and thinner shell plating due to the increased ability to resist water pressure – the Author's note.

[16] http://www.transport-research.info [accessed on 27th July 2017].

400 TEU per hour at reduced costs.[17] As the IPSI terminal is supposed to be a link in a well-controlled intermodal transport chain, it is required that all export cargo should be loaded onto the first IPSI vessel arriving. The import cargo from an IPSI vessel should be handled and processed to inland modes before the arrival of the next IPSI vessel. This becomes a difficult challenge to the IPSI terminal operators but also to service providers on land. It also determines the use of dedicated telematics solutions. For hinterland transport, three modalities are included:[18]

- barges for inland shipping – only IPSI barges are considered there, an IPSI barge carries cassettes only, which are handled exclusively by the AGV system,
- trains for the rail modality, establishing "Rail Service Centre" (RSC),
- trucks for the road modality, establishing "Truck Service Centre" (TSC).

A schematic view of the IPSI terminal layout is given in Figure 4-8. On the left side of the illustration, TSC and RSC are drawn. Both TSC and RSC use the same pool of straddle carriers.

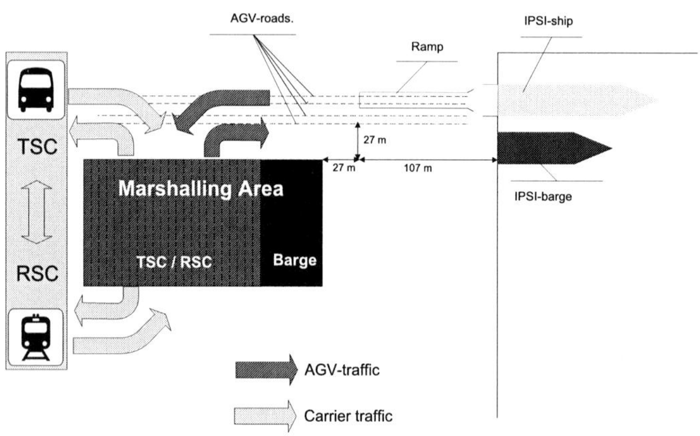

Figure 4-8. A schematic view of the IPSI terminal layout

Source: adapted from Improved Port/ship Interface (IPSI), Deliverable D 6005 IPSI WP6200..., op. cit., p. 39.

[17] H. Lindstad, E. Uthaug, *Innovative technology: Ro-Ro vessel, terminal, and barge design that will improve the cost position and lead-time for the logistic chain.* IMDC 03 Conference, Athens, Greece, 3–5 June, 2003.

[18] *Improved Port/ship Interface* (IPSI), Deliverable D 6005 IPSI WP6200..., op. cit., p. 40.

The marshalling area is divided into rows. The left part of the marshalling area contains the rows of containers for TSC and/or RSC. The right part is reserved for containers to be delivered to or coming from the barges. The AGVs enter and leave the marshalling area only on one side, so the straddle carriers can always use the other side.

The AGV system should cooperate closely with the system of automated gantry cranes in the marshalling area. The operational principles of automated gantry cranes are presented in Figure 4-9.

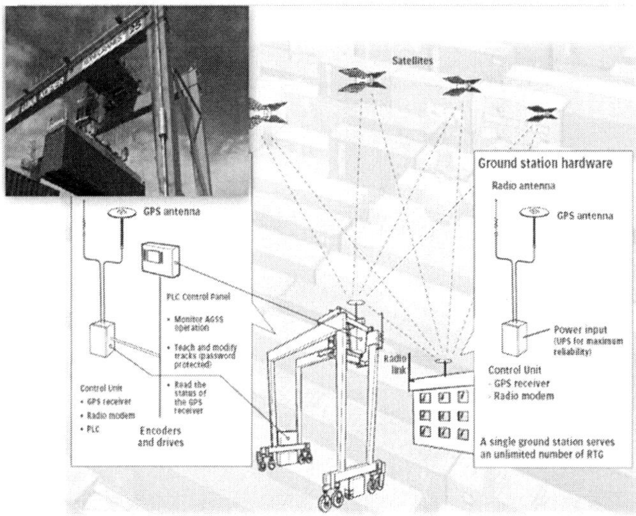

Figure 4-9. The system of automated gantry cranes

Source: adapted from materials provided by Solvo at http://www.solvo-europe.com/en/solutions/solvo-tos-container-en [accessed on 2nd August 2018].

Such understanding of operational processes which are performed in inland waterway transport clearly indicates areas and fields of application for telematics solutions. Hence, they will be applied in processes which are related to inland navigation handling and transportation of cargo (also in the context of the IPSI system application).

Generally, telematics solutions in inland waterway transport are most often applied to collect information about quantitative and qualitative conditions of transportation processes with regard to the capabilities of using such information in decision-making processes referring to the choice of optimal economic solutions, undertaking proper action when too much risk referring to life and property is involved, or when there is a threat to natural environment. They also provide a possibility of starting remedy actions, elimination of consequences resulting from the current situation e.g. congestion, accidents (failures) or ecological disasters.[19] Considering the aspect of innovative technical and operational solutions, some broader application of the RIS system in the process of management optimisation in inland waterway transport gains special significance. However, before that integrated system is discussed here, it is advisable to have a closer look at the taxonomic division of telematics systems already existing in inland waterway transport.

4.1.2. Taxonomy of telematics and vessel monitoring systems in inland waterway transport

The demand for the specific information from the field of inland waterway transport, defined in the previous sub-chapters, has resulted in the development and implementation of numerous systems for acquiring, collecting, processing and transmitting information in inland transport practice. Considering different specificity of that transportation mode and its character, which is complementary to maritime transport, a considerable number of telematics system come as an extension of some other systems which have been already used in maritime transport (it mainly refers to navigation and terminal systems, systems of vessel/barge and container monitoring).

Similarly to maritime transport and considering the variety of systems and their key purposes, the following systems can be distinguished in inland waterway transport:[20]

- applications (including telematics ones) in regional and national systems of inland waterway navigation monitoring, used at the strategic level by government institutions and agencies (based on the RIS components);

[19] Bujak A., P. Zając, *Monitoring of Cargo in Logistic Systems of Transport and Storage* [in:] J. Mikulski (ed.), Activities of Transport Telematics, Springer, Berlin Heidelberg 2013, p. 362.

[20] Based on the following accessible sources: *Improved Port/ship Interface* (IPSI), Deliverable D 6005 IPSI WP6200, Final Report for Publication; L.M. Punter, W.J. Hofman, *Digital Inland Waterway Area. Towards a Digital Inland Waterway Area and Digital Multimodal Nodes*, EUROPEAN COMMISSION, Directorate-General for Mobility and Transport Directorate D Waterborne Unit D.3 Ports and Inland Navigation, Luxembourg: Publications Office of the European Union, 2017 – the Author's note.

- commercial applications based on the available Inland-AIS signal, dedicated to entities of inland water economy, including logistical operators at sea and sea-river ports in order to streamline management at the integrated level;
- auxiliary applications providing support to the processes in land-sea, land-river and sea-river logistical chains which optimise operational processes;
- ship-owner applications (accessible on-board on barges, resembling an integrated bridge on seagoing vessels).

Considering the position taken by the particular systems, the discussed division can be defined as a division by a vertical criterion. Again, similarly to maritime navigation, the second division criterion applied in inland waterway navigation is a horizontal criterion which indicates a significant diversification of current systems, with regard to the distinguished homogeneous factors. Based on such a division, it is possible to distinguish the following telematics systems of inland waterway transport:

- considering the subject of the monitoring:
  - systems for monitoring river barge traffic (systems of inland waterway navigation monitoring),
  - systems for monitoring loading units (e.g. containers),
  - systems for monitoring parameters defining the operation of the river barge systems (including the drive system and emission control),
  - systems for monitoring port facilities (which partially result from the IPSI regulations for maritime navigation and in this case applied at sea-river ports);
- considering the object collecting and processing information provided by monitoring systems (data administrators):
  - national (domestic), and international systems (e.g. the EU systems),
  - public (of inland waterway administration) and private (commercial) systems;
- considering the sensory factor:
  - systems based on the Inland-AIS signal,
  - systems based on visual and radar technologies (including CCTV cameras and infrared cameras),
  - systems of satellite photographs (most often limited to the monitoring of estuaries);

- considering the purpose of collected information:
  - systems displaying the situation along the inland waterways (monitoring function only for the traffic management purpose),
  - analytical and decisional systems (effector systems, telematics, automation and autonomation systems);
- considering the range (of the field of data collection):
  - local,
  - regional;
- considering the accessibility of databases and the character of collected information:
  - public (with unlimited access through the Internet and the cloud) and requiring authorisation (user registration),
  - confidential, with access to commercial sensitive data provided only to the previously defined entities (e.g. ship-owner systems, systems for groups of users, national systems);
- considering the type of information and its significance to safety/security processes:
  - safety systems (navigational including river nautical information, such as water levels, monitoring barge parameters and operational regimes to control the drive and navigation systems, monitoring systems of cargo, spills, etc., port terminal and ship-owner systems),
  - security systems (informational, analytical and decisional systems used by services responsible for state security in the inland waters domain – usually in the field of safety of sea-river ports).

Table 4-1 presents the division of telematics systems of inland waterway transport, according to the horizontal criterion in a synthetic (tabular) approach.

Table 4-1. Horizontal classification of telematics systems of inland waterway transport

Division criteria	**Categories of telematics systems of inland waterway transport**
the subject of the monitoring	• systems for monitoring inland waterway navigation, • systems for monitoring loading units (e.g. containers), • systems for monitoring parameters defining the operation of the river barge systems (including the drive system and river barge navigation system), • systems for monitoring port facilities at sea-river ports

Division criteria	Categories of telematics systems of inland waterway transport
an entity collecting and processing information provided by the monitoring systems (data administrators)	• national (domestic), sub-regional and international systems (e.g. the EU systems), • public (of inland waterway administration) and commercial systems
the sensory factor	• systems based on the Inland-AIS signal, • systems based on visual technologies
the purpose of collected information	• systems displaying situation along the inland waterways, lagoons and estuaries, • analytical and decisional systems (mixed effector systems)
the range (the field of data collection)	• local systems, • regional systems
the accessibility of databases and the character of collected information	• public (with unlimited access through the Internet) and requiring authorisation (user registration), • confidential, with access to commercial sensitive data provided only to the previously defined entities (e.g. ship-owner systems)
the type of information and its significance for safety/security processes	• safety systems (navigational including river nautical information, including the water levels, monitoring barge parameters and operational regimes to control the drive and navigation systems, systems for monitoring cargo, spills, etc., port terminal and ship-owner systems), • security systems (informational, analytical and decisional systems used by services responsible for state security in the inland waters domain – usually in the field of safety of sea-river ports)

Source: the Author's own study.

However, considering the above-mentioned specificity of inland waterway transport, it is possible to distinguish systems which – under the taxonomy presented above – are typical exclusively to this mode of transport. The general level of digitalisation observed in the processes of acquiring (sensory), processing (analytical and decisional) and implementation (effector) of information in inland waterway transport is presented in Figure 4-10.

Applied in the pragmatics of the processes related to management of a logistical supply chain, the SCM-class systems (Supply Chain Management) are usually integrated with the ERP-type systems (Enterprise Resource Planning, for procurement, production planning and sales) and MRP I and II systems (Materials Requirements Planning, Manufacturing Resource Planning).

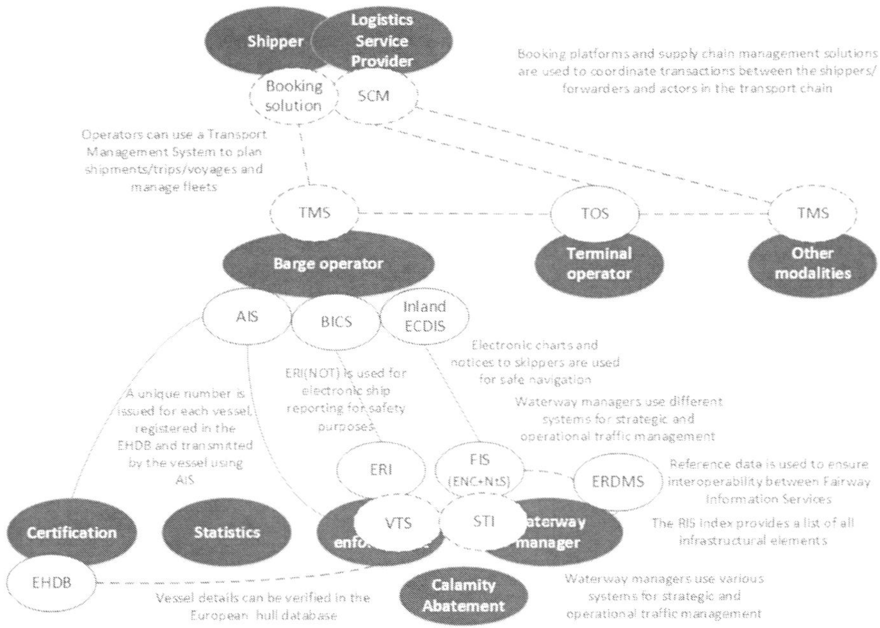

Figure 4-10. The digitalisation level of information processes in inland waterway transport

Source: adapted from L.M. Punter, W.J. Hofman, *Digital Inland Waterway Area. Towards a Digital Inland Waterway Area and Digital Multimodal Nodes*, EUROPEAN COMMISSION, Directorate-General for Mobility and Transport Directorate D Waterborne Unit D.3 Ports and Inland Navigation, Luxembourg: Publications Office of the European Union, 2017, p. 28.

However, in many cases there is no real-time link between the SCM systems of shippers and various systems used by logistics actors (logistics service providers and transport operators). Hence, systems based on the operational information of inland waterway transport become more and more significant. They include the following:[21]

- freight exchange (IWT cargo booking solutions),
- the TMS-class systems (Transport Management Systems),
- the TOS-class systems (Terminal Operating Systems),
- navigation systems based on the AIS transponders (Inland-AIS, Vessel Tracking and Tracing – VTT as a part of the RIS system – River Information Service),
- the Inland ECDIS with the ENC system of electronic charts,
- the ERI-class systems of electronic reporting ERI (Electronic Reporting International with the ESR standard – Electronic Ship Reporting and the BICS[22]),
- the FIS-class systems (Fairway Information Services),
- traffic management systems – Traffic Management Services (VTS – Vessel Traffic Services and TTI/STI – Tactical/Strategic Traffic Information as a part of the RIS system),
- database systems/centralised databases – (ERDMS – The European Reference Data Management System, EHDB – The European Hull Database, ECQD The European Crew Qualifications Database) – compatible with the RIS system.

The efficiency and economics of inland waterway transport are achieved through the efficient use of transport capabilities and optimisation of barge carrying capacities. A tool which proves to be very useful to achieve such aims is the freight exchange. Inland waterway transport operators and ship-owners (barge owners) use systemic software available on-line (e.g.: GT Nexus, INTRRA – electronic transaction platform for maritime shipping), which – in fact – is dedicated to maritime transport, and they use a platform accessible under the systems of logistical supply chains, based

[21] L.M. Punter, W.J. Hofman, *Digital Inland Waterway Area. Towards a Digital Inland Waterway Area and Digital Multimodal Nodes*, EUROPEAN COMMISSION, Directorate-General for Mobility and Transport Directorate D Waterborne Unit D.3 Ports and Inland Navigation, Luxembourg: Publications Office of the European Union, 2017, p. 34.

[22] BICS was developed by *Bureau Telematica Binnenvaart and Rijkswaterstaat* (Netherlands) providing skippers with a light-weight environment for electronic reporting (https://bics.com [accessed on 24th June 2017]) – the Author's note.

on the EDI format – Electronic Data Interchange (EDIFACT).[23] There are also platforms broadly used for the exchange of information about transportation processes (cargo, barges), dedicated exclusively to inland waterway transport, such as, for example, the Bargelink systemic software. Figure 4-11 presents the interface of the Bargelink platform.

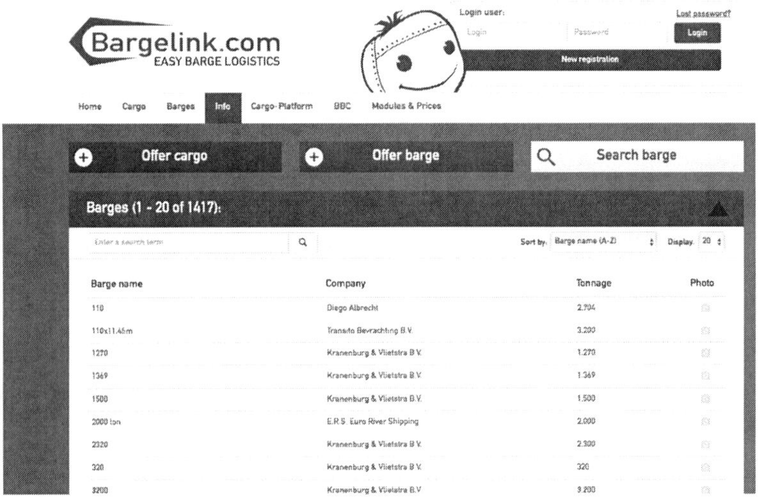

Figure 4-11. Bargelink platform

Source: adapted from materials provided by Bargelink at http://direct.bargelink.com/Index [accessed on 20th July 2018].

The application offers some considerable advantages (access to key transportation data – cargo, barges, available transport, directions of shipments, a tool for selecting barges in accordance with specified parameters) however, considering relatively high costs (a subscription that allows the user to enter the freight exchange) and a necessity of having the proper hardware and on-

[23] The message standard currently in use is the UN/EDIFACT in accordance with the syntax rules for the message structure (ISO 9735-1). A more recent developed syntax, specially designed for personal computers and based on HTML, is XML (Extended Markup Language) that is flexible and independent from the data format. Both, EDIFACT and XML use the same data structures and code tables. Presently, the UNECE together with ISO and other standardisation bodies have finalised the first version of the XML standards and messages: this is called ebXML. To avoid ambiguous interpretation, this standard uses only the data and codes of the UN/EDIFACT directories and the TDED (Trade Data Elements Directory) – the Author's note.

line connection with the Internet on the board of a barge, it is accessible only to major carriers (with regard to profitability).[24]

The TMS-class systems are used for route planning and cargo management, especially when intermodal transport is involved with the partial use of road transport (when a road transport operator acts as an integrator of the shipping). The TMS-class systems offer a number of functions which provide the monitoring of transportation means (sometimes also the monitoring of the containerised cargo) and anomaly alerts (previously defined in the system), such as delays resulting from congestion, etc. Using the TMS-class systems for barge operators depends on the access to the Internet and the quantitative level of the operations that would justify the purchase of such software (so again, the most interested entities would be major operators/ship-owners).

The TOS-class systems are applied in the operations performed at port terminals (sea and sea-river ports) to handle river barges, especially their containerised cargo. It requires a combination of the internal system of barge operators with the TOS system, in order to obtain current information about the number of containers to be handled, the updated estimated time of arrival of the barge to the port (ETA). Some major sea and sea-river ports (the Rhine Ports and PSA Antwerp) offer such a function as a part of the PCS packet (Port Community System).[25]

The ERI[26] standard provides standardised electronic messages for Electronic Ship Reporting and other reporting purposes. Other reporting purposes (passenger lists, voyages and berth management) are not covered by the directive and underlying regulations. The standards specify the syntax to be used. Barge operators use the BICS to create and to exchange the required messages according to the ERI standard. Frequently, data need to be filled-in manually in the BICS tool. Figure 4-12 presents the interface of the BICS system.

The application also includes some elements which may cause some inconveniences to barge operators. The most problematic ones refer to the fact that:

- reporting systems are not connected to the underlying data sources – as a result, skippers need to re-enter data,
- reporting standards should be worked out for the areas beyond four types of messages covered by the current standard.

---

[24] L.M. Punter, W.J. Hofman, op. cit., p. 36.

[25] Ibidem, p. 36–37.

[26] The RIS directive requires fairway authorities to use this standard – the Author's note.

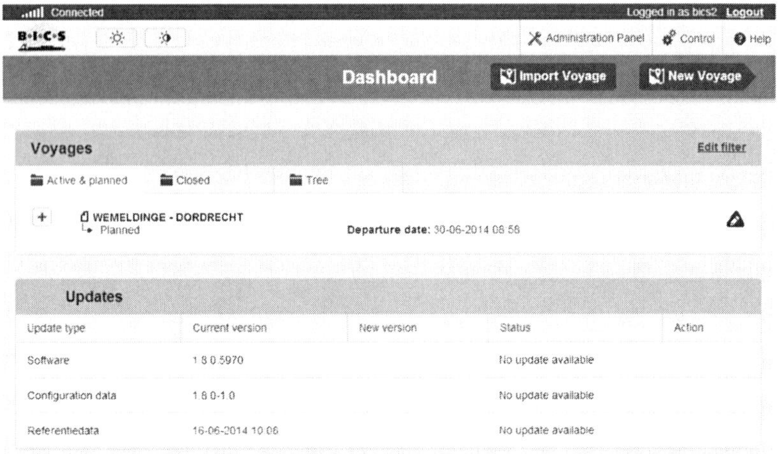

Figure 4-12. The interface of the BICS system

Source: adapted from L.M. Punter, W.J. Hofman, *Digital Inland Waterway Area. Towards a Digital Inland Waterway Area and Digital Multimodal Nodes*, op. cit., p. 45.

Other systemic solutions for telematics of inland waterway transport have been integrated under the RIS-class system (River Information Service).

4.2. The RIS (River Information Service) as a telematics system of inland waterway transport management

4.2.1. The assumptions and tasks of the RIS

The intensification of all the operations aimed at the development of inland waterway transport capabilities found its expression in the European Integrated Programme: Navigation and Inland Waterway Action and Development in Europe – NAIADES, presented by the European Commission for Transport in 2006. It was focused on five strategic areas, including the development of a concept, coordination of the development and implementation of the RIS system (River Information Services) in Europe.[27]

[27] Under the development of the proper infrastructure for inland waterway transport, in accordance with the current concept included in the EU PLATINA project – the Author's note.

The above-mentioned activities were preceded by the adoption of the Directive 2005/44/WE on harmonised River Information Services (RIS) for inland waterways in the European Community.[28] The RIS Directive defines a legal framework for the distribution and conditions which must be met to use the service in the EU inland waterway transport, in order to increase safety, efficiency and to improve the character of an impact exerted on natural environment, including improved cooperation with other modes of transport. The obligation of the RIS implementation is imposed on all the inland waterways of international importance which are connected with other waterways of the same standards,[29] starting from the IV class of waterways, in accordance with the UNECE classification of 1992.[30]

As a result, most of the EU member countries have already implemented the service of river integrated information, the RIS, on their inland waterways. The area in which the RIS Directive is applicable in Europe (European waterways covered by the RIS Directive) is presented in Figure 4-13.

It is related to a necessity of developing the system, of providing an informational service for vessels (their captains, skippers and managers) and of implementing an integrated system for monitoring and control in the field of traffic management on the EU inland waterways.[31] The establishment of the RIS technical means involves the implementation of the following:[32]

---

[28] (Journal of Laws L255/152 of 30th Sept. 2005) Directive 2005/44/WE of the European Parliament and the European Council of 7th Sept. 2005 on the synchronised River Information Services (RIS) for inland waterways in the European Community refers to the implementation and operation of the RIS exclusively on inland waterways of at least the IV class, in the EU member countries, which are connected by inland waterways of at least the IV class with other waterways of at least the same class in other EU member countries and ports located along such waterways, defined in the Decision no. 1346/2001/WE. Brussels 2005 – the Author's note.

[29] http://hydrogeo.am.szczecin.pl/projekty/ris/4-ris [accessed on 27th July 2017].

[30] Listed in the Directive, a waterway of the IV class should be defined in accordance with the principles of the European inland waterway classification established by the Resolution no. 30 of the UNECE of 12th November 1992. According to these principles, a IV class waterway is a waterway which allows a vessel or a barge up to 80–85 m long and up to 9.5 m wide to navigate safely – the Author's note.

[31] Wawruch R., *Usługa Informacji Rzecznej (RIS) na śródlądowych drogach wodnych Unii Europejskiej i zasadność jej ustanowienia w Polsce*, Prace Wydziału Nawigacyjnego Akademii Morskiej w Gdyni, no. 24, 2010, pp. 125–135.

[32] Regulation of the Commission (EC) no. 414/2007 of 13th March 2007 on the guidelines for the River Information Services (RIS).

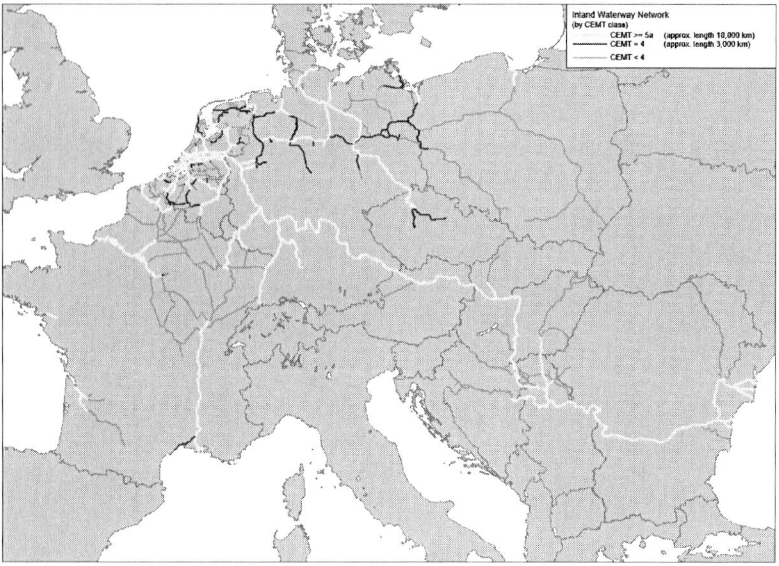

Figure 4-13. European waterways covered by the RIS Directive [2014]

Source: adapted from *Evaluation of RIS Implementation for the period 2006–2011*, Main Report. Zoetermeer, July 2014, p. 14. [c]

- the display system of electronic charts and information in inland waterway navigation (inland ECDIS),
- the electronic reporting system for vessels used via the inland automatic identification system (Inland AIS),
- radio information service,
- vessel positioning systems with the relevant level of precision and integrity,
- radar systems of traffic control and systems for vessel monitoring.

Figure 4-14 presents a scheme of the system which includes sensory, infrastructural and effector elements.

The framework structure of the harmonised river information system (RIS) should provide transport operators and other users of the system with a possibility of more efficient fleet management and transport operation management, based on the access to the information collected and transmitted in real time, actually forms the service. The scope of the RIS system is increased by the EU postulate, which states that [...] *at the discretion of the EU member*

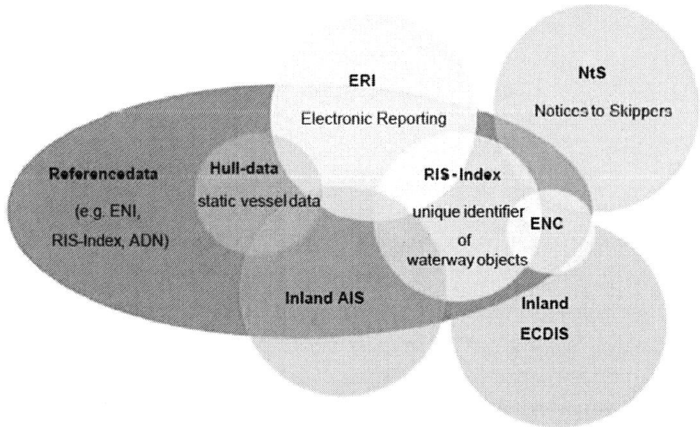

Figure 4-14. The scheme of the RIS systemic structure

Source: adapted from *Evaluation of RIS Implementation for the period 2006–2011…*, op. cit., p. 16.

countries, the resolutions of the RIS Directive can be applied also on inland waterways which are not under the obligation stated by those resolutions. National inland waterways under the above-mentioned obligation, which are not related with an inland waterway network of any other member country, do not have to meet the requirements and technical specifications defined by the RIS Directive for the harmonised system or river information, however they are recommended to do so.[33]

Therefore, it can be clearly stated that in each EU member country, the RIS system is developed on the basis of the local (national) systems with regard to their interoperability. The systems are based on open and public technical standards which are available without any limitation for all system providers and users. Most of these standards meet the criteria of data stream compatibility and, by definition, they have features attributed to telematics and monitoring systems.

Therefore, the RIS is, by definition, a harmonised (and integrated) system which has been developed as a combination of local (that is: national, in the context of the EU) informational (sub-)systems, including human

[33] Commission Regulation (EC) no. 414/2007 of 13th March 2007 on the guidelines for the river information services (RIS).

resources, equipment, communication and software, along with relevant and applicable legal regulations intended to perform tasks related to processing information about inland waterway navigation.[34]

The main aim of the European RIS system is to turn inland waterway navigation into a reliable, flexible, economically and ecologically efficient (low emission, low anthropopressure and low external costs) mode of transport which will be able to successfully compete with other modes of transport and to efficiently enter inter- and multimodal structures and relations. Considering the main objectives of the RIS systems, the European Commission has defined four main strategic lines for the implementation of that system:[35]

- intermodal competitiveness of inland waterway navigation,
- optimisation of the use of the existing infrastructure and superstructure,
- higher safety of cargo and navigation,
- higher protection of natural environment in inland waterway transport, achieved through further decrease in its anthropopressure.

In order to achieve the strategic aims defined in the above-stated way, the RIS (as a system) uses the services which enable it to develop an interoperable, intelligent system for vessel (barges and pusher tugs) traffic management, transport processes (inland waterway navigation and river port operations) and safety management through the implementation of systems applied for monitoring, analysis and optimisation of processes and support of river navigation.[36] The RIS should integrate all the newly-developed systems and services of river information of the various (local, regional and national/domestic) range into one common (shared) operational system of the European range.[37]

Such a configuration of the RIS systemic relations is implemented on two basic levels (display and navigation management and support of transport process management). It has generated eight main categories of services: four (FIS, TTI, TM, CAS) for the level referring to the vessel traffic, and four

---

[34] M. Kaup, op. cit., pp. 5279–81.

[35] Directive 2005/44/EC.

[36] *Sprawozdanie z inicjatyw i opinii Komisji Europejskiej 09.2013 r.*, transport, natural environment protection, tourism, sport, industry, scientific research, innovation, digital agendas, health care, employment policy (a report on the initiatives and an opinion of the European Commission 2013), Brussels, September 2013, pp. 46–48.

[37] *River Information Services, Modernizing inland shipping through advanced information technologies.* (2010), Transport Research Knowledge Centre (TRKC), European Commission's Directorate-General for mobility and transport, Belgium 2010, p. 58.

(ITL, ILE, STAT, WCHD) for the level referring to the transport characteristics.[38] Table 4-2 presents the main categories of the RIS services broken into the levels discussed above.

Table 4-2. The main categories of the RIS for the level of vessel traffic information and the level of transport characteristics

No.	The RIS categories referring to the level of vessel traffic
1.	FIS – Fairway Information Services – it provides geographical, hydrological, administrative information which is used by captains and fleet managers for planning and vessel monitoring
2	TTI – Tactical Traffic Information – it provides information supporting captains in making quick decisions referring to vessel movements in emergency situations
3.	TM – Traffic Management Services: it is used for optimal use of the infrastructure, safe navigation and protection of natural environment
4.	CAS – Calamity Abatement Services: it is used for mitigating the consequences of disasters, providing information about accidents on waterways to vessels, police boats and rescue units; it is used for the initiation of search and rescue operations.
The RIS categories referring to the level of transport characteristics	
5.	ITL – Information for Transport Logistics: it provides support to inland waterway transport and logistical operations; it is ultimately intended for planning sailings, port and fleet management.
6.	ILE – Information for Law Enforcement – it fosters law enforcement through providing information about legal requirements referring to border crossing, immigration and customs regulations, etc.
7.	ST – Statistics – it is applied to collect information about inland waterway traffic
8.	WCHD – Waterway Charges and Harbour Dues: it is applied to support planning of expenditures related to sailing and staying at ports.

Source: adapted from Kaup M., *Znaczenie systemów telematycznych w funkcjonowaniu transportu wodnego śródlądowego na przykładzie Systemu Informacji Rzecznej (RIS)*, "Logistyka" no. 6/2014, pp. 5279–5287.

As it has been already mentioned, viewed from the perspective of the EU and understood as a systemic integration solution, the RIS is based on the national (domestic) solutions which have been developed in the particular member countries. As a result, numerous solutions have been provided which are specific for the particular areas. Table 4-3 presents the current RIS-type systems which have been developed at the national (domestic) levels.

[38] Eight main systemic categories of the RIS include about 75 various services at a higher level of task specification – the Author's note.

Table 4-3. National River Information Systems (RIS) in Europe

No.	Country	Name of the information system
1.	Germany	ELWIS
2.	Bulgaria	BulRIS (Bulgarian River Information Service)
3.	Austria	DoRIS (Donau River Information System)
4.	Slovakia	SlovRIs (Slovak River Information System)
5.	France	VNF
6.	Belgium	RIS Flanders II
7.	The Netherlands	LIVRA (Logistic Information Fairway Rotterdam-Antwerpen)
8.	Romania	RoRIS (Romanian River Information System)
9.	Serbia	Plovput
10.	Croatia	CroRIS (Croatian River Information System)
11.	Ukraine	UkrRIS (The River Information System of Ukraine)
12.	Poland	RIS of the Lower Oder River
13.	Hungary	Pannon RIS
14.	The Czech Republic	LAVDIS

Source: adapted from M. Kaup, op. cit., pp. 5279–5287.

Considering the requirement referring to the interoperability of the particular RIS systems at the national level (among others, possible integration and data compatibility), and the basic safety standards (referring to transport processes and information, including monitoring), it has been decided that the requirements and technical specifications of the RIS systemic infrastructure (sensors, data transmission systems, process standards, etc.) must be based on the solutions already developed in the field of inland waterway transport by international organisations, such as:[39]

- The Permanent International Association of Navigation Congresses (PIANC),
- Central Commission for Navigation on the Rhine (CCNR),
- United Nations Economic Commission for Europe (UNECE).

At the European level, the RIS has been conceptually developed on the basis of the PLATINA 2, a project of the EU which provides access to the key functions of the system services, such as the following:[40]

[39] *Sprawozdanie z inicjatyw i opinia Komisji Europejskiej 2013*, op. cit., p. 68

[40] R. Miler, *Systemy telematyki i monitoring w zarządzaniu transportem wodnym śródlądowym (RIS)* [in:] R. Miler, M. Gogołkiewicz, B. Pac (eds), *Wymiary Logistyki – Ujęcie systemowe*, Prace Naukowe WSB Gdańsk, vol. 51, CeDeWu Warszawa 2017, pp. 161–197.

- European Hull Database – EHDB: similarly to maritime navigation, inland navigation vessels require technical inspections that can certify their serviceability in inland waterway navigation. The minimal information provided in that sub-system includes the *Unique European Vessel Identification Number* data: the name of a vessel, its dimensions, draft and current certificate status. In accordance with the Rhine Shipping Examination Regulations (RheinSchUO), EC Regulation 164/2010 and the Directive 2013/49/EU, it comes as a set of data which should be exchanged between certification institutions and users of the RIS system (according to the EU PLATINA). The system provides the characteristics of over 13 800 vessels, that is approximately 80% of the total inland navigation fleet providing transportation services along inland waterways of the EU. In March 2016, the European Commission took control over the EHDB, and now it requires regular reports on all changes from the relevant institutions of the member countries (up-to-date maintenance of the system).
- European Reference Data Management System-ERDMS: data provided by this sub-system are used in various RIS applications; they include information about the location of the main elements of the linear (rivers, channels, lakes with waterways) and nodal (ports, weirs, locks) infrastructure with their individual ISRS location codes. It allows interested entities to identify an object immediately and to compare the data from the compatible (including national) RIS sub-systems, giving static data (e.g. location of a bridge) according to ENC charts and dynamic data (e.g. limited clearance of a bridge resulting from some maintenance technical work) which are provided to skippers/vessel captains (NtS Notices to Skippers) by the national RIS systems. Therefore, advanced standardisation of data is necessary as well as the development of interfaces that would facilitate comparison and analysis. The ERDMS also provides access to the code libraries of the RIS objects and reference tables. It organises, unifies and standardises codes in the following groups:
  - RIS index: information about the infrastructure (RIS objects) and their location,
  - ADN codes (The European Agreement concerning the International Carriage of Dangerous Goods by Inland Waterways) providing information about hazardous cargo in inland waterway transport,
  - HS codes (harmonised system codes from the customs organisation – non-dangerous goods): providing information and codes referring to non-hazardous cargo to the system of customs services,

  - ERI Location (electronic reporting stated as the SRS codes): it uses the ERINOT format, and it is replaced by the RIS index in the area of the EU, however it may include the SRS indices for some locations outside Europe,[41]
  - container codes: identification of the container types used in inland waterway transport (ISO 6364),
  - national (domestic) codes: they allow interested entities to identify the origin country (of an infrastructural object, vessel registration, etc., in accordance with the ISO 3166),
  - cargo codes (Codes for Types of Cargo, Packages and Packaging Materials, in accordance with the UN Rec 21),
  - identification code of the vessel type (in accordance with the UN Rec 28),
  - NtS codes (Notices to Skippers), Figure 4-15 presents the scope of the ERDMS operation in the RIS system in a schematic way (starting from the RIS data administrator, *RIS Data Admin Focal Point* and ending up with the RIS users).
- The Electronic Navigational Chart Register and the digital parts of the Inland ENC standard – Inland ENC Register – a type of electronic maps for the Inland ECDIS standard (*Electronic Chart Display and Information System for Inland Navigation*), which is based on the S-57 solution/standard, implemented by the International Hydrographic Organisation (IHO). It is composed of an electronic catalogue of objects formed as libraries with relevant symbols. This standard is successively replaced with new solutions, such as S-100 and S-101 standards (Universal Hydrographic Data Model) which allow users to use Inland ENC charts in

[41] The ERI format for the dangerous goods notification is the UN/EDIFACT "International Forwarding and Transport Dangerous Goods Notification (IFTDGN) message". The port authorities of Antwerp, Bremen, Felixstowe, Hamburg, Le Havre and Rotterdam have derived the PROTECT message from the IFTDGN message. In close consultation and in co-operation with the PROTECT group of ports, the ERI notification and the ERI response messages have been derived for inland navigation; moreover, the BERMAN message has been defined ensuring similar procedures for inland and maritime vessels in the respective ports. By means of the Memorandum of Understanding between the PROTECT group of ports and the ERI Expert Group, it has been agreed to mutually develop and use the respective messages, data and codes which will lead to the required level of conformity. This procedure ensures the conformity between maritime and inland navigation for dangerous, polluting goods and the SPS requirements where applicable – the Author's note.

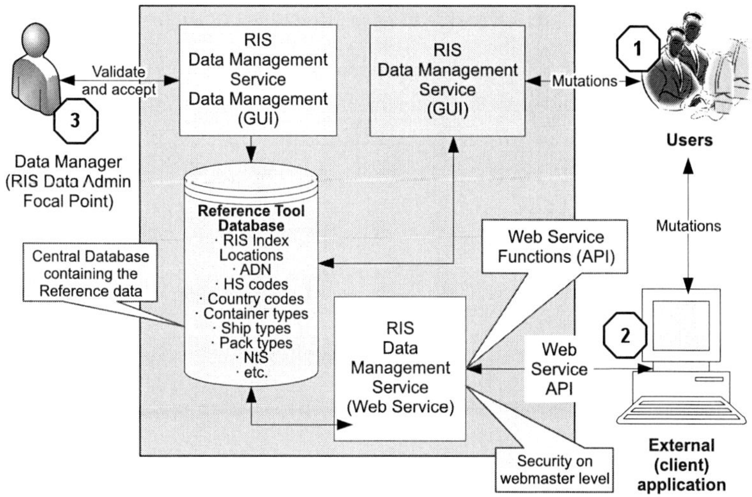

Figure 4-15. The functional scope of static data management in the RIS system (ERDMS)

Source: adapted from http://www.ris.eu/services/european_reference_data_management_system__erdms [accessed on 20th June 2018].

a broader and automated way in the systems of automated navigation compatible with the ECDIS database; the S-100 standard (Geospatial Information Registry) takes the following domains into consideration:[42]
- Hydro Domain,
- ICE Domain – referring to inland waterways where ice may occur,
- Inland ECDIS Domain,
- Marine Information Objects (MIO) Domain – databases providing information about objects affecting navigation safety,
- Port ENC Domain – databases providing information about ports.

- The RIS portal has been developed in order to collect information about inland waterway navigation and – more broadly – inland waterway transport (including port operations). It provides on-line access to river information (www.ris.eu, which has been operating since 2009).

Presented above, the functional structure of the RIS system considerably determines the organisational structure and operational capabilities of the system.

[42] R. Miler, *Systemy telematyki i monitoring w zarządzaniu transportem wodnym śródlądowym (RIS)*, op. cit., p. 168.

4.2.2. The organisational and operational structure of the RIS

Similarly to all systems based on acquisition, processing and transmission of information, the RIS consists of four basic segments:[43]

- the sensor segment – the source of acquiring information,
- the data transmission segment,
- the RIS Centre – a place where information is processed (analytical and decisional modules),
- the RIS user segment – addressees of information (the use of information with application of the effector elements).

Sensors which acquire spatial information (Geospatial-GIS) and support information supply from the current databases in the RIS system include the following:[44]

- the ARPA-type radars,
- AIS and Inland AIS transponders,
- GPS/ DGPS navigation receivers,
- water gauges, gyro-compasses, logs,
- intelligent elements of waterway marking,
- CCTV systems.

Information is processed and displayed in the electronic chart systems of the Inland ECDIS and the Integrated Navigation System (INS) which comes as a part of the Integrated Bridge System installed on vessels and some inland navigation vessels (the number of which is still scarce). When viewed from the operational perspective, the RIS must provide its users with undisturbed access to at least three levels:[45]

- electronic navigation charts for inland navigation in the Inland ECDIS standard,
- communication for vessel skippers/captains (NtS – Notices to Skippers),
- the system of vessel traffic control (VTT-Vessel Tracking and Tracing) which closely cooperates with the VTS-class systems (in maritime navigation in fairway areas).

---

[43] Ibidem, p. 170.

[44] Miciuła I., *Model systemu wspomagania decyzji w oparciu o dane geoprzestrzenne na potrzeby rzecznego systemu informacyjnego z wykorzystaniem dynamicznej domeny 3D*, Zachodniopomorski Uniwersytet Technologiczny, Szczecin 2010., pp. 26–28.

[45] R. Miler, *Systemy telematyki i monitoring w zarządzaniu transportem wodnym śródlądowym (RIS)*, op. cit., pp. 171–172.

Applied in the RIS, the Electronic Chart Display and Information System (Inland ECDIS) is an informational and navigational system which integrates an electronic navigation chart and information received from navigation measurement sensors on one display screen in order to support navigators in planning and monitoring their routes. Figure 4-16 presents a display under the Inland ECDIS which uses on-board inland electronic navigation charts (IENC).

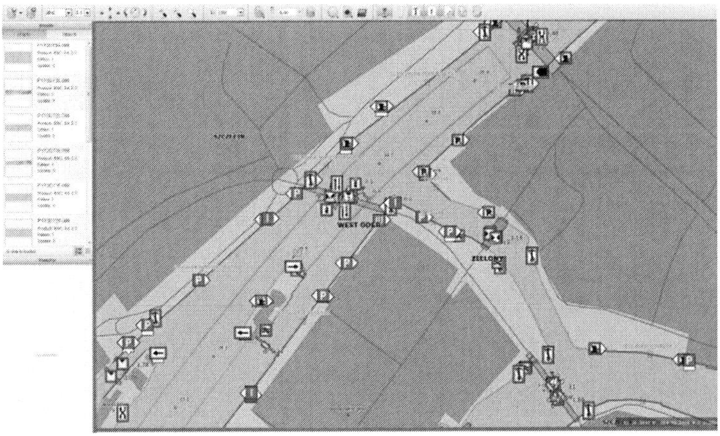

Figure 4-16. An image presenting the use of the Inland ECDIS system in the RIS system: a display that can be observed in a screen installed on a vessel (for the Oder River, in the vicinity of Szczecin)

Source: adapted from materials provided by UZS at https://szczecin.uzs.gov.pl/zsic7_elektroniczne_mapy_nawigacyjne.html [accessed on 2nd April 2018]. [c]

The IENC (Inland Electronic Navigational Charts) is a standardised product in terms of its structure, format and content. Additionally, users of the RIS system in some EU countries can use an interactive chart of the RIS area; this function is included in the display and traffic monitoring services (Figure 4-17).

Such a diversified and multi-layered display of the situation is presented in a graphical way in the ECDIS system. The advantage offered by charts of this type involves a possibility of supporting the chart content with a radar overlay. It considerably increases the efficiency and safety of the navigation process (and the whole transportation process).[46] Figure 4-18 presents such a situation.

[46] R. Miler, *Systemy telematyki i monitoring w zarządzaniu transportem wodnym śródlądowym (RIS)*, op. cit., p. 172.

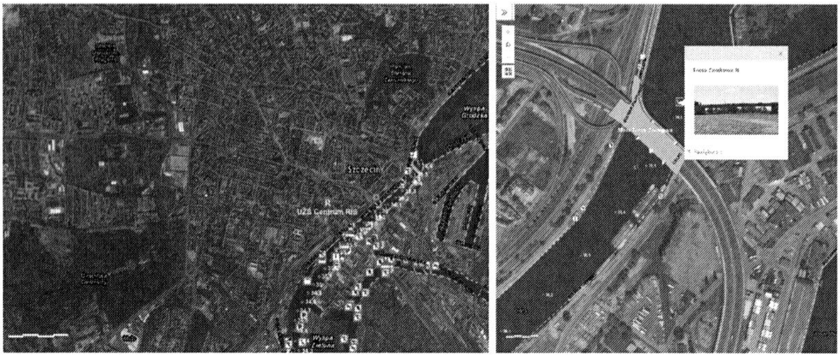

Figure 4-17. An interactive chart of the Oder River area

Source: adapted from materials provided by Żegluga Szczecińska at http://zegluga.home.pl/zdjecia_ris/index2.html [accessed on 2nd April 2018]. [c]

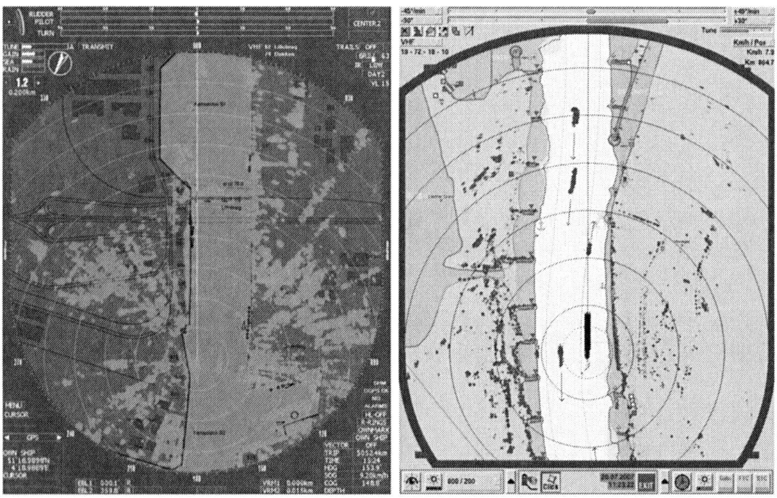

Figure 4-18. A display of the situation in a water area presented with the use of the ARPA-type radar (left) and a radar display overlaid on the ENC chart (right)

Source: adapted from materials provided by Central Commission for the Navigation of the Rhine (CCNR) Leaflet Inland ECDIS, Edition 2014, p. 15. [c]

Another feature in the field of operational application of the RIS (in the ECDIS function) is constant monitoring and display of traffic with regard to restrictions entered into the system (depth, navigational obstacles, predefined zones and water areas). Figure 4-19 presents an example of a defined display (with regard to safety isobaths for the inland waterway (river) traffic of vessels characterised by a particular draft).

Figure 4-19. Safety isobaths for a vessel navigating along an inland waterway

Source: adapted from www.caris.com. [in:] I. Miciuła, op. cit., p. 80. [c]

Another component of the RIS system is the NtS communication service for captains/skippers. NtS communication is a standardised, coded and ready to be downloaded message which contains at least two of the following pieces of information indispensable for safe navigation:[47]

- Fairway and Traffic related Message (FTM),
- Water Related Message (WRM),
- Weather Related Message (WERM),
- Ice Message (ICEM).

The application provides a possibility of selecting messages with regard to the area, expiry date or type. Users can also select a language of the message, and they can register their e-mail address into the service on the same website in order to receive messages to the stated address.[48] Figure 4-20 presents a display of the NtS application for the RIS system.

---

[47] *Pilotażowe wdrożenie RIS Dolnej Odry Studium wykonalności*, vol. 1. The main section. The study provided by a specialist consortium of the Maritime University of Szczecin and Lemtech Konsulting Ltd., Szczecin 2011, p. 38.

[48] R. Miler, *Systemy telematyki i monitoring w zarządzaniu transportem wodnym śródlądowym (RIS)*, op. cit., p. 174.

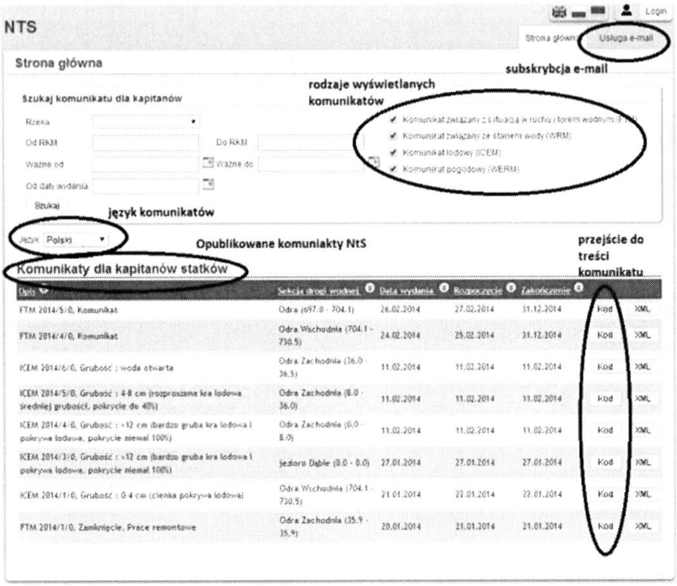

Figure 4-20. The NtS application for the RIS system, the Oder River, Poland

Source: adapted from materials provided by UZS at https://szczecin.uzs.gov.pl/zahsf_komunikaty_dla_kapitanow_statkow_nts_notices_to_skippers.htm [accessed on 2nd April 2018].

The data are provided through the web portals operated by fairway authorities. In addition to these portals, NtS messages can be downloaded in various ways (e.g. using subscriptions). Through various RIS projects, these portals have been improved over the recent years providing a more integrated view on data. Some examples are depicted in Figure 4-21.

The VTT technology is one of the key elements of the RIS system. It is applied in the surveillance and monitoring of vessel traffic, with the use of the most advanced technologies such as:

- Inland AIS (Inland Automatic Identification System) with the characteristics similar to the AIS for maritime navigation,
- DGPS (Differential Global Positioning System),
- radars including the ARPA-type radars (Automatic Radar Plotting Aids),
- CCTV (Closed-Circuit TeleVision),
- IALA DGNSS for inland waterways – Differential GNSS (Global Navigation Satellite System).

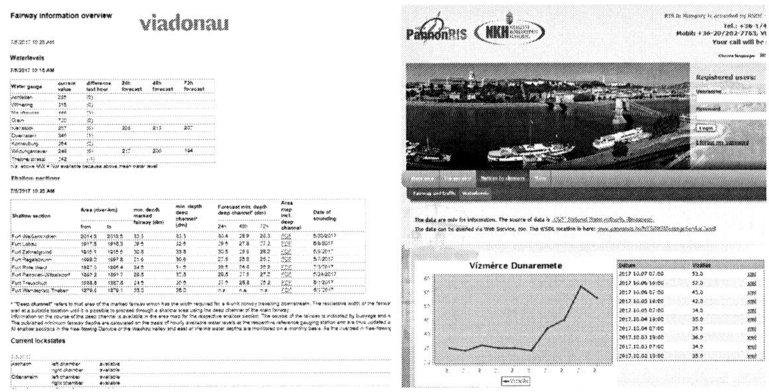

Left: Integrated Fairway information provided in the PDF-format (Via Donau)
Right: PannonRIS providing water level information in Hungary

Figure 4-21. An example of information available in the NtS RIS system

Source: adapted from materials provided by L.M. Punter, W.J. Hofman, *Digital Inland Waterway Area. Towards a Digital Inland Waterway Area and Digital Multimodal Nodes*, EUROPEAN COMMISSION, Directorate-General for Mobility and Transport Directorate D Waterborne Unit D.3 Ports and Inland Navigation, Luxembourg: Publications Office of the European Union, 2017, pp. 17–18.

Figure 4-22 presents a model of the support provided by the RIS to decisions related to the safety, efficiency and effectiveness of inland waterway transport management processes.

Considering the above-mentioned model solution and the main aims of the RIS system development, it is possible to distinguish the fundamental areas of information that form the display at the following levels in the architecture of the information flows:[49]

- tactical – Tactical Traffic Image – provides information generated by the analysis of a traffic image. It includes everything which results from the traffic image, among others: vessel course and speed. Such information affects decisions made by the users, and their decisions directly affect the current safety of navigation;
- strategic – Strategic Traffic Image – it provides information which refers to water transport, such as current positions of vessels, traffic volume. Such information generates relevant changes in decisions which affect the current traffic planning.

[49] I. Miciuła, op. cit., p. 26–28.

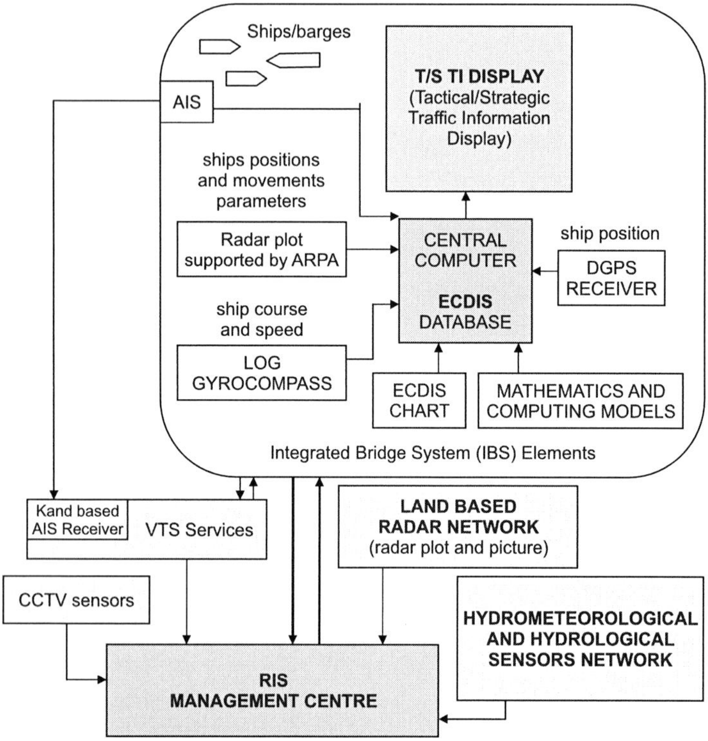

Figure 4-22. An organisational and process (operational) model of the RIS system operation

Source: adapted from I. Miciuła, op. cit., p. 59.

Tactical and strategic display is connected with the Fairway Information System which provides information related to the planning and organisation of vessel traffic.[50] It presents information about conditions along waterways which directly affects possibilities of route planning for sailings along the particular waterway sections. Such information includes data referring to the current water level, currents, ice sheet or meteorological and hydrological forecasts.

Information for the levels identified above is acquired from the current databases (Figure 4-23).

[50] R. Miler, *Systemy telematyki i monitoring w zarządzaniu transportem wodnym śródlądowym (RIS)*, op. cit., p. 175.

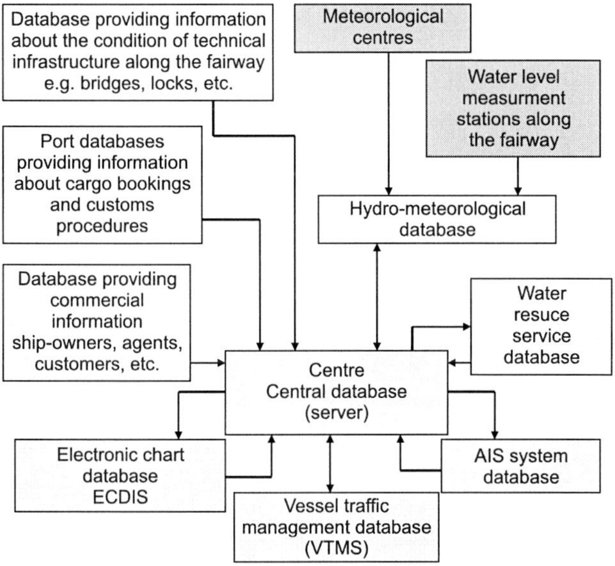

Figure 4-23. The RIS database sub-systems

Source: adapted from I. Miciuła, op. cit., p. 56.

In order to exchange computerised data without direct human interference between the RIS users and the RIS services, the use of codes and references is strongly advised and in a number of cases required to ensure unambiguous information. The indicated classification and the defined code lists shall be used to the highest possible extent. The following classifications shall be used in inland ship reporting:[51]

- vessel and convoy type (UN Recommendation 28),
- official ship number (OFS),
- IMO ship identification number (IMO),[52]
- ERI ship identification number,
- ENI unique European vessel identification number,
- harmonised commodity description and coding system 2002 (HS, goods),
- combined nomenclature goods (CN goods),

[51] Central Commission for the Navigation of the Rhine (CCNR) Leaflet Electronic Reporting in Inland Navigation, Edition 2015, p. 13.

[52] IMO ID is the Lloyds Register as published for every seagoing vessel number without the letters LR – the Author's note.

- standard goods classification for transport statistics /Revised (NST/R goods),
- UN dangerous goods number (UNDG),
- international maritime dangerous goods code (IMDG),
- ADN,
- UN code for country and nationality,
- UN code for trade and transport locations (UNLOCODE),
- fairway section code,
- terminal code,
- freight container size and type code,
- container identification code,
- package type code,
- handling instructions,
- purpose of call
- nature of cargo.

The above presentation of the technical, organisational and operational aspects of the RIS system makes it possible for the Author to take the next step in the analysis of the RIS system usefulness for inland waterway transport, namely identification of the main functions of the system.

4.3. Identification of the main functions of the RIS in inland waterway transport

4.3.1. Application of the Inland ECDIS system (Electronic Chart Display and Information System for Inland Navigation)

A European concept focused on the harmonisation of information services aimed at providing surveillance and management of inland vessel traffic, the River Information Service – by definition – should provide two-direction information flows in relations between:[53]

- onshore objects and the RIS Centre,
- onshore objects and navigating vessels,
- navigating vessels and the RIS Centre.

The systemic integration of such information flows in the RIS makes it possible to indicate some fundamental functions of the system which consider-

[53] Pilotażowe…, op. cit., p. 57.

ably determine the advantages it offers.[54] As a system, the RIS is based on the software available on the IT market, which has been offered by numerous software developers. In order to present the functions of the system, one example of the RIS software will be discussed: the ORCA™ River offered by SevenCs. Understood as the software for the RIS system, the product provides the following:[55]

- in the sensory layer:
  - identification of the quality and precision of data obtained from the onshore sensors in order to determine the precision of the position stated (and risk calculation),
  - additional systemic support for the sensors which monitor the course, engine speed, depth and wind (speed, direction);
- in the operational layer (monitoring scheduled sailings):
  - river port interfaces which provide information concerning the closest secondary port, in case there are any problems with loading/unloading at the intended destination port (a fallback option with a secondary port),
  - a display of a silent alert, in case there are any deviations from the predefined parameters of the sailing;
- in the operational layer (towing and pushing):
  - a special TowBuilder function which provides a graphical display of a barge train and precise management of the process,
  - selection of barges from the database (among others: European Hull Database – EHDB) according to the selected criteria;
- in the operational layer (a display of the situation on the water with the use of the AIS and information provided by the Centre):
  - manual or automatic object tracking,
  - multi-criteria filter for the requirements of the automatic tracking,
  - a system of graphical presentation (colour code) of objects which are highly significant to the safety of navigation (e.g. vessels on a collision course, ways to pass and to overtake other vessels), a display of such situations with CPA-type solutions (closest point of approach), with the calculation of TCPA (time to CPA). It is presented in Figure 4-24;

---

[54] I. Broeke, *River Information Services – An envisaged Vessel Traffic Management System on European Inland waterways*, International Conference, Singapore, 2000., p. 201.

[55] Based on different available sources incl. Kaup op. cit.,; Miciuła op. cit.; https://www.sevencs.com, R. Miler, *Systemy telematyki i monitoring w zarządzaniu transportem wodnym śródlądowym (RIS)*, op. cit., p. 175–176.

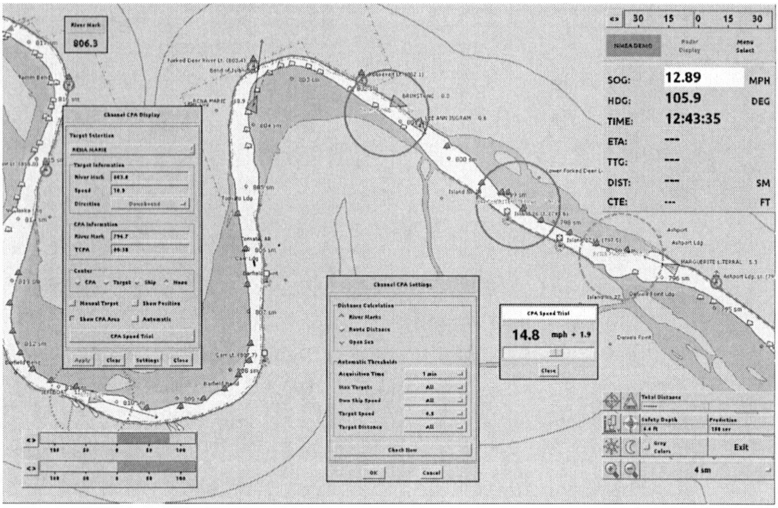

Figure 4-24. The interface of the ORCA™ River user in the CPA/TCPA function

Source: adapted from materials provided by SevenCs at https://www.sevencs.com/navigation-software/orca-river/ [accessed on 15th May 2017]. [c]

- in the navigational layer (with the use of the Inland ECDIS):
  - automatic download of charts during the sailing/operation,
  - possibility of using Inland ENCs, SENCs, BSB raster charts and charts which have been individually prepared for the requirements of the particular operator with high precision and chart coding (S-63 protected ENCs),
  - possibility of using charts provided by the ChartWorld producer in the DNV (S)ENC Approval standard, in accordance with the PRIMAR Standards.[56]

Another set of functions offered by the RIS system results from the use of ENC and ECDIS (Inland-ECDIS). An example of such software is PC-Navigo which offers functions in the following fields:[57]

- voyage planning in the Voyage Planner which is compatible with the Inland-ECDIS and provides planning and displaying scheduled routes in two modes:

---

56 http://www.chartworld.com [accessed on 27th July 2017].

57 http://www.pcnavigo.com [accessed on 26th July 2017].

- PLANNING and LOGISTICS general mode (a general display of the route for the visual, logistical and other purposes, without any irrelevant details which may generate problems with the reading resulting from too many unnecessary details),
- NAVIGATION and MANOEUVRING detailed mode (a display of the route section with a large number of details indispensable for navigation, such as buoy location, depth of the water area, the location of navigational obstacles, etc.); the system can automatically switch the display mode according to the decision made by the operator (when vessel enters the shorter-range area, the chart automatically switches to its navigation mode[58]). Both display modes (planning mode and navigation mode) are presented in Figure 4-25;

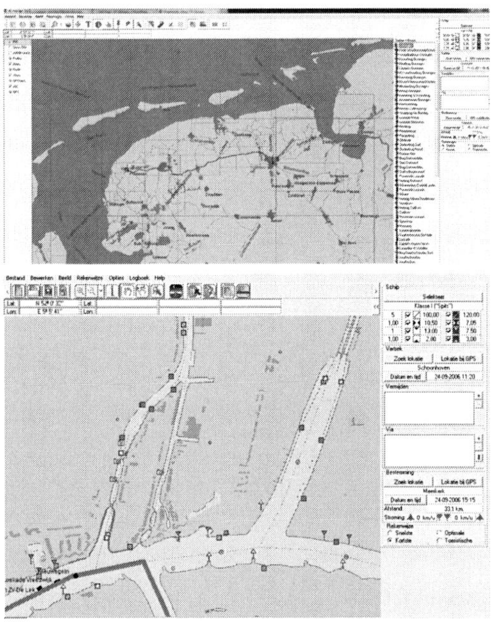

Figure 4-25. A display of a voyage in the PC-Navigo for the Planning mode (top) and the Navigation mode (bottom)

Source: adapted from materials provided by PC-Navigo at http://www.pcnavigo.com/charts/enc-and-ecdis/?lang=en [accessed on 28th June 2017]. [c]

[58] R. Miler, *Systemy telematyki i monitoring w zarządzaniu transportem wodnym śródlądowym (RIS)*, op. cit., pp. 175–176.

- automatic supply of information packets to ENC charts under the Inland-ECDIS, which allows the users to navigate precisely, especially with the use of the GPS positions. It also makes it possible to eliminate the effect resulting from the different precision levels of charts and positions. Figure 4-26 presents such a situation;

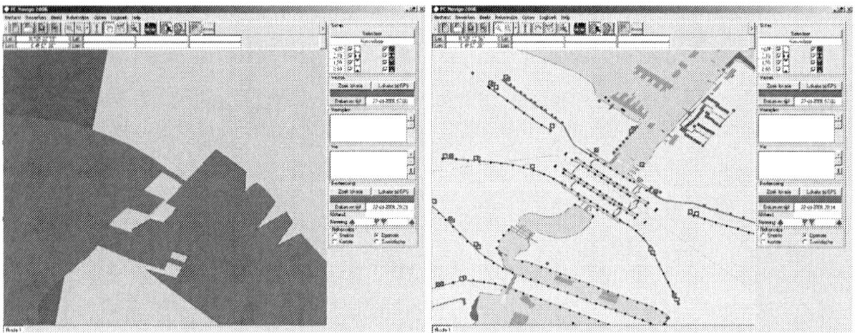

Figure 4-26. A display of a navigational chart in the PC-Navigo without the information from the Inland-ECDIS database system (left) and with such information provided (right)

Source: adapted from materials provided by PC-Navigo at http://www.pcnavigo.com/charts/enc-and-ecdis/?lang=en [accessed on 28th June 2017]. [c]

- automatic supply of information acquired from the AIS system to the ENC display (optionally, when an AIS receiver is available); it provides presentation of such data as the name of a vessel, its course, speed, destination port and the estimated time of arrival (ETA) of all the vessels which are operating within the range of the AIS receiver installed on the vessel for which the voyage is planned and monitored;
- automatic supply of data referring to the vessel (its length, beam, draft, speed, manoeuvre capabilities) for which the voyage is planned to the system of route planning. It makes it possible for users to take such data into consideration during the process of route planning and to use the available functions of the system (e.g. safety isobaths);
- automatic download of information about any difficulties in vessel traffic and necessary stops, e.g. at locks; such information is taken into consideration during route planning (optionally, when the access to the Internet is available – an on-line service);

- export of data referring to the voyage to the Google Earth application; it makes it possible to transfer the voyage route and its reading to all the users of the software which handles the *.kml format (information sharing).

In order to identify the position of a ship and, thus, for the positioning of the electronic river chart, a satellite-based navigation system is required. At the moment, a GPS navigation system is available throughout Europe. Its accuracy is sufficient for strategic navigation and for the information mode of Inland ECDIS equipment. The accuracy of chart positioning is subject to more stringent requirements in the navigation mode, which cannot be complied with without a differential service. Therefore, the IALA DGNSS service is provided. IALA DGNSS reference stations along major inland waterways in Europe are planned or already provided (Figure 4-27).

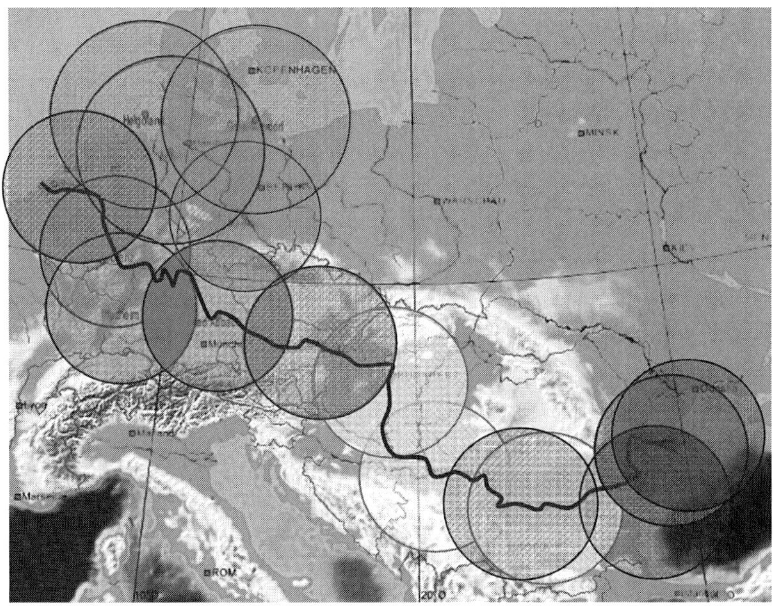

Blue circles (dark) – IALA DGPS Reference stations already in operation; red circles – IALA DGPS Reference stations planned

Figure 4-27. IALA DGNSS beacon coverage (planned or already in service) along major inland waterways in Europe [2014]

Source: adapted from materials provided by Central Commission for the Navigation of the Rhine (CCNR) Leaflet Inland ECDIS, Edition 2014, p. 4. [c]

The identification of the RIS functions based on the representative ORCA™ River and PC-Navigo software makes it possible to make an attempt at the analysis and systematisation of advantages offered by the RIS applied in the management of inland waterway transport.

4.3.2. Examples of benefits resulting from the application of the RIS system in inland waterway transport management

Generally, is should be observed that at present the development of transport is focused on spatial and temporal integration of all its modes, co-modality and optimisation in the field of integrated management. However, structural differentiation in technical, technological, legal, administrative, financial and organisational fields results in the fact that there are considerable disproportions between the basic modes. It comes as a barrier to the development of an integrated transport system.[59]

Hence, the questions referring to the incorporation of inland waterway transport into integrated transport and logistical chains become highly important. Undoubtedly, it requires application of some integrated telematics solutions which can optimise the processes of management, monitoring, risk assessment and safety in transport. By definition, such systems are aimed at an efficient use of the existing linear, nodal, informational and IT infrastructure (including systems of navigational support – AtoN Aid to Navigation) and efficient transmission of information to all the stakeholders of the system (active participants of transportation processes).[60]

After the implementation of the RIS systems, it has been concluded that they contribute to the improvement in the safety of inland waterway navigation, increasing its efficiency. They also increase the level of protection provided to transported goods and employees of this mode of transport. Furthermore, they lower the level of anthropopressure, and they contribute to the protection of natural environment.[61] The harmonisation of vessel traffic systems (VTS), vessel traffic management and information systems (VTMIS) and river information service systems (RIS) which cover more and more European inland waterways, increases the level of efficiency in the whole system, and some advantageous economies-of-scale effects can be observed.

---

[59] I. Miciuła, op. cit., p. 36–38.

[60] M. Kaup, op. cit., pp. 5279–81.

[61] *RIS, Modernizing…*, op. cit., p. 86.

In order to exemplify the advantages resulting from the RIS system application, the area where such advantages can be observed can be divided into relations which define the determinants of their taxonomic division:[62]

- national and international: local and regional effects,
- intra- and inter-modal: mono- and co-modal effects,
- operational and managerial: managerial and operational effects.

The effects of the RIS system application at the national and international level undoubtedly include the following:[63]

- rationalisation and optimisation of the use of the existing infrastructure of inland waterway transport, which increase the capacity, transport capabilities and economisation of transportation processes;
- increasing managerial capabilities through the monitoring of processes and application of optimisation algorithms;
- increasing the extent to which inland waterway navigation is used in the transportation system and increasing the integration level;
- increasing the level of safety (e.g. more efficient information management);
- decreasing a detrimental impact exerted by transportation processes on natural environment – lowering the anthropopressure coefficient (directly – through an increased participation of inland waterway transport in shipping; indirectly – through the optimisation of inland waterway transport processes).

Considering the analysis of intra- and intermodal effects, understood as mono- and co-modal outcomes, it is possible to draw a conclusion that the main line to be followed should be a more significant role played by inland waterway transport in the general volume of shipping. Such effects can be divided into:[64]

- intramodal effects:
  - improved quality of provided services, which is manifested by the growing level of indices related to the customer service (timeliness, quality, safety),

[62] R. Miler, *Systemy telematyki i monitoring w zarządzaniu transportem wodnym śródlądowym (RIS),* op. cit., pp. 176–177.

[63] M. Kaup, op. cit., pp. 5279–81.

[64] Based on different sources incl. M. Kaup, op. cit; I. Miciuła, op. cit., *Inland Waterway Transport...*, op. cit., *Evaluation of RIS Implementation for the period 2006–2011*, Main Report. Zoetermeer, July 2014 – the Author's note.

  - improved availability of inland waterway transport services (development of nodal infrastructure in particular),
  - increased efficiency of particular operations in transport (optimisation of transportation processes) and cargo handling operations (optimisation of port operations in river ports);
- intermodal effects:
  - development of networks involving intermodal relations with the use of integrated systems and common transportation technologies (a higher level of integration results in some increased pressure put on the use of inland waterway navigation as an alternative mode of transport to land road and rail transport),
  - further integration of management systems (relations of the RIS with the VTMIS systems of maritime navigation management and systems of management applied to other modes of transport will result in further optimising premises),
  - the effect of economies of scale resulting from the standardisation of processes, procedures and technologies.

Considering the effects generated by the implementation of the RIS systems in the category of managerial and operational outcomes, it is necessary to mention the following:[65]

- in the category of managerial effects:
  - improved efficiency of managerial processes achieved through the implementation of the systems which provide generation, presentation and information exchange (Inland-ECDIS) among all the participants of transportation processes in real time (e.g. recognition of the situation and traffic in the vicinity of a particular vessel),
  - increased safety of transportation processes (support provided to navigation on a display screen on a vessel) and safety of transported cargo (e.g. through the use of anti-collision systems);
- in the category of operational effects:
  - optimal voyage planning (e.g. Voyage Planning in the PC-Navigo) which provides more efficient competition with other transport operators,
  - optimal use of the existing capabilities (e.g. barge carrying capacity, availability of port operations in relation to the voyage itinerary and the data dynamically imported from other systems, e.g. NtS).

---

[65] Based on different sources incl. M. Kaup, op. cit; I. Miciuła, op. cit., *Inland Waterway Transport…*, op. cit., *Evaluation of RIS Implementation for the period 2006–2011*, Main Report. Zoetermeer, July 2014 – the Author's note.

Identification of advantages which result from the implementation of the RIS systems makes it possible to draw some more general conclusions which can be useful for the holistic analysis of the RIS as a telematics system in inland waterway transport:[66]

- telematics systems find their broad and common application in all the modes of transport, and their practical telecommunication, informational, IT and analytical solutions result in the fact that the system of transport albo has evolved into an intelligent system;
- the implementation of the RIS systems is a key line for innovative changes and further development of inland waterway transport (it refers to the local scale as well as to the national and international scale, e.g. European scale);
- providing adequate correlations and coordination of information flows among all the participants of transportation processes, the RIS systems contribute to increased efficiency, organisational and economic efficiency which leads to improved competitiveness of inland waterway transport in terms of the volume of shipping work performed in the fields of their implementation;

Presentation of the main functions and advantages of the implementation of the RIS system clearly indicates that the key feature of this system is the possibility (or rather an imperative) of its further integration with systems of maritime navigation in particular.

4.4. Integration of telematics systems in inland waterway transport

4.4.1. The concept of the Navigation and Inland Waterway Action and Development in Europe (NAIADES II) and the PLATINA supporting platform

If we consider some simple facts mentioned in this publication, which indicate that:

[66] Based on different sources incl. A. Dziubińska, A. Weintrit, *Śródlądowe drogi wodne w Polsce i ich klasyfikacja*, "Logistyka" No. 3/2014, L. Filina-Dawidowicz, M. Kaup, A. Wiktorowska-Jasik, *Zintegrowany transport wodny i lądowy*, Wydawnictwo Uczelniane Zachodniopomorskiego Uniwersytetu Technologicznego w Szczecinie, Szczecin 2014, Trojanowski 2010, M. Kaup, op. cit., I. Miciuła, op. cit. – the Author's note.

- each of five largest sea ports in the European Union is connected with inland waterways,
- every year approximately 140 billion tonne-kilometers are made along inland waterways of the European Union,
- the network of TEN-T corridors includes 230 inland ports, 75 of which are incorporated into the base network, 38 of these ports are river and sea ports,
- the level of CO_2 emission and fuel consumption on large inland navigating vessels is only 1/3 of the same values indicated in road transport,
- in 2010, about 30% of all the cargo was shipped by inland waterways from Rotterdam, the largest sea port in the European Union,

it is possible to state clearly that inland waterway transport has a number of considerable capabilities, and it plays a more and more significant role in transportation processes performed in logistical supply chains in Europe. However, the discussed potential has not been fully used yet.

Hence, in 2006 some intensive work was commenced in the EU in order to successively increase the use of inland waterway navigation in trans-European transportation processes. The NAIADES programme (Navigation and Inland Waterway Action and Development in Europe) was initiated by the European Commission in order to increase the use of inland waterway navigation as one of the solutions applied in the field of inter-modal cargo transport to create a permanent, competitive and environmentally-friendly network of transport in Europe. It includes the guidelines for transport policy to achieve a common approach, supporting inland waterway transport and the implementation of actions recommended in five fields of intervention:[67]

- market,
- fleet,
- employment and qualifications,
- image,
- infrastructure (including the implementation of the RIS).

[67] The Commission Communication on the promotion of inland waterway navigation NAIADES – the Navigation and Inland Waterway Action and Development in Europe COM(2006) 6 final, pp. 23–27.

The NAIADES programme[68] was followed by the NAIADES II programme,[69] developed in 2013. The actions undertaken under the NAIADES II (for the years 2014–2020) should ultimately contribute to the development of a stable, long-term framework for investment in inland waterway navigation and intensification of actions undertaken for the innovation of that transportation mode. The NAIADES II programme is aimed at providing proper conditions for the development of inland waterway navigation which should become a high-quality transport means. It should be well-managed, efficient, safe, integrated into a chain of inter-modal transport and offering high-quality positions to qualified employees. It should also comply with high standards in the field of natural environment protection.

The pursuit of high quality along with maintaining profitability should allow the discussed sector to break out of the vicious circle in which insufficient capabilities, insufficient investment, lack of innovation and blocked markets prevent it from generating profits. Providing high quality in the sector becomes indispensable in order to achieve aims such as a considerable increase in the participation of inland waterway transport of cargo in the EU and lower emission.

Considering such priorities, the European Commission has suggested undertaking activities in the following fields:[70]

- the quality of infrastructure and the efficient functioning of the market through elimination of barriers to the development of the sector and so called bottlenecks, such as improper dimensions of locks, bridges or

---

[68] The evaluation of the NAIADES indicates that it has considerably mobilised the EU member countries and other interested parties to undertake relevant actions. Nevertheless, the prospects for inland waterway navigation in the field of economic development and protection of natural environment have been still worsening and the advance in the field of elimination of so called bottlenecks at the key points of the infrastructure is still limited. Moreover, the IWT sector has been in a difficult situation, facing the problem of carrying overcapacity in some segments and dispersion of market participants. Carrying overcapacity in the sector of liquid bulk cargo results from the obligation of double-hull construction. The obligation has resulted in a situation in which double-hull vessels are launched to service but old single-hull crafts have not been withdrawn from service yet. It is expected that carrying overcapacity will start to shrink in 2018, when single-hull vessels are going to be out of service. Hence, the continuation of the programme as the NAIADES II seems to be well justified – the Author's note.

[69] 8 COM(2013) 623 final of 10th Sept. 2013 *Towards the high quality of inland waterway transport – NAIADES II.*

[70] Ibidem.

fairways, and the development of some missing links, such as the connection of the Seine and the Scheldt river systems;
- protection of natural environment through the development of ecological and innovative market sector, based on such solutions as new standards for engines that can encourage investment in low-emission technologies, including LNG, and significant support provided to the development of research and innovation;
- high-quality positions offered to qualified employees;
- integration of the European river systems with the European transportation field;
- incorporation of inland waterway navigation into the multi-modal logistical chain through better connections with other modes of transport.

The infrastructure of the EU inland waterways (despite the investments which have been already made) does not have a sufficiently developed network of connections, and it is insufficiently integrated with other modes of transport. Frequently, inland waterways are not connected to any logistical centres. Some considerable difficulties in a smooth traffic flow (bottlenecks) are caused by improper dimensions of locks, bridges or fairways, and the missing links become obstacles to the development of the full potential which has been lying in this sector.[71] An example of such a missing link is the Seine-Scheldt connection between France and Belgium, which is currently being addressed by the Member States concerned and by the TEN-T (Figure 4-28).

Furthermore, inland waterways come as an important element of six (out of nine) corridors of the TEN-T[72] basic network. A consistent approach

[71] Bridge clearance and the width of the passages between the supports determine the size of inland vessels and the number of container layers they can transport. The bridge vertical clearance diminishes with high water levels and increases with low water levels. The width and shape of the fairway determine whether vessels navigating upstream and downstream can pass simultaneously and they also determine the speed of navigation. The water depth of the fairway determines how many tonnes of goods (cargo) may be carried on an inland vessel. The draught of loaded vessels has a decisive influence on the cost-effectiveness of inland waterway transport. Lock capacity can lead to prolonged travel times because of waiting times due to the size of the vessel or convoys that can pass through the chamber(s). Single chamber locks risk blocking inland navigation on the entire river if just one is closed for maintenance. Missing links are the parts of the future network of inland waterways of international importance that do not exist at present – the Author's note.

[72] The corridors the North Sea – the Baltic Sea, the Mediterranean Sea, the Near East countries/the eastern coasts of the Mediterranean Sea, the Rhine – the Alps region, the North Sea – the Mediterranean Sea and the Rhine – the Danube – the Author's note.

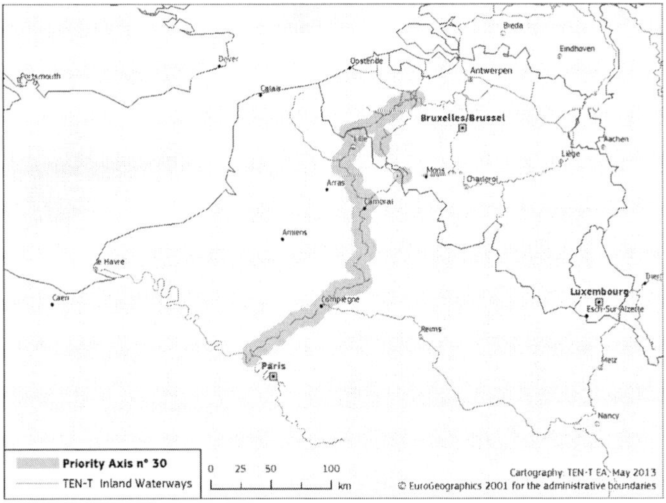

Figure 4-28. The missing link between the Seine and the Scheldt

Source: adapted from materials provided by European Court of Auditors, Special Report, *Inland Waterway Transport in Europe: No significant improvements in modal share and navigability conditions since 2001*, Luxembourg: Publications Office of the European Union, 2015, p. 16.

towards the problems specific for this sector applied in all the corridors will facilitate the planning and monitoring of the implementation of the TEN-T corridors project with regard to the requirements of inland waterway navigation. A representative example is the implementation of navigation systems and the integration of telematics systems understood in a broader sense. It provides the achievement of considerable operational advantages.[73] The implementation of more precise navigation systems and their influence on the fairway parameters are presented in Figure 4-29.

When compared to other modes of transport, the pace of innovation implementation in the sector of inland waterway transport is rather slow, due to the longtime vessel exploitation (the exploitation time of a river barge may be as long as 50 years), fragmentation of the market, the lack of innovation culture and a worsening financial situation of inland navigation operators. The task of the NAIADES II programme is to implement efficient encouragement

[73] Special Report, *Inland Waterway Transport in Europe: No significant improvements in modal share and navigability conditions since 2001…*, op. cit., p. 16

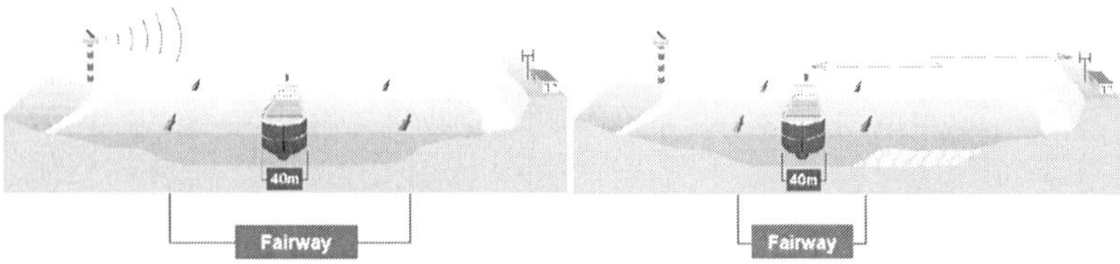

Fairways vary according to adopted navigation control systems, PIANC recommends a fairway 5× the width of the largest ship when guided by a radar and only 2.8× the width of the largest ship when guided by more precise DGNSS/DGPS navigation systems.

Figure 4-29. The influence of modern and precise navigation systems and barge positioning to the required width of the fairway

Source: adapted from http://www.inlandnavigation.eu [accessed on 23rd March 2018].

that will foster the commencement of work on research, development and innovation (R&D&L). It is addressed to such entities as ship-owners, shipyard companies, equipment producers, authorities of sea and inland ports, terminal operators, forwarders and integrators of logistical processes.[74]

The market of inland waterway transport is fully liberalised. However, some regulations – for example, regulations on vessel crews or crew working hours are still decided at the regional level or at the level of the EU member countries (in the Rhine River basin in particular). Moreover, some problems can be observed in relation with the lack of a uniform methodology for external cost internalisation.[75] The lack of harmonisation in this respect can become an obstacle to the efficient functioning of the internal market. Therefore, some actions have been undertaken under the NAIADES II in order to eliminate such malfunctions.

[74] Communication From The Commission To The European Parliament, The Council, The European Economic And Social Committee And The Committee Of The Regions on the quality of inland waterway transport, NAIADES II, Brussels, 10.9.2013 COM(2013) 623 final.

[75] At present, some possibilities of using the charges collected for the use of infrastructure have been considered in order to facilitate the achievement of external cost internalisation in inland waterway transport. Contrary to other sectors of transport, there have not been any regulations implemented for inland waterway transport to limit emission. The strict regulations concerning NOx and particulate matter emission generated by river vessels and barges will be provided with the implementation of a new EU Directive on Non-Road Mobile Machinery (NRMM) in 2020. In this way, it will also constitute the internalisation of transport costs in inland navigation – the Author's note.

Inland waterway transport (despite its strong ecological advantages, according to the SWOT matrix presented above) does not follow a positive tendency towards the limitation of GHG emission. The acceptable emission values, which are less strict for inland waterway navigation, and a long time of engine exploitation are the main reasons for that serious deficiency (when compared to restrictions for vessel engines in maritime navigation, for example). At present, the implementation of ecological drive systems (hybrid, electric, solar and LNG) in river barges faces a number of financial, technical and infrastructural difficulties. The NAIADES II is aimed at the implementation of the EU strategy concerning alternative fuels, including technical standards of the LNG bunker stations for inland waterway navigation and the use of LNG as fuel.[76] A possible solution might be a small-scale terminal or a small satellite terminal. The concept of an inland LNG terminal (for bunkering river vessels and barges) is presented in Figure 4-30.

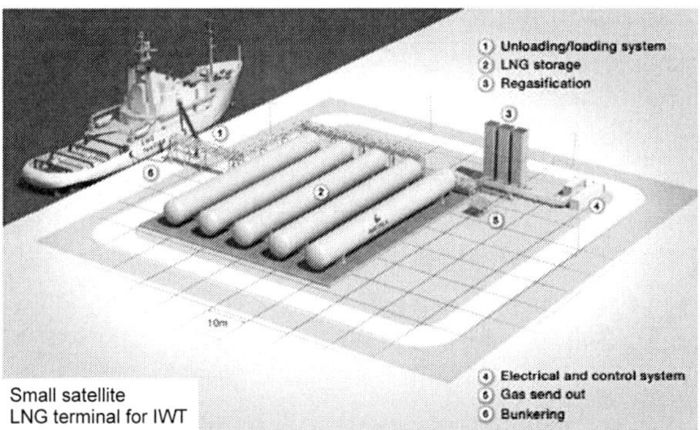

Figure 4-30. The concept of an LNG terminal for IWT

Source: adapted from www.vartsila.com [in:] M. Hapanionek, P. Sobkowicz, W. Ślączka, *Śródlądowe terminale LNG i ich możliwa lokalizacja na rzece Odrze,* 6/2017 "AUTOBUSY", p. 1383.

[76] Alternative fuel which contributes most to the limitation of emission is liquefied natural gas (LNG). It reduces the content of nitrogen monoxide (NOx) by 90% and the emission of particulate matter by 95%. On 22nd October 2014, the Directive of the European Parliament and the European Council 2014/94/EU was adopted. It refers to the development of alternative fuel infrastructure. Due to the updated ADN agreement on international inland waterway transport of hazardous cargo, it is possible to transport LNG in tankers. Hence, LNG as fuel for vessels is more and more broadly applied. Despite the fact that some investment is necessary to provide the common use of LNG for inland waterway navigation, it will be quickly returned through the lower cost of that fuel. The price of LNG is considerably lower than the price of diesel fuel – the Author's note.

Despite quite numerous pro-access activities, admission to the profession and recognition of the qualifications of people employed in inland waterway transport still face some obstacles. Going through all the procedures necessary to prove the experience and qualifications requires a lot of effort from a candidate. On the other hand, however, imprecise formulations make it also possible to ignore such procedures. Legal regulations referring to the requirements which must be met by employees of the IWT sector in the field of skills and qualifications lack consistency. Therefore, a new legal framework presented in the NAIADES II programme is aimed at a considerable decrease in the number of obstacles to the admission to the professions and mobility in that sector; it is also aimed at the appreciation of professional qualifications and careers and also at equal-opportunity treatment. At present, some electronic tools for registering and monitoring navigation time of vessels and crews and their qualifications are being implemented.[77]

Integration of inland waterway transport with the multi-modal logistical chain requires integration of infrastructure, services and information flows:

- integration of infrastructure is included in the first group of activities in the field of the infrastructure discussed above;
- integration of services requires, first of all, more commitment from the representatives of the sector, which is emphasized very clearly in the NAIADES II. In accordance with the recommendations, in some relevant cases public authorities should provide or – through the spatial development and planning – make sufficient areas along the rivers available in order to foster logistical operations in the field of inland waterway navigation (e.g. construction of new terminals, including LNG terminals). Municipal authorities should fully consider transportation of cargo and waste by inland waterways in their plans of sustainable mobility and in their strategies referring to urban logistics.

Integration must also make progress in telematics systems. The main tool, namely the RIS, implemented with the use of the IRIS Europe platform,[78] is now

[77] Communication from The Commission to The European Parliament, The Council, The European Economic and Social Committee and The Committee of the Regions, *Towards quality inland waterway transport*, NAIADES II, Brussels, 10.9.2013 COM (2013) 623 final.

[78] The IRIS Europe initiative focuses on the coordinated Implementation of RIS in Europe. For that purpose, the multinational projects IRIS Europe I (2006–08) and its follow-up IRIS Europe II (2009–11) – co-funded by the European Union within the TEN-T

applied mainly to manage traffic, to manage infrastructure or to perform activities in the field of safety improvement. In accordance with the NAIADES II, the current data exchange will be extended and integrated with the information flow in other sectors of transport, with the consideration of the e-Freight concept and the implementation of the RISING project (RIS Services for Improving the Integration of Inland Waterway Transports into Inter-modal Chains).[79]

The implementation of the NAIADES II programme received technical and organisational support in the form of the PLATINA 2 (2013–2016) platform. The PLATINA 2 brought together all its stakeholders in a multidisciplinary knowledge network in order to foster the development of inland navigation towards a more sustainable and competitive part of the European transport network. The PLATINA 2 was based on the results of the PLATINA project (2008–2012). In accordance with the NAIADES II, the PLATINA 2 was focused on four thematic fields:

- Markets and Awareness,
- Fleet and Innovation,
- Jobs and Skills,
- Infrastructure.

The PLATINA 2 provided technical and organisational assistance for the European Commission, member and neighbouring countries, River Commissions and entrepreneurs. Thereby, the PLATINA 2 significantly supported the development and the performance of targeted activities for the implementation of the NAIADES II Action Programme.

Despite some considerable efforts made by the EU under the above-mentioned NAIADES/PLATINA programmes, the level of integration in the field of operational procedures and telematics systems under the framework of intermodality of transportation processes taking place between the particular modes of transport and IWT is still far from satisfying. Hence, an imperative has clearly appeared to provide a high level of interoperability under the framework of complementary transportation processes between

Programme – were initiated with the main objective of starting and proceeding with the harmonised implementation of the RIS at the European level (harmonised pan-European RIS – ERIS) – the Author's note.

[79] The RISING has the objective of developing and testing new River Information Services for the European IWT sector and its enterprises. In contrast to other RIS projects, the RISING focuses exclusively on the present and future needs of the European transport and the logistics sector – the Author's note.

the main telematics systems of maritime transport (VTIMIS) and inland waterway transport (RIS).

4.4.2. The imperative for the cooperation of the VTMIS (maritime) and RIS (inland waterway) transport systems under the concept of a Digital Inland Waterway Area (DINA)

Considering the main entities involved in the processes of maritime and inland waterway economy, which have been identified in the previous sub-chapters along with their mutual interactions, it is necessary to outline a foundation of the modern and innovative architecture of a system integrating both those environments in order to achieve synergy and optimisation of all the procedures and processes supporting activities of telematics systems. Under the framework of the EfficientSea 2[80] and DINA projects, conceptual basis for such systemic integration has been developed. The DINA concept assumes that integration processes will take place between the entities presented in Figure 4-31, namely ship-owners, logistics operators, owners of river barges, operators of other (land) modes of transport, (river and sea-river) port operators, maritime and inland waterway administration, (local) managers of inland waterway transport and all other stakeholders of transportation processes (in a multi-modal approach).

In accordance with the accepted assumptions, integration should take place in the following fields:[81]

- vertical (A)[82] – aiming at full synchronisation of the processes taking place among ship-owners, maritime logistics operators and owners (operators) of river barges;
- horizontal (B) – aiming at synchronisation of transportation processes taking place among owners (operators) of river barges, river and sea-river ports and logistics centres (terminals and hubs)[83];

---

[80] EfficienSea 2 has gathered a unique level of competence in a consortium of 32 partners from ten countries representing authorities, academia, international organisations as well as equipment manufacturers combining all the right capacities for the effective achievement of the objectives e.g. development of the Maritime Cloud – a communication framework for both e-maritime and e-navigation – enabling efficient sharing of information between all maritime stakeholders – the Author's note.

[81] L.M. Punter, W.J. Hofman, *Digital Inland Waterway Area. Towards a Digital Inland Waterway Area and Digital Multimodal Nodes*, op. cit., pp. 60–62.

[82] According to the numbering in Figure 4-31 – the Author's note.

[83] According to the *hub and spokes concept* – the Author's note.

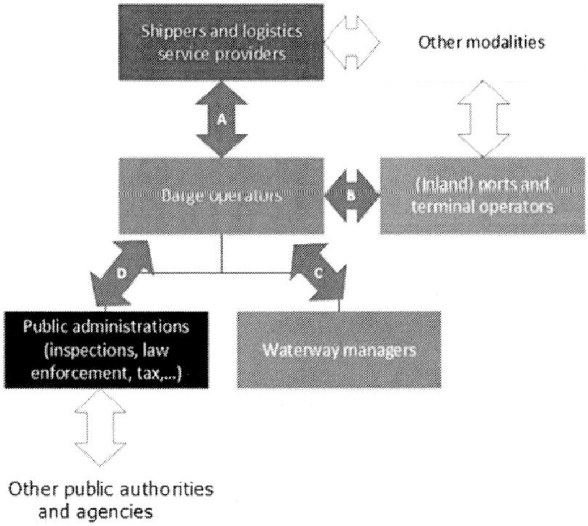

Figure 4-31. The high-level framework for the digital inland waterway area

Source: adapted from L.M. Punter, W.J. Hofman, *Digital Inland Waterway Area. Towards a Digital Inland Waterway Area and Digital Multimodal Nodes*, op. cit., p. 59.

- operational (C) – including (mainly information) exchange between transportation entities and administration of the infrastructure in order to improve safety, efficiency and economy (also in the ecological aspect – lowered anthropopressure);
- administrative (D) – including the development of standards for information exchange in order to provide the efficient functioning of inspection systems, government statistics, maritime and river safety and other systems (in accordance with local, national and European legal regulations).

Vertical integration (A) refers to the field of activities related to the course of transportation processes and the flow of transportation documents. Hence, it should include such information as freight exchange, available carrying capacities (referring to inland waterway operators who operate in the 1–2PL and 3–4PL regimes[84]), the monitoring of transportation processes in real time (including the position of a vessel, given in the ETA function – Estimated

[84] In logistics terms: barge operators currently act as first or second party logistics providers (1PL and 2PL – offering just transportation services). Increasingly, barge operators can also act as customers to other barge operators and possibly even carriers of other modalities, thus becoming 3PL (providing outsourced services) or 4PL (outsourced services and supply chain solutions) – the Author's note.

Time of Arrival). Hence, the specific character of that field of activities requires some development of new integration services (based on telematics solutions) which should include:[85]

- marketplace services – based on generally available freight exchanges which optimise shipping in the IWT sector through the matching of information about the demand and the supply for inland waterway transport services;
- booking and ordering services – providing a possibility of making transactions (including planning and estimation of transport costs, transport terms and general terms of agreement);
- visibility and payment triggering services – based on systems of financial telematics (invoices, confirmations and notifications) with automatically generated financial documents at the end of each step of the transportation process.

Horizontal integration (B), that is namely the synchronisation of the processes which take place between owners (operators) of river barges, river and sea-river ports and logistics centres (terminals and hubs), requires precise information about at least two aspects of transportation processes (within IWT), such as information about ETA and the scope of logistical and transport operations (cargo handling operations, loading, unloading) which must be performed at a particular port (and/or terminal). It generates the demand for a telematics system which will provide unlimited exchange of information of this kind among the stakeholders of transportation processes understood in such a way. It particularly refers to integrated information concerning:[86]

- precisely estimated ETA (based on the river AIS systems, under the function of the RIS),
- information about the cargo distribution (based on the vessel cargo planning system, stowage plan) on a vessel (river barge) and the number of operations which must be performed at the next river or sea-river port/terminal of call (NPoC),

[85] In modern and intelligent telematics solutions applied in other sectors, e.g. in freight exchange for road transport, all these functions can be integrated under one exchange system; the largest exchanges, such as Trans.eu, TimoCom and some others have been already operating in this system – the Author's note.

[86] L.M. Punter, W.J. Hofman, *Digital Inland Waterway Area. Towards a Digital Inland Waterway Area and Digital Multimodal Nodes*, op. cit., pp. 62–64.

- cargo handling capabilities of a terminal and its temporal and spatial optimisation (based on an efficiently operating TOS-class system).

The key information resource which provides optimisation of transportation processes in the field of horizontal integration is the ability of the system to determine ETA precisely, which becomes a key function of modern RIS-class systems (ETA is a derivative of information acquired from numerous RIS sub-systems discussed in previous sub-chapters). This aspect is inseparably related to another field of recommended integration – namely to operational integration.

Operational integration (C) combines two optimisation aspects. The first aspect is related to optimisation of navigation (the navigation aspect viewed from the perspective of an operator/skipper of a river barge); the second one is related to optimisation of the accessible transport infrastructure (the aspect of using the river infrastructure, viewed from the perspective of the fairway authorities). Considering the perspective of a river barge operator, a telematics system which provides integration, understood in such a way, is the Inland-ECDIS (developed as an integral part of the RIS), the system of ENC charts available during the whole navigation process (along all the sections of navigable waterways) and providing an easy and undisturbed access to the NtS (through the broadband Internet in 4G/5G technology and an access to the application in the cloud). In the future, further development of technology may result in semi-autonomic or fully autonomic inland waterway navigation, which will be also shared by fairway authorities (in the context of automated operation of infrastructural facilities of inland waterway navigation, such as locks, inclined planes, lifts, etc.).[87]

Resulting from the necessity of sharing information about river barge traffic displayed in telematics (and monitoring) systems of inland waterway transport with other stakeholders (e.g. administration, reporting, economic statistics, customs services, operators of other modes of transport) of broadly understood economic and transport processes (e.g. in logistical supply chains), administrative integration (D) should be developed on the basis of dedicated access to the environment of the RIS-type systems (with regard to the relevant levels of access matched with the relevant levels of precision and detailed information (e.g. ETA, current position, stowage plan). It should provide a quick definition of barge parameters, transport and technical

---

[87] Ibidem, pp. 64–68.

documentation, required certificates, any possible violation of legal regulations (e.g calamity abatement, law enforcement) and access to archive data (e.g. a histogram of passings through a particular infrastructural object in order to update the plans of its extension, etc.). Due to the implementation of the RIS in the EU, some administrative information in the field of IWT has already been undergoing integration processes (e.g. EHDB – European Hull Database), it has already entered its initial/preliminary stage of integration (e.g. ECQD – European IWT Workers' Qualifications Database), which is spread under the BICS standard. However, there is still a number of fields which require full integration.

This challenge is closely related to decisions about the principles for future integration/standardisation of new regulations, procedures and systems (including telematics systems) which will soon provide support to IWT operations. The problems related to such decisions refer, first of all, to the following aspects:[88]

- decisional level (public/regional/national/local versus private implementation) which involves an answer to the question whether the institution responsible for the establishment of new standards should be, for example, the European Commission, national navigation authorities or software companies (as private enterprises which provide some particular telematics solutions); a clear dichotomic division can be observed in decisions made at the European level (e.g. the RIS standards) and at the local private level, e.g. a selection of a TOS system that meets the requirements of a particular river terminal;
- integration level (shared solutions versus components) which involves an answer to the question to which extent it is possible to implement acceptable uniform standards under the framework of already operating Port Community Systems, considering the specificity of the IWT sector, and how strong an impact forcing high variety and systemic freedom will be to generate problems with the ultimate systemic compatibility (especially data compatibility);
- open standard application level, referring to commonly acceptable systemic solutions which should include at least the following elements (in order to guarantee the minimal effect of economies of scale):

---

[88] L.M. Punter, W.J. Hofman, *Digital Inland Waterway Area. Towards a Digital Inland Waterway Area and Digital Multimodal Nodes*, op. cit., pp. 64–68.

- uniform communication protocols in common standards of information transmission (Internet, 4/5G in mobile telephony),
- uniform message standards (syntax) sharing a uniform and common structure of information in the exchange among the computer systems, e.g. XML, EDIFACT (the architecture must be open for the compatibility with other standards which are still being developed),
- a uniform technical paradigm which defines technical framework for the development of telematics systems with regard to the acquisition (sensors) and transmission of information (messaging-based paradigm), the Internet database services (web services paradigm) or services of active data search (on request, linked data paradigm),
- uniform semantic/data dictionaries which provide a basis for the restructuring of information (and terminology) and which attribute it with the same meaning that will be interpreted in the same way by all telematics systems;[89]

- level of the local solution application with regard to the achievable level of centralisation (local development versus cloud-based deployment) which refers to a defined level of the application of technical paradigms and the tendency towards storing the information in the cloud (despite the current and very local approach towards generating and storing the information within the framework of transport management in the IWT domain). On the other hand, there is a possibility of applying new blockchain technologies which have become a counterbalance to the processes of centralisation, storage and access to information in the domain of IWT (e-IWT implementation with all natural consequences of that

---

[89] Semantics can be represented in a syntax (e.g. an XML Schema Definition or XSD – see ISO standards – or an EDIFACT message structures like those developed by the WCO and UN/CEFACT) and/or a data model (e.g. the WCO data model for Business-to-Government data sharing). Such data models are based on dictionaries like the United Nations Trade Data Elements Directory (UNTDED). They may also have data groups in common, e.g. dangerous goods specifications, but not all the data values of these groups are identical for all modalities. A semantic model and/or dictionary for terminology shared amongst stakeholders in inland waterways, ERINOT, contains a number of terms, whereas ERDMS provides valid code lists. Both are the components of a dictionary. It should be noticed that harmonisation of data dictionaries still needs to be done; different modalities each have their data dictionary and IT systems utilised by stakeholders built upon these dictionaries. Terminology of inland waterways is (partly) aligned with that of sea transport, since messages like ERINOT are based on the agreed components which are also utilised in sea transport – the Author's note.

> process).[90] Moreover, the controlled process of implementing solutions which are aimed at full autonomation of water (maritime and inland waterway) transport processes shall affect the structure of the information flow (through the remote-control rooms/centres).[91]

The concept of a data cloud, applied as a telematics solution in new integrated systems of maritime and inland waterway transport (maritime cloud) is presented in Figure 4-32.

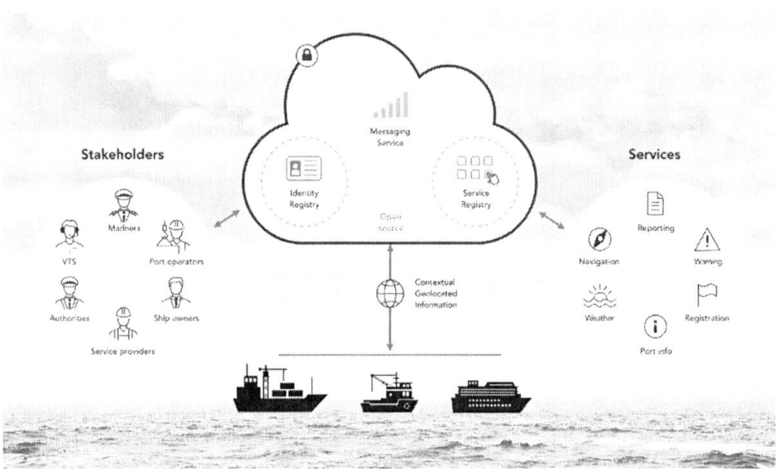

Figure 4-32. A concept of a data cloud for water transport (maritime cloud)

Source: adapted from L.M. Punter, W.J. Hofman, *Digital Inland Waterway Area. Towards a Digital Inland Waterway Area and Digital Multimodal Nodes*, op. cit., p. 68.

---

[90] These can be used to register transactions (transport orders, arrivals at a port, members of crew coming onboard, etc.) in a secure way without the need for a centralised repository. While the technology stems from digital currencies, its TRL (Technology Readiness Level) is relatively low for logistics applications. Nevertheless, such technology can be considered for future applications, e.g. for registering data from an Electronic Service Record Book for implementing e-IWT – the Author's note.

[91] Autonomous barging requires development of on-board applications that can be controlled remotely (for instance to monitor autonomous sailing of deep sea vessels from a remote-control room). As such, a barge becomes a sailing computer with computational power and storage on which applications can be installed, and which is linked to a cloud-based environment. A similar trend can be observed in the automotive domain where, for instance, Tesla cars (partly) rely on cloud services for implementing autonomous navigation – the Author's note.

Based on the facts which have been established so far, it is possible to assume that further systemic integration in the field related to telematics of inland waterway transport, under the framework of the DINA concept, should be developed in three key direction lines:

- further development of the RIS towards the implementation of new functions and on-board and e-IWT applications (Figure 4-33),

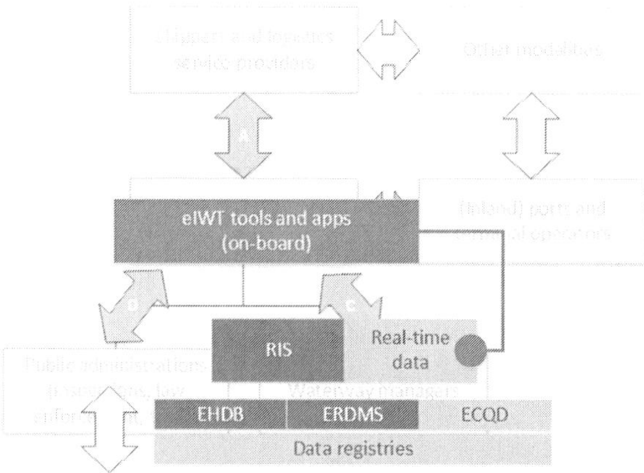

Figure 4-33. A functional model of the RIS development in the function of on-board and e-IWT services

Source: adapted from L.M. Punter, W.J. Hofman, *Digital Inland Waterway Area. Towards a Digital Inland Waterway Area and Digital Multimodal Nodes*, op. cit., p. 70.

- development of a platform for information exchange (cloud) addressed to river barge operators (more broadly, for all IWT operators) (Figure 4-34),
- deep integration of IWT information with booking applications and freight exchanges in all complementary modes of transport (Figure 4-35).

Presented above, the three-level concept of further integration of water (maritime and inland waterway) transport still faces a number of problems of technical, organisational, legislative and operational nature. The most important problems with implementation (in the context of a potential route map) which refer to the new architecture of the integrated system of water transport telematics (according to the DINA concept) are presented in Table 4-4.

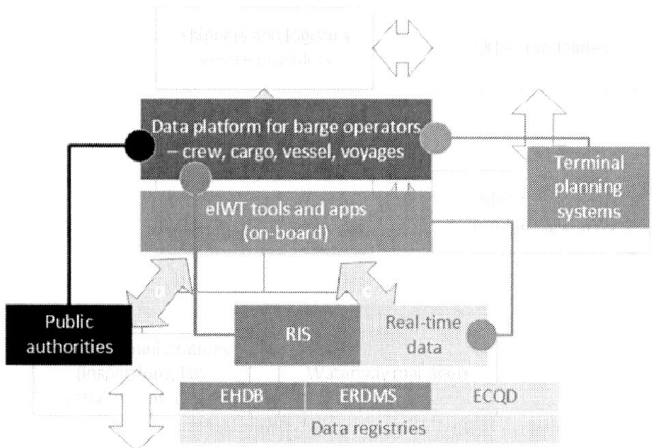

Figure 4-34. A functional model of a platform for information exchange dedicated to river barge operators

Source: adapted from L.M. Punter, W.J. Hofman, *Digital Inland Waterway Area. Towards a Digital Inland Waterway Area and Digital Multimodal Nodes*, op. cit., p. 73.

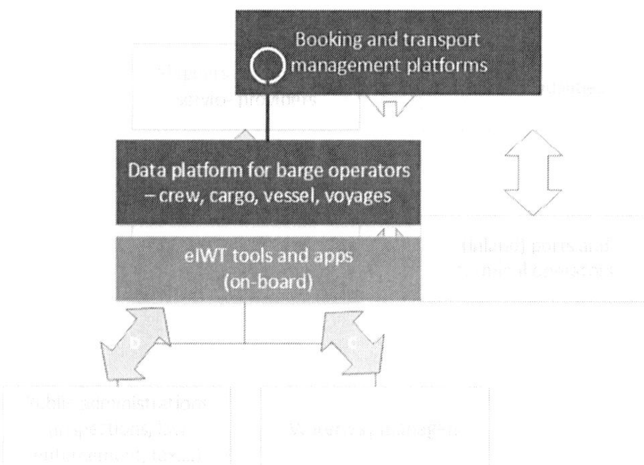

Figure 4-35. A functional model of development of booking platforms for IWT

Source: adapted from L.M. Punter, W.J. Hofman, *Digital Inland Waterway Area. Towards a Digital Inland Waterway Area and Digital Multimodal Nodes*, op. cit., p. 73.

Table 4-4. Implementation problems related to the integrated system of water transport telematics (according to the DINA concept)

Problem driver	Addressed in the architecture	Implication for the roadmap implementation
River Information Services have not been developed with continuous and controlled data sharing in mind	New features proposed (e.g. data registrics) to enable real-time data sharing on the infrastructure, vessel, cargo and crew	• Standardisation • Governance • Legislation (updating of the RIS directive)
Legal and commercial bottlenecks for sharing data	Controlled data sharing is proposed through a platform for barge operators. This enables the barge operator to control who can access his data	• Governance • (Potentially) legislation
IT-setup of barge operators which is not yet capable of continuous interaction with third parties	New on-board e-IWT tools are proposed to be used with a cloud-based platform for barge operators	• Standardisation • Public-private collaboration needed
The IWT sector being limited in size and fragmented, making it difficult to achieve economies of scale for new solutions	It is proposed to work with shared solutions and cloud-based approaches, providing the required scalability	• Standardisation • Public-private collaboration needed
The lack of an overall framework for electronic reporting covering different purposes, making it difficult to address the administrative burden	The architecture supports new reporting mechanisms and facilitates the re-use of data	• Future reporting requirements and administrative formalities should be transformed in electronic way

Source: adapted from L.M. Punter, W.J. Hofman, *Digital Inland Waterway Area. Towards a Digital Inland Waterway Area and Digital Multimodal Nodes*, op. cit., p. 78.

Overcoming the above-mentioned problems and other implementation challenges, applying and following technological regimes and the suggested implementation schedule (Figure 4-36) will have eventually allowed the interested entities to achieve all functions of the new integrated system of water transport telematics (Figure 4-37) in the next decade (by 2024 at the latest).

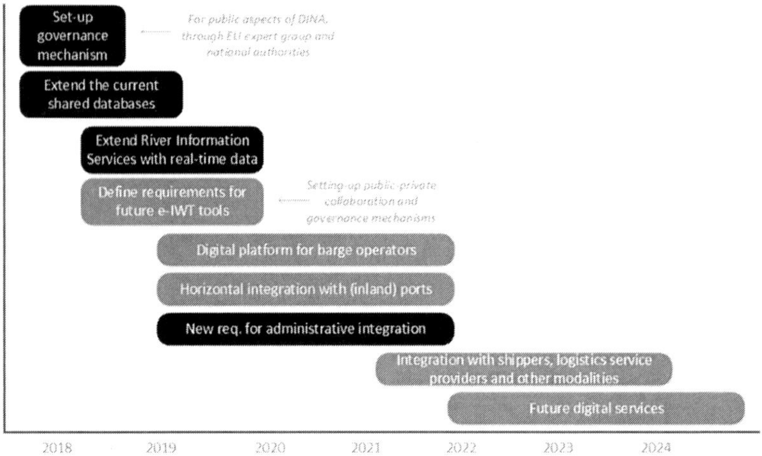

Gray blocks – public-private partnership; black blocks – public initiatives

Figure 4-36. The schedule for the implementation of integration solutions suggested by the DINA for the system of water transport telematics (building blocks and timeline)

Source: adapted from L.M. Punter, W.J. Hofman, *Digital Inland Waterway Area. Towards a Digital Inland Waterway Area and Digital Multimodal Nodes*, op. cit., p. 79.

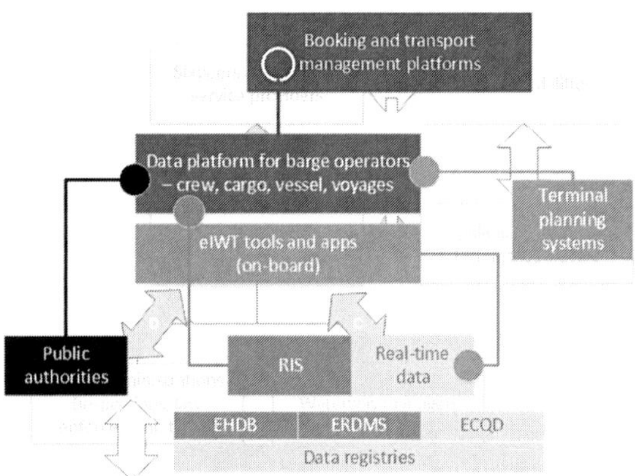

Figure 4-37. The ultimate model presenting the architecture of the system of water transport telematics, in accordance with the DINA concept

Source: adapted from L.M. Punter, W.J. Hofman, *Digital Inland Waterway Area. Towards a Digital Inland Waterway Area and Digital Multimodal Nodes*, op. cit., p. 77.

The achievement of interoperability of the particular solutions suggested in the DINA concept will be possible due to a secular approach which assumes five steps which must be made to actually achieve interoperability (Figure 4-38):[92]

- identification and standardisation of data referring to logistic operations (based on the reported demand and supply, with an assumption that the criteria of technical unification are met);
- identification of immanent features attributed to logistical and transport operations, which will lead to the development of unique measures for their evaluation and presentation;
- attribution of technical standards and the syntax type to the specific features and measures which have been identified;

Figure 4-38. The concept presenting the achievement of interoperability in the integrated system of water transport telematics (in accordance with the DINA assumptions)

Source: adapted from L.M. Punter, W.J. Hofman, *Digital Inland Waterway Area. Towards a Digital Inland Waterway Area and Digital Multimodal Nodes*, op. cit., p. 77.

---

92 L.M. Punter, W.J. Hofman, *Digital Inland Waterway Area. Towards a Digital Inland Waterway Area and Digital Multimodal Nodes*, op. cit., p. 102.

- registration of stakeholders in one database system and specification of information standards;
- operational exchange of logistical and transport data in real time among the stakeholders of the system.

The implementation of the integrated system should involve a systematic implementation of new applications and services dedicated to cooperation among all the stakeholders of that system. Therefore, some applications and services must appear, which will be especially dedicated to logistic, transport, navigation and optimisation processes in the field of transport and logistic infrastructure. It is also necessary to provide technical and complementary services (e.g. in the field of transaction, financial and accounting operations). The structure of services organised in such a way (by their types and significance to the advance of IWT digitalisation) is presented in Figure 4-39.

The last problem which is significant when viewed from the perspective of the potential implementation of the telematics system, in accordance with the DINA concept, refers to its costs (the detailed specification of costs is presented in Appendix 2). It is estimated that the total investment costs can reach the level of EUR 300 million up to EUR 350 million, broken down into:[93]

- ca. 70–80 million EUR for digitalisation of barges,
- ca. 20–30 million EUR for improvement in the integration of IWT in the field of logistics processes of shippers and logistics service providers,
- ca. 90 million EUR for digitalisation of (inland) ports and terminals,
- ca. 50 million EUR for digitalisation of public authorities (fairway managers, inspection agencies, law enforcement).

Additionally, it is necessary to estimate annual costs of service and maintenance which should not exceed the following amounts:[94]

- maintenance of digital tools, e.g. as a result of service level agreements and license fees – approximately 20% of the initial technology investment – ca. EUR 35 million/year;
- governance of the required standards and the coordination of the implementation – approximately 5% of the initial technology investment – ca. EUR 1 million/year.

---

[93] L.M. Punter, W.J. Hofman, *Digital Inland Waterway Area. Towards a Digital Inland Waterway Area and Digital Multimodal Nodes*, op. cit., p. 89.

[94] Ibidem, pp. 89–90.

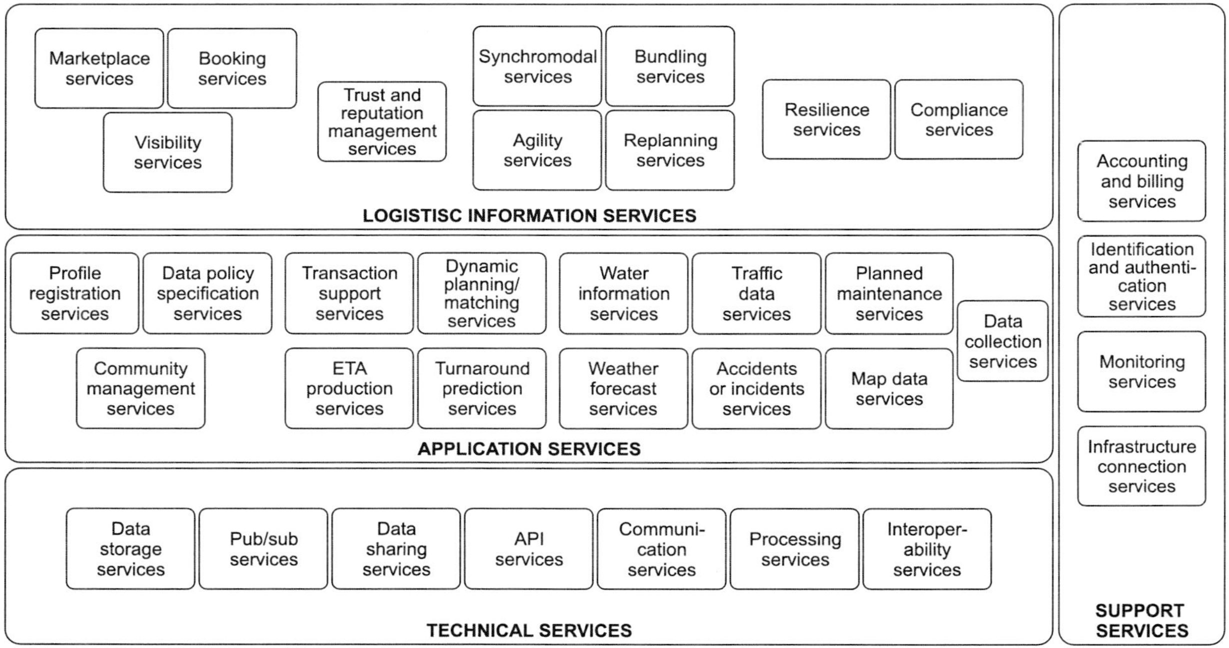

Figure 4-39. The concept of services under the integrated system of water transport telematics (in accordance with the DINA)

Source: adapted from L.M. Punter, W.J. Hofman, *Digital Inland Waterway Area. Towards a Digital Inland Waterway Area and Digital Multimodal Nodes*, op. cit., p. 104.

Presented above, the information about the implementation of systemic and integrated solutions applied in water transport telematics sums up the considerations in this chapter.

The chapter provides an attempt made at the identification of telematics systems applied in inland waterway transport. The classification of river vessel and barge monitoring systems has initiated a discussion on the role and significance of telematics systems in inland waterway transport; the specific character of this mode of transport, identified in the previous chapters, proves to be particularly important here (with regard to the scale of transportation processes, the type of the infrastructure and the specificity of vessels, which are translated into the scale and less complication of operational processes). The chapter also presents the original classification of telematics systems implemented to inland waterway transport, with consideration of the IPSI-type terminal system (Improved Port Ship Interface), which imposes systemic and operational compatibility on port systems and vessels which are operated in maritime and inland waterway navigation. However, the chapter is mainly focused on the RIS-class systems (River Information Services) which come as an original European telematics solution in the field of inland waterway transport. Next to the identification of the main systemic assumptions of the RIS, the chapter presents its organisatonal and operational structure as well as the main functions of the system, which determine its potential applications and advantages resulting from its implementation. The chapter provides some examples of the key functions offered by the RIS system, which are based on the systems of intelligent navigation and Inland ECDIS charts (Electronic Chart Display and Information System for Inland Navigation). They allow ship-owners and vessel/barge managers to succeed in safe navigation. In order to avoid a dispersion of telematics systems, which would be similar to a dispersion of such systems in maritime transport, the compatibility of the standards and systems has been emphasized since the very beginning in the domain of inland waterway transport in the EU. Therefore, the concept of the Navigation and Inland Waterway Action and Development in Europe (NAIADES II) and the PLATINA 2 platform for the systemic support of recommended solutions is presented as well. Considering such a context, the cooperation of telematics systems applied in both modes of water transport (maritime VTMIS and inland RIS) becomes particularly significant under the uniform DINA concept (Digital Inland Waterway

Area), which is discussed at the end of the chapter. Summing up, as specialists from Rolls-Royce say "In this digital age, data is like oxygen – crucial for survival and prosperity".[95]

[95] https://www.rolls-royce.com/products-and-services/marine/ship-intelligence.aspx#section-intelligent-asset-management [accessed on 24th March 2018].

SUMMARY

Being inseparable from the advance of our civilisation, maritime transport and inland waterway navigation undoubtedly come as integral elements of modern economic processes. At present, more and more cargo is transported under the framework of regional and global transportation processes, and such processes themselves face numerous challenges. A number of such challenges refer to stricter economic and operational regimes, including time and cost optimisation, a high level of reliability and safety, support provided to management processes through monitoring and tracking, tendencies towards automation and, eventually, towards autonomation. The challenges have already resulted in the fact that maritime transport and inland waterway navigation have to be equipped with systemic telematics solutions. Drawing some conclusions from this monograph, it is possible to assume that maritime transport and inland waterway navigation have already achieved a considerably high level of saturation in terms of ICT tools. Unfortunately, considering the lack of coordination, the architecture of the systems proves to be of dispersed and incoherent nature. The multiplication of the systems can be observed, and the problem of information entropy appears (especially in the context of diversified and differentiated national systems implemented in various countries).

Hence, the first attempts have been already made towards the integration of systems and the achievement of the economies of scale (compatibility of data and systems, and eventually their systemic integration in the future). Common environment (approach fairways towards sea and river ports located in estuaries) in which transportation processes are performed generates the necessity of further functional integration of telematics solutions, which are dedicated to both modes of water transport. This fact becomes highly significant. The discussed integration should bring about the effect of the economies of scale and even higher usefulness of integrated systems in the European and global scale. The most advanced telematics systems of maritime transport and inland waterway navigation also take external effects into consideration, providing a possibility of precise control over fuel consumption and emission. This fact contributes to lower anthropopressure generated by broadly understood water transport.

As a result, coming as a contribution to the full identification and systematisation of current telematics solutions applied in maritime and inland waterway transport, this monograph presents a review of the implemented systems and their taxonomy with the use of the methodology which has been specifically developed for this purpose. It is the first time most telematics systems applied in the modes of water transport have been described in detail, with the indication of their usefulness in the optimisation of transportation processes and the increase in their safety and efficiency of operational management. The scope of the application of telematics systems should be also emphasized, particularly in maritime transport, as it includes simple systems of navigation and port operation monitoring as well as autonomic navigation and unmanned drive systems which have resulted in the first trials of fully autonomic vessels. As it has been already mentioned, the full integration of maritime and inland waterway transport systems and other modes of multimodal transport becomes a serious challenge.

LIST OF ABBREVIATIONS AND GLOSSARY OF DEFINITIONS

5G – the fifth generation standard for cellular (wireless) communication; focusing on improved speeds, more reliable connections and better support for all sorts of wireless data applications

AIS – (*Automatic Identification System*); a digital radio-beacon used on vessels to broadcast data on their position, heading and vessel characteristics

ALARP (*As Low As Resonably Practicable*) – the scope of acceptable risk in the assessment of maritime transport safety

AMOS (*Asset Management Operating System*) – a system providing support to the management of resources performed by ship-owners (the counterpart of the ERP-class software – Enterprise Resources Management)

AMS (*Advanced Manifest Surcharge*) – an additional charge collected in maritime freight for an advanced manifest

AMVER (*Automated Mutual-assistance Vessel Rescue system*)

ARPA (*Automatic Radar Plotting System*) – one of the standards applied in the radar display of the situation on the water

ATA – Actual Time of Arrival

AtoN (*Aid to Navigation*) – systems providing hydrographic and navigational aid

BAF (*Bunker Adjustment Factor*) – one of freight adjustment types in maritime navigation (bunker adjustment)

BICS – (*Binnenvaart Informatieen CommunicatieSysteem*) – a tool for electronic reporting in IWT

BPMC (*border procedures monitoring and control*) – procedures facilitating border control in maritime transport and correlating information with systems of navigation display

BRITE (*Baseline for Rapid Iterative Transformation Experimentation*) – an experimental system of navigation monitoring

BSC (*balanced score card*) – a method applied for the assessment of the project implementation level

BSR – Baltic Sea Region

C/S or PCS – a congestion surcharge in maritime freight

C2 (*command and control*) – one of the military systems providing support to decision making processes

CAF – the currency adjustment factor

CBA (*Cost-Benefit Analysis*) – a type of efficiency analysis in the cost-outcome relation

CBM (*condition-based monitoring*) – the monitoring of parameters related to vessel drive operation

CBR (*commodity box rate*) – a way to present freight prices for containerised units in maritime transport

CCNR – Central Commission for the Navigation of the Rhine

CDM (*Clean Development Mechanism*) – adopted by the Kyoto Protocol signatories, a mechanism of sustainable development which limits the anthropopressure of economic processes,

CEF – *Connecting Europe Facility*; the European funding scheme for infrastructure investments

CESNI – *Comité européen pour l'élaboration de standards dans le domaine de la navigation intérieure*; a standardisation body initiated by the CCNR

CG – Coast Guard

CISE (*Common Information Sharing Environment*) – a concept referring to the integration of current national monitoring systems of maritime transport safety, defined as a platform of integrated maritime surveillance with an experimental tool IMDatE

CL – Catch Logbook

Cloud – software and data storage deployed using online service managed, by a service provider, instead of deploying it on-premises in a data center

COBO (*Coal-Ore-Bulk-Oil*) – a type of oceangoing vessels dedicated to transportation of bulk cargo (a universal bulk carrier to transport four main types of bulk cargo in maritime transport)

CoC (*Compliant Licenses and Certificates of Competency*) – a module applied to monitor the qualifications of crew members to perform their duties, certificates and diplomas

COI (*Contact of Interest*) – a vessel identified by the monitoring system of maritime safety as a contact of interest

COMCIS – Collaborative Information Services for Container Management – the European FP7 research project
COMEX – RIS Corridor Management Execution – the CEF funded project
CORE – Consistently Optimised REsilient ecosystem – the FP7 research project on supply chains and information exchange
CoRISMa – RIS0-enabled Inland waterway Transport Corrdidor Management – an innovation project
Corridor management – traffic management in a transport corridor
CSN (*CleanSeaNet*) – a system applied to monitor ecological threats in maritime transport
CSO (*Company Security* Officer) – an element of the ISPS regulations system at ports and oceangoing vessels
CSS – Coastal States Security system
DATEX II – a model of standardised information for road traffic and travel information in Europe
DGPS – Differential Global Positioning System
DINA – Digital Inland Waterway Area
DMN – Digital Multimodal Nodes
DOS (*Declaration of Security*) – an element of the ISPS regulation system applied at ports and oceangoing vessels
DSM – Digital Single Market strategy of the European Commission
DTLF – Digtal Transport and Logistics Forum of the European Commission
e-IWT – an electronic toolkit for IWT
ECDIS – Electronic Chart Display Information System applied to provide support to navigation
ECQD – European Crew Qualifications Database, currently being set up as a shared European database to register skills certificates
ECTS (*Electronic Container Tracking Service*) – a module applied to monitor containers (position, internal conditions, integrity) on vessels
EDI (*Electronic Data Interchange*) – a format which provides the exchange of communication and documents under the EDI standard, which takes place among the entities involved in the operations of a logistical supply chain
EDIFACT – Electronic Data Interchange for Administration, Commerce and Transport
EEDI – Enegry Efficiency Design Index of a vessel
EEOI – Energy Efficiency Operational Indicator
EEZ – Exclusive Economic Zone; a part of the maritime territory of a coastal state

EGNOS (*European Geostationary Navigation Overlay Service*) – a European system of satellite navigation

EHDB – European Hull Database; a shared European database containing unique identifiers and basic characteristics of all vessels

EIS – Equipment Imbalance Surcharge

EMC (External Marginal Costs) – a category of external marginal costs which is applied in the methodology of the European Commission to calculate external costs of maritime transport

EMCIP – European Marine Casualty Information Platform

EMSA – European Maritime Safety Agency

ENC (*Electronic Navigational Chart*) – a maritime map used for e-navigation

EPIRB – Emergency Position Indicated Radio Beacon

ERDMS – European Reference Data management system; a shared database containing code-lists and unique identifies of infrastructural objects

ERI – Electronic Reporting International; a standard for electronic reporting

ERP – Enterprise Resource Planning; software for financial and operational planning of organisations

ERS (*Electronic Reporting System*) – a system applied to collect data on fishing for the Electronic Reporting System

ERS (*equipment re-positioning surcharge*) – a special surcharge applied to eliminate the ship-owner's loss resulting from non-balanced container availability

ETA – Estimated Time of Arrival

ETD – Estimated Time of Departure

EU NAVFOR – European Union Naval Force

EUMSS – European Union Maritime Safety Strategy

EUSBSR – European Union Strategy for the Baltic Sea Region

FCL (*Full Container Load*) – a container which does not undergo any stripping operations

FIS – Fairway Information Service; an online information portal for providing information on fairway conditions

FoN – the Freedom of Navigation principle

FP7 – the 7th Framework Programme; a European programme for innovation and R&D, succeeded by Horizon 2020

FSC – Flag State Control

GCAF (*Gross Cost of Averting Fatality*) – a total cost of fatality prevention applied in one of the methods used for the assessment of risk at sea

GHG – greenhouse gas

GIS (*Geographic Information System*) – a database system applied in intelligent systems of maritime telematics

GISIS – Global Integrated Shipping Information System

GL (*Gear Logbook*) – a system applied to monitor legal use of fishing nets

GLONASS – Global Navigation Satelite System

GMDSS (*Global Maritime Distress and Safety System*) – a global maritime distress and safety system which is operated on the basis of approved safety procedures

GPS – Global Positioning System

GSOP – Geoinformatics Port Security System

GT – Gross Tonnage

H2020 – Horizon 2020; a European programme for innovation and R&D

HAZMAT – Hazardous Materials

HELCOM – Helsinki Commission

HOA (*Horn of Africa*) – a region of the Somalian coast

HQ MC Northwood – Allied Maritime Command Headquarters Northwood

HVG – high value goods

IACS (*International Association of Classification Societies*) – an association of classification societites which evaluate technical conditions of oceangoing vessels

IALA – International Association of Lighthouse Authorities

IBS (*Integrated Bridge System*) – an integrated bridge system applied on oceangoing vessels

ICAF (*Implied Cost of Averting a Fatality*) – an indicator used in one of the methods applied to evaluate risk at sea

ICF (*International Compensation Fund for GHG emissions from ships*) – an international fund applied to compensate external costs of maritime transport resulting from the GHG emission

IEM – Information Exchange Model

IMAP (*Integrated Maritime Awareness Picture*) – a type of maritime situation display

IMB PRC (*International Maritime Bureau – Piracy Reporting Centre*) an institution which monitors and fights against maritime piracy

IMDatE (*Integrated Maritime Data Environment*) – an experimental monitoring system of navigation safety in Europe

IMDG Code – International Maritime Dangerous Goods Code

IMO – International Maritime Organisation

INS (*Integrated Navigation System*) – an integrated navigation system commonly applied on oceangoing vessels

IoT – Internet of Things; a technological trend regarding the connection of all sorts of devices and sensors to the internet, not just traditional computers and handheld devices

IRR – Internal Rate of Return

IRTC (*Internationally Reccomended Transit Corridor*) – a safe protected transit corridor across the waters of the Gulf of Aden threatened by piracy

ISM (Code) (*International Safety Management Code*) – legislation which implements uniform international standards and procedures of maritime transport safety

ISO – International Standards Organization; a formal standardisation body

ISP (*Integrated Ship Profile*) – navigation monitoring with the use of many sources, integration of information about a vessel

ISPS Code – International Ship and Port Facility Security Code

ISSC (*International Ship Security Certificate*) – an element of the ISPS regulations system at ports and oceangoing vessels

IT – Information Technology

ITOPF – The International Tanker Owners Pollution Federation

IVEF (*Inter-VTS Exchange Format*) – a standard developed by the EMSA to incorporate a display of vessel traffic, relevant static information and information about the voyage

IVU IWC – In-Vessel Unit, a part of the e-IWT concept,

IWRAP (*IALA Waterway Risk Assessment Programme*) – a programme considered as a quantitative method of risk analysis, assessment and control along fairways

IWT – Inland Waterway Transport

LASH (*Lighter Aboard Ship Vessels*) – a barge carrier

LCL (*less container load*) – a container that can undergo stripping operations

LEO (*Low Earth Orbital*) – satellite systems applied in telematics of transport and logistics

LNG – Liquefied Natural Gas

LQI (*life quality index*) – an indicator of life quality applied in one of the methods applied to assess risk at sea

LRFPR – Lloyd's Register – Fairplay Research

LRIT (*Long Range Identification and Tacking*) – a long-range system for monitoring oceangoing vessels

LSP – Logistics Service Provider; a company providing logistics services such as planning, coordination, transport and warehousing.

MARPOL 73/78 – International Convention for the Prevention Pollution from Ships

MAS (*Maritime Assistance Services*) – a system providing support to operations at sea during collisions and failures

MDA (*Maritime Domain Awareness*) – an initiative of the United States of America which refers to maritime safety; it has been developed in Europe as the MSA

MEPC – IMO *Marine Environment Protection Committee*

MOOTW (*Maritime Operations other than War*) – military operations other than operations performed at war

MRP I, II – Materials Resource Planning, Materials Requirement Planning; a supply chain management concept relating to the expected amount of required materials in a chain of supply (MRP I) and production (MRP II),

MRV of CO_2 – monitoring, reporting and verification of CO_2 emission

MS (*Member States*) – participating countries

MSA (*Maritime Situational Awareness*) – a system providing warnings about threats at sea

MSC (*Member States Costs*) – costs incurred by the participating countries

MSCHOA – Maritime Security Centre Horn of Africa

MSO (*Maritime Security Operations*) – a civil and military concept of providing security to maritime operations and warnings about threats at sea

MSSIS (*Maritime Safety and Security Information System*) – an international system of navigation display based on the AIS signal

Multimodality – the usage of multiple modalities (waterways, road and/or rail) for arranging transport

NAF (*North Atlantic Format*) – a standard applied in cooperation of the VMS-type systems which are used for monitoring fishing vessels [the ISSCFV International Standard Classification of Fishing Vessel sub-standards – identification of the fishing vessel type; the ISSCFG International Standard Classification of Fish Gear – identification of fishing nets (catching type); ISSCAAP International Standard Classification of Aquatic Animals and Plants – identification of fish species which are caught]

NAIADES – a European action programme aimed at moving more freight transport onto inland waterways

NtS – Notices-to-skippers; the RIS technology to inform skippers about fairway conditions (e.g. blockages)

NAMESIS (*Naval Merchant Shipping Information System*) – a system developed for the requirements of the Royal Navy; at present, it is applied in the military cooperation and counselling provided to navigation by the NATO Navy forces

NAVTEX (*NAVigational TEXt Messages*) – a part of the integrated GMDSS system which provides information important for navigation safety with the use of telex transmission from dedicated on-shore stations to vessels which are equipped with proper receivers

NCAF – Net Cost of Averting Fatality applied in one of the methods used for the assessment of risk at sea

NCAGS (*Naval Co-operation and Guidance for Shipping*) – a department of civil and military cooperation

IUU fishing – illegal, unreported and unregulated fishing at sea

NPOC – Next Port of Call

NPV – Net Present Value

NSA – National Shipping Authorities

NSC – NATO Shipping Centre at the Regional Maritime Command in Great Britain

NSSN – the national *SafeSeaNet* system

NSW (*National Single Window*) – a mechanism developed for the comprehensive exchange of information and FAL documents

OAE – the NATO operation *Active Endeavour*

OECD – Organisation for Economic Co-operation and Development

OPRF (*Ocean Policy Research Foundation*) – a foundation for the sustainable research and development of the ocean (Japan)

Paris MOU – Paris Memorandum of Understanding

PAWSA (*Ports and Waterways Safety Assessment*) – a qualitative method applied to provide the analysis, assessment and control of risk at ports and along fairways

PCMS – *Propulsion Condition Monitoring Service* – the monitoring of the operational parameters related to drive systems at oceangoing vessels combined with maintenance service planning

PFSA (*Port Facility Security Assessment*) – the analysis of the security level of a port facility; an element of the ISPS regulations system at ports and oceangoing vessels

PFSO – Port Facility Security Officer, an element of the ISPS regulation system at ports and oceangoing vessels

PFSP – Port Facility Security Plan; an element of the ISPS regulation system at ports and oceangoing

PHICS (*Polish Harbours Information and Control System*) – a system applied to verify the authenticity of Polish seamen's documents

Physical Internet – a future logistics concept based on extensive use of digital technologies, smart hubs and self-organisation

PIANC – Permanent International Association of Navigation Congresses; the world's association for waterborne transport infrastructure

PLL (*Potential Loss of Life*) – the rate of potential loss of life applied in one of the methods used for the assessment of risk at sea

PMC (*Private Marginal Cost*) – a category of private marginal costs in the methodology of the European Commission which is applied to calculate external costs of maritime transport

PSC – Port State Control

PSSA – Particularly Sensitive Sea Area

RAA – Restricted Access Area

RECAAP – Regional Cooperation Agreement on Combating Piracy and Armed Robbery against ships in Asia

RF (*Radiative Forcing*) – a risk factor (radiation/emission) applied in the analysis of the anthropopressure of maritime transport

RFID (*Radio Frequency Identification*) – a technique of automatic identification in the processes of transport and logistics

RFT (*Retail and Fishstock Traceability*) – a system applied to monitor fish processing and marketing

RIS – River Information Services

RMP (*Recognised Maritime Picture*) – a navigation display in its military version for the NATO, EU NAVFOR, the national Navy forces of the member states, with consideration of classified information

RNC (*Raster Navigational Chart*) – a type of electronic charts applied in maritime e-navigation

RSO (*Recognized Security Organization*) – an element of the ISPS regulation system at ports and oceangoing vessels

RTW (*Round-the-World*) – a maritime round-the-globe connection

SAR – Search and Rescue

SCM – Supply Chain Management

SCS (*Suez Canal Surcharge*) – an additional charge for each container passing the Suez Canal

SECA – Sulphur Emission Control Area

SEEMP (*Ship Energy Efficiency Management Plan*) – a plan referring to the principles of energy efficiency applied at newly constructed vessels
SMC – Social Marginal Cost
SME – Small and Medium-Sized Enterprises
SMS – Safety Management System
SOLAS 1974 – the Convention for the Safety of Life at Sea
SRES – Special Report on Emissions Scenarios
SRS (*Ship Reporting System*) – a system defining ship reporting procedures
SSA (*Ship Security Assessement*) – an element of the ISPS regulations system at ports and oceangoing vessels
SSC (*Security (Sur)Charge*) – maritime transport charge under the ISPS
SSN (*SafeSeaNet*) – a computer system of navigation data exchange – the EU EMSA
SSO (*Ship Security Officer*) – an element of the ISPS regulation system at ports and oceangoing vessels
SSP (*Ship Security Plan*) – an element of the ISPS regulation system at ports and oceangoing vessels
SSS – Short Sea Shipping
STI – Strategic Traffic Information; a possible RIS service
STIRES (SafeSeaNet *Traffic Information Relay and Exchange System*) – a system of vessel traffic information exchange
Synchromodality – logistics concepts whereby logistics choices are made based upon real-time data of cargo destinations, infrastructure conditions in multiple modalities and availability of logistics services.
TEN-T – Trans-European Transport Network
TCO – Total Cost of Ownership
TEU (*Twenty-Foot Equivalent Unit*) – a capacity unit often applied in reference to ports and vessels. It is equal to the capacity of a 20 ft container
THC – Terminal Handling Charge in maritime freight
TMS – Transport Management System; software used by a transport operator to plan and monitor his operations
TOGAF – The Open Group Architecture Framework; a framework used for designing and managing information technology based on the principles of Enterprise Architecture.
TOS – Terminal Operating System; software used by terminal operators for planning and monitoring their operations
TTI – Tactical Traffic Information; a possible RIS service
TSB – Total Social Benefit

TSC – Total Social Cost
TTW (*Territorial Waters*) – parts of the territories under the exclusive jurisdiction of a coastal state
TWIC – Transportation Worker Identification Card
UMS – Unattended Machinery Space
UN/CEFACT – United Nations Centre for Trade Facilitation and Electronic Business; an organisation focusing on trade facilitation procedures and electronic business
UN/ECE – United Nations Economic Commission for Europe; an organisation to encourage economic cooperation
UN/TDED – United Nations Trade Data Element Directory; a model of ISO-standardised information for trade documents
VHF – Very High Frequency radio; voice communication between vessels and on-shore actors using wireless equipment
UNCLOS – United Nations Convention on the Law of the Sea
UNCTAD United Nations Conference on Trade and Development
UNEP – United Nations Environment Programme
UNFCCC – United Nations Framework Convention on Climate Change
USD (*User Specific Data*) – the resources of data in independent national acquisition systems and navigation safety information processing
VDS (*Vessel Detection System*) – a system of satellite positioning and detection of fishing vessels at sea and at ports
VMS – Vessel Monitoring System
VTMIS – Vessel Traffic Monitoring and Information System
VTS – Vessel Traffic Services; possible RIS service
VTS (*Vessel Traffic Service*) – a system for vessel traffic surveillance
VTT – Vessel Tracking and Tracing; a possible RIS service based on AIS (positioning) data
WECDIS – Warship Electronic Chart Display and Information System
WIC (*Worker Identification Card*) – active biometric documents of an employee
WRP (*War Risk Premium*) – an additional freight surcharge in maritime transport for the risk related to war
XML – Extensible Markup Language; a technology providing a syntax/structure for electronic data exchange

APPENDICES

Appendix No. 1

The list of states (flags) of the highest number of registered vessels (as for 1st January 2013; data in DWT)

Flag of registration	Number of vessels	Share of world total, vessels	Deadweight tonnage (thousands DWT)	Share of world total (percentage DWT)	Cumulated share (percentage DWT)	National ownership (percentage)[a]	DWT growth 2013/2012 (percentage)
Panama	8 580	9.87	350 506	21.52	21.52	0.14	5.03
Liberia	3 144	3.62	198 032	12.16	33.68	0.01	5.83
Marshall Islands	2 064	2.37	140 016	8.60	42.27	0.11	11.08
Hong Kong (China)	2 221	2.55	129 806	7.97	50.24	12.15	16.87
Singapore	3 339	3.84	89 697	5.51	55.75	36.60	16.62
Greece	1 551	1.78	75 424	4.63	60.38	92.60	5.13
Bahamas	1 446	1.66	73 702	4.52	64.91	1.18	1.44
Malta	1 794	2.06	68 831	4.23	69.13	0.35	8.18
China	3 727	4.29	68 642	4.21	73.35	98.18	9.83
Cyprus	1 030	1.18	31 706	1.95	75.29	19.51	7.61
Isle of Man	422	0.49	22 629	1.39	76.68	0.00	9.32
UK	1 343	1.54	21 095	1.30	77.98	49.88	6.99
Italy	1 506	1.73	20 612	1.27	79.24	93.46	2.44
Japan	5 379	6.19	20 409	1.25	80.50	99.32	11.04
Norway (NIS)	536	0.62	18 093	1.11	81.61	82.33	5.37

Flag of registration	Number of vessels	Share of world total, vessels	Deadweight tonnage (thousands DWT)	Share of world total (percentage DWT)	Cumulated share (percentage DWT)	National ownership (percentage)[a]	DWT growth 2013/2012 (percentage)
Republic of Korea	1 894	2.18	17 720	1.09	82.69	96.47	−10.74
Germany	781	0.90	17 128	1.05	83.75	97.59	2.30
India	1 385	1.59	15 876	0.97	84.72	96.16	−3.45
Indonesia	6 293	7.24	14 267	0.88	85.60	90.28	0.17
Antigua and Barbuda	1 302	1.50	14 142	0.87	86.47	0.00	4.27
Denmark (DIS)	482	0.55	13 739	0.84	87.31	92.53	1.24
Bermuda	168	0.19	12 378	0.76	88.07	1.69	0.45
US	3 452	3.97	12 321	0.76	88.83	73.93	−1.18
Malaysia	1 539	1.77	10 508	0.65	89.47	92.82	−3.15
Turkey	1 365	1.57	10 215	0.63	90.10	96.94	3.30
United Republic of Tanzania	198	0.23	8 815	0.54	90.64	0.30	10.45
The Netherlands	1 250	1.44	8 712	0.53	91.17	70.90	6.73
France	543	0.62	7 431	0.46	91.63	52.40	−0.22
Vietnam	1 772	2.04	7 284	0.45	92.08	97.55	1.52
Belgium	216	0.25	6 913	0.42	92.50	58.35	0.46
Russian Federation	2 324	2.67	6 784	0.42	92.92	84.57	−2.14
Philippines	1 383	1.59	6 417	0.39	93.31	26.36	−2.41
St. Vincent and the Grenadines	1 046	1.20	4 919	0.30	93.61	0.08	−18.09
Thailand	755	0.87	4 811	0.30	93.91	97.95	−6.63
Cayman Islands	174	0.20	4 310	0.26	94.17	0.00	2.12
Top 35 total	66 404	76.38	1 533 889	94.17	94.17	24.30	6.71
World total	86 942	100.00	1 628 783	100.00	100.00	23.00	5.98

[a] The estimate of national ownership is based on available information of commercial seagoing vessels of 1,000 GT and above.

Source: adapted from Review of Maritime Transport 2013, UNCTAD/RMT/2013 (data supplied by Clarkson Research Services), New York–Geneva 2013, p. 56.

Appendix No. 2

A simplified model of maritime accident consequences at sea

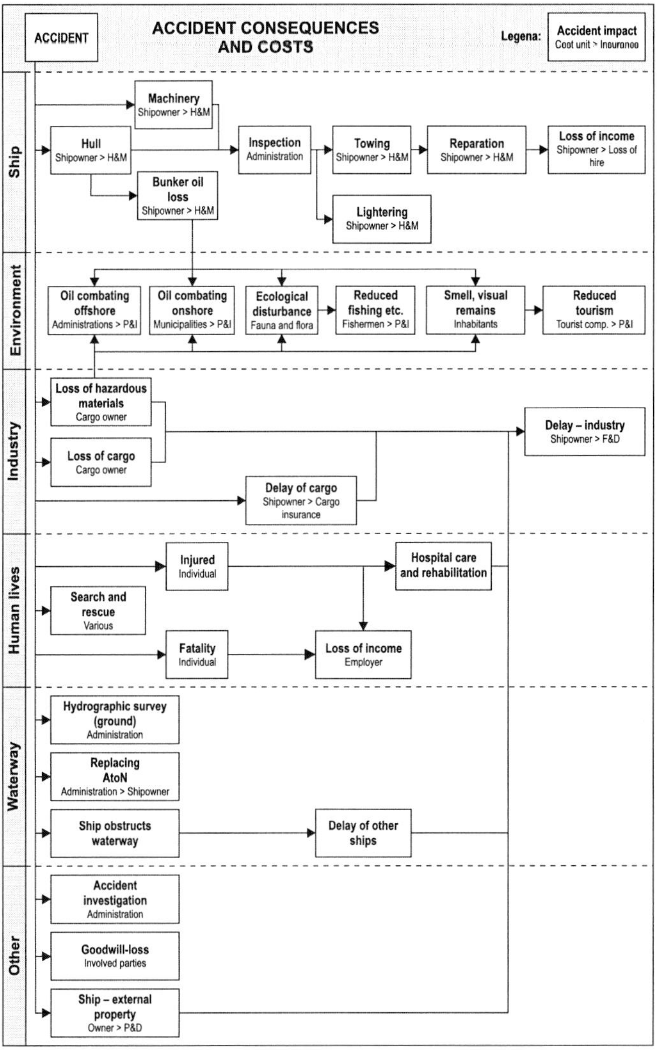

H&M – hull and machinery; P&I – protection and insurance; F&D – freight and dispatch

Source: adapted from Lundkvist, M. 2010. Riskvärdering av sjötrafikinformation. Draft report 2010-12-30. Swedish Maritime Administration.

Appendix No. 3

The classification of the DINA costs broken into the main groups of stakeholders

Barges / skippers

Description	Number	Cost / unit	Total cost
New cloud platforms for managing data of barges (crew, vessel, cargo, voyages)			€ 15.000.000
On-board multifunctional digital infrastructure ('eIWT tools')	12.500	€ 2.000	€ 25.000.000
Apps for compliance / law enforcement	12.500	€ 500	€ 6.250.000
Apps for smart navigation	12.500	€ 500	€ 6.250.000
Apps for communicating with terminals	12.500	€ 500	€ 6.250.000
Total investment in technology			€ 43.750.000
Design, set-up, and organisation implementation (75%)			€ 32.812.500
Total required investment			**€ 76.562.500**

Shippers / logistics service providers

Description	Number	Cost / unit	Total cost
Integration with booking platforms and visibility tools of customers			€ 15.000.000
Total investment in technology			€ 15.000.000
Design, set-up, and organisation implementation (75%)			€ 11.250.000
Total required investment			**€ 26.250.000**

Terminals and ports

Description	Number	Cost / unit	Total cost
Equipment and apps for small terminals and inland ports	180	€ 5.000	€ 900.000
Digital platforms for large terminals (e.g. Duisburg, maritime ports, inland hubs)	20	€ 2.500.000	€ 50.000.000
Total investment in technology			€ 50.900.000
Design, set-up, and organisation implementation (75%)			€ 38.175.000
Total required investment			**€ 89.075.000**

Authorities

Description	Number	Cost / unit	Total cost
Adaptation of the existing River Information Services, extension towards corridor management. Including shared European facilities	30	€ 500.000	€ 15.000.000
Dashboards and digital infrastructure for law enforcement and inspection agencies	30	€ 500.000	€ 15.000.000
Total investment in technology			€ 30.000.000
Design, set-up, and organisation implementation (75%)			€ 22.500.000
Total required investment			**€ 52.500.000**

Source: adapted from L.M. Punter, W.J. Hofman, *Digital Inland Waterway Area. Towards a Digital Inland Waterway Area and Digital Multimodal Nodes*, op. cit., pp. 87–88.

Appendix No. 4

An example of the cost calculation in the CISE model

Excerpt Consolidated Cost Estimates

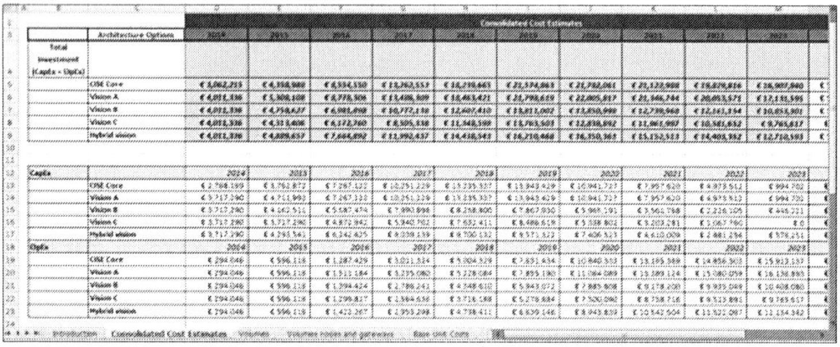

Example illustration of Volume Estimates

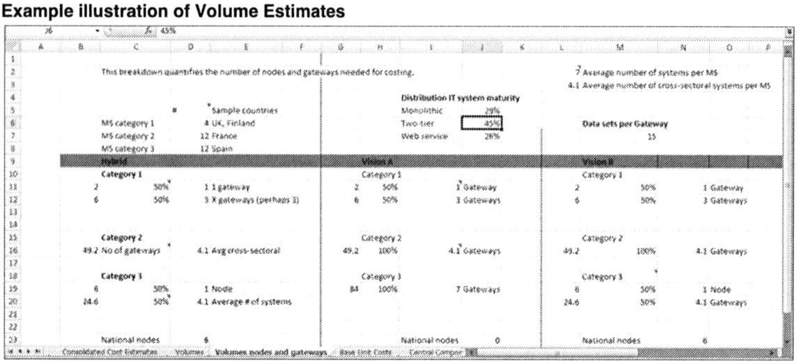

Example illustration of Base Unit Cost

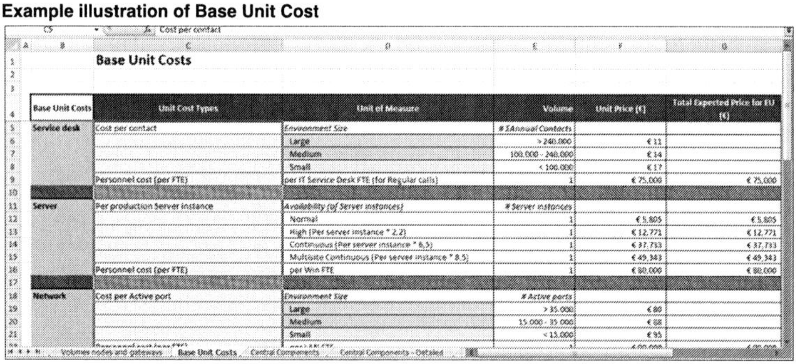

Base Unit Costs

Base Unit Costs	Unit Cost Types	Unit of Measure	Volume	Unit Price (€)	Total Expected Price for EU (€)
Service desk	Cost per contact	Environment Size	# ΣAnnual Contacts		
		Large	> 240.000	€ 11	
		Medium	100.000 - 240.000	€ 14	
		Small	< 100.000	€ 17	
	Personnel cost (per FTE)	per IT Service Desk FTE (for Regular calls)	1	€ 75,000	€ 75,000
Server	Per production Server instance	Availability (of Server instances)	# Server instances		
		Normal	1	€ 5,805	€ 5,805
		High (Per server instance * 2,2)	1	€ 12,771	€ 12,771
		Continuous (Per server instance * 6,5)	1	€ 37,733	€ 37,733
		Multisite Continuous (Per server instance * 8,5)	1	€ 49,343	€ 49,343
	Personnel cost (per FTE)	per Win FTE	1	€ 80,000	€ 80,000
Network	Cost per Active port	Environment Size	# Active ports		
		Large	> 35.000	€ 80	
		Medium	15.000 - 35.000	€ 88	
		Small	< 15.000	€ 95	

Source: adapted from Sustainability and Efficiency of Visions for CISE, Report for European Commission – Directorate General Maritime Affairs, 330012909 – Version 1.0.

Appendix No. 5

The potential benefits resulting from the integration of the monitoring systems of navigation safety for the CISE [MEUR]

	Output	Impact	
	Value and unit	Unit values	MEUR
Economic			
Cost-savings in information gathering and use of assets	26.6 MEUR	Greater efficiency and general improvement of maritime surveillance	26.6
Value of seized goods	4.2 MEUR	Estimate for all sea basins based on an assessment for the Mediterranean sea. Impact is assumed to amount to a similar value of foregone legal sales	4.2
Interception of counterfeit goods	4.1 Cases	Average value of sea-borne counterfeit interception case is EUR 190,000 (2010 to 2012). Similar interception increase assumed for other relevant sea basins. Value of foregone legal sales	4.7
Social			
Saved lives of irregular immigrants	1502 Number of people	Assumption that 0.5% of the additionally saved irregular immigrants would have died – using the value 1.5 MEUR per life	11.3
Irregular immigrants refrained from entering the EU waters	225 Number of people	Saved interception costs per immigrant: EUR 60,000	13.5
Maritime accidents	–65 Number of accidents	Material (and environmental) damage at MEUR 0.5 per accident	32.6
Maritime accidents (saved HELO costs)	–0.03 MEUR	Benefit = saved costs	0.03
Maritime accidents (casualties at sea)	–3.5 Number of people	Health and other related costs of EUR 50,000 per casualty	0.18
Maritime accidents (lives lost at sea)	–0.8 Number of lives lost	MEUR 1.5 per life	1.2
Value of seized drugs (sea)	313 MEUR	Socioeconomic benefits – e.g. lower health and crime costs	n.a.
Value of seized drugs (ports)	27 MEUR	Socioeconomic benefits – e.g. lower health and crime costs	n.a.

	Output	Impact	
	Value and unit	**Unit values**	**MEUR**
Smuggling of small arms/light weapons	–3 Number of weapons	Socioeconomic benefits – e.g. lower crime costs	n.a.
Search And Rescue	–424 Cases	EUR 3,350 per SAR case	1.4
Environmental			
Interceptions of IUU fishing	0.2 Cases	MEUR 0.1 per case	0.02
Total amount of illegal oil discharges	–3 Cases	MEUR 0.1 per cas	0.3
Accidental oil spills	–1 Cases	MEUR 20 per case.	20
Number of positive validations of suspected oil spills	8 Cases	Socioeconomic benefits have here been infeasible to put values on	n.a.
Number of solved pollution cases of confirmed oil spills	8.5 Cases	MEUR 5 per case	42.5
Detection of pollution	–43 Cases	MEUR 0.1 per case	4.3
		Total cost savings	40.1
		Total other impacts	122.6
		GRAND TOTAL	162.7

Source: adapted from *The development of the CISE for the surveillance of the EU maritime domain and the related Impact, Assessment*, PART 2: Combined Analysis, European Commission and DG Mare, Lyngby 2014, p. 115.

BIBLIOGRAPHY

2006 Baltic Maritime Outlook, Risbergs Media AB, Uddevalla 2006

ABB Shore to Ship Power Supplies, by K. Henderson, *Schematic of shore to ship power supply*, Image credit: ABB – Maritime Propulsion Environmental State aid guidelines 18.01.2012 [accessed on 12th October 2015]

Abramowicz-Gerigk T., *Bezpieczeństwo manewrów krytycznych statków w systemie transportowym autostrady morskiej*, Oficyna Wydawnicza Politechniki Warszawskiej, Warszawa 2012

Abramowicz-Gerigk T., Z. Burciu, P. Kamiński, *Kryteria akceptowalności ryzyka w żegludze morskiej*, Prace Naukowe Politechniki Warszawskiej – Transport (96), Warszawa 2013

Abramowicz-Gerigk T., Z. Burciu, *Systemy bezpieczeństwa i obsługi autostrady morskiej*, Prace Naukowe Politechniki Warszawskiej, Transport no. 70/2009

Andritsos F., European Commission Joint Research Center – *eIWT Electronic tool for Inland Waterway Transport* – https://ec.europa.eu/transport/sites/transport/files/2017-03-14-study-eiwt-d4-postws-rev7.pdf [accessed on 12th October 2017]

Andrzejewski L., Fechner I., *Uwarunkowania lokalizacyjne i funkcjonale terminali kontenerowych w Polsce,* Logistyka 1/2014

Bąbczyńska-Jelonek Z., *Na straży morskiego ładu – pakiet dyrektyw KE "Erika III"*, on-line edition: http://sophico. onet.pl/2006/10/28/na-strazy-morskiego-ladu-pakiet-dyrektyw-ke-erika-iii/ [accessed on 12th October 2015]

BAS-WASGiPU/WAPEiM-1909/13 Opinia w sprawie komunikatu Komisji do Parlamentu Europejskiego, Rady, Europejskiego Komitetu Ekonomiczno-Społecznego i Komitetu Regionów: Włączenie emisji z transportu morskiego do unijnej polityki w zakresie ograniczenia emisji gazów cieplarnianych (COM 2013 479 final) – materiały niepublikowane Biura

Analiz Sejmowych (an opinion of the communication of the European Commission to the European Parliament, European Council, European Economic and Social Committee and the European Committee for the Regions: including emissions generated by maritime transport to the EU policy on the limitation of GHG emission (COM 2013 479 final) – non-published materials of the Bureau of Research at the Polish Parliament Chancellery

Bertalanffy von L., *Ogólna teoria systemów*, PWN, Warszawa 1984

Biniek Z., *Elementy teorii systemów modelowania i symulacji*, Wydawnictwo Internetowe INFOPLAN, Warszawa 2017

Bluhm B., *Risks of Pollution by Shipping – How to Improve Preparedness to Oil Spills*, EMSA Publication, Lisbon 2011

Bogalecka M., *Bezpieczeństwo transportu morskiego w regionie Morza Bałtyckiego*, Prace Naukowe Wydziału Przedsiębiorczości i Towaroznawstwa, Gdynia Maritime Academy, Gdynia 2011

Bogalecka M., M. Borucki, *Analysis of Sea Accidents in 2004–2007*, "Polish Journal of Environmental Studies", no. 17(3C) 2008

Broeke, I., *River Information Services – An envisaged Vessel Traffic Management System on European Inland waterways*, International Conference, Singapore, 2000

BSR Innoship, (Project Seminar: *Limitation of toxic compounds emission generated after fuel combustion in vessels in the Baltic Sea region and at ports;* non-published materials of the Polish Register of Ships), Gdańsk 2012

Buck Consultants International, "Supply Chain Structures of the Future: An Industry Sector Approach", Nijmegen 2006

Bujak A., A. Orzeł, R. Miler, *Telematics of Supply Chain – Areas, Opportunities, Challenges* [in:] J. Mikulski (ed.), *Telematics – Support for Transport*, Springer, Berlin Heidelberg 2014

Bujak A., K. Szczyrbowski, R. Miler, *Implikacje wprowadzenia kodeksu ISPS dla gospodarki morskiej*, "Logistyka" no. 6/2014

Bujak A., M. Smolarek, A. Orzeł, *Telematic Innovations in Transport Safety* [in:] J. Mikulski (ed.), *Activities of Transport Telematics*, Springer, Berlin Heidelberg 2013

Bujak A., P. Zając, *Monitoring of Cargo in Logistic Systems of Transport and Storage* [in:] J. Mikulski (ed.), *Activities of Transport Telematics*, Springer, Berlin Heidelberg 2013

Burciu Z., *Niezawodność akcji ratowniczej w transporcie morskim*, Oficyna Wydawnicza Politechnika Warszawska, Warszawa 2012

Burnewicz J., *Nowoczesna wizja transportu i jej potencjalny wpływ na zagospodarowanie przestrzenne*. Konferencja "Koncepcja zagospodarowania kraju a wizje i perspektywy rozwoju przestrzennego Europy", Jachranka, 25–26 września 2008, (the proceedings of the Conference on the Development of the Country and the Visions and Prospects of the Spatial Development of Europe; Jachranka, 25th–26th Sept. 2008)

Cempel C., *Teoria i inżynieria systemów – zasady i zastosowania myślenia systemowego*, Instytut Technologii Eksploatacji, Radom 2008

Cempel C., *Teoria Inżynierii Systemów*, Wyd. Politechniki Poznańskiej, Poznań 2000

Chrzanowski A.R., I. Głażewska, *Wpływ rewolucji technologicznej na usprawnienie procesów biznesowych* [in:] Zeszyty Naukowe POU "Zarządzanie Zmianami" no. 2, Warszawa 2012

Ciekanowski Z., *Działania asymetryczne jako źródło zagrożeń bezpieczeństwa*, „Bezpieczeństwo i Technika Pożarnicza" no. 3, Warszawa 2009

Cimek G., *W labiryncie pojęć – globalizacja jako kategoria teoretyczna*, "Zeszyty Naukowe AMW", 2011, no. 4

Collaboration Concepts for Co-Modality Legal Framework (2011) http://www.co3-project.eu/wo3/wp-content/uploads/2011/12/CO3-D-2-9-Legal-Framework-excl-contr.-august-2014.pdf [accessed on 12th October 2017]

COMCIS Collaborative Information Services for Container Management http://comcis.eu [accessed on 12th October 2017]

Corbett J.J., J.J. Winebrake, E.H. Green, P. Kasibhatla, V. Eyring, A. Lauer, *Mortality from ship emissions: A global assessment*, Environmental Science & Technology, 2007

Czermański E., *Rola i znaczenie polskich armatorów w światowych przewozach ładunków drogą morską* [in:] Zeszyty Naukowe Uniwersytetu Gdańskiego – Ekonomika Transportu Morskiego "Handel Morski i Turystyka", Gdańsk 2003

Czermański E., *Żegluga bliskiego zasięgu w rejonie Morza Bałtyckiego*, on-line edition https://www.portalmorski.pl/referaty/2004/23.pdf [accessed on 12th October 2017]

Daszkiewicz A., *Kontenery i konteneryzacja w transporcie morskim*, Wirtualny Leksykon Transportu i Logistyki, on-line publication: logistyka.edu.pl, p. 9 [accessed on 15th April 2018]

Demczuk R., M. Kościelski, R. Miler, *Miejsce, rola i znaczenie Centrum Operacji Morskich w tworzeniu Krajowego Systemu Bezpieczeństwa Morskiego oraz w procesie wdrażania natowskiej koncepcji MSA*, Zeszyty Naukowe AMW, 172 K/1 2008

Dering J., Biuletyn informacyjny PRS S.A. no. 3/259 June 2006, Gdańsk

Digital Transport and Logistics Forum – Background, https://ec.europa.eu/transport/sites/transport/files/media/news/doc/2015-04-15-setting-up-dtlf/background.pdf [accessed on 12th October 2017]

Directive 2005/44/EC of the European Parliament and of the Council of 7 September 2005 on harmonised river information services (RIS) on inland waterways in the Community – http://eur-lex.europa.eu/legal-content/EN/TXT/?qid=1490188053423&uri=CELEX:32005L0044 [accessed on 12th October 2017]

Distributed ledger technology: Blackett review UK Government Office for Science, https://www.gov.uk/government/publications/distributed-ledger-technology-blackett-review [accessed on 12th October 2017]

Drozd W., M. Dziewicki, M. Waraksa, *Status polskiego systemu automatycznej identyfikacji statków (AIS)*, materials provided by the Maritime Office of Gdynia, Gdynia 2007

Dulęba D. (ed.), *Spedycja,* Wydawnictwo Difin, Warszawa 2011

Dura D., D. Szulc, *Nowe standardy bezpieczeństwa morskiego i jego ochrony. Międzynarodowy kodeks ochrony statków i obiektów portowych (kodeks ISPS),* Prace Wydziału Nawigacyjnego Akademii Morskiej w Gdyni, Gdynia 2004

Dura D., D. Szulc, *Nowe standardy bezpieczeństwa na morzu i w portach*, "Czas Morza" no. 1 (20) 2004

Dura M., *Koniec wolności pływania,* "Przegląd Morski" no. 1(019) / 2009

Dyrcz Cz., *Koncepcja bazy danych nawigacyjno-hydrograficznego zabezpieczenia (NHZ) na polskich obszarach morskich*, ZN Akademii Marynarki Wojennej, rok LII no. 3 (186), Gdynia 2011

Dyrcz Cz., *Wybrane problemy terroryzmu XXI wieku* (1), "Przegląd Morski" no. 10/2005

Dyrcz Cz., *Wybrane problemy terroryzmu XXI wieku* (2). "Przegląd Morski" no. 11/2005

Dyrektywa 2005/44/WE Parlamentu Europejskiego i Rady z dnia 7 września 2005 r. w sprawie zharmonizowanych usług informacji rzecznej (RIS) na śródlądowych drogach wodnych we Wspólnocie (Dz. U. L255/152 z dnia 30.09.2005 r.); (The Directive of the European Parliament and European Council 2005/44/EC of 7th Sept. 2005 on harmonised services of river information RIS along inland waterways in the European Community (Journal of Laws L25/152 of 30th Sept. 2005))

Dziubińska A., A. Weintrit, *Śródlądowe drogi wodne w Polsce i ich klasyfikacja*, "Logistyka" no. 3/2014

Economic Impacts Of Accidents On The Marine Industry, US dept. Of Commerce Report, Springfield VA 1997

EU FP7 CORE project, http://www.coreproject.eu/media/16662/16-12-22-core-leaflet-2.pdf [accessed on 12th October 2017]

EUNAVFOR http://eunavfor.eu/key-facts-and-figures [accessed on 12th October 2017]

European Commission, *A Digital Single Market Strategy for Europe* – Communication from the Commission to the European Parliament, the Council, the European Economic and Social Committee and the Committee of the Regions – http://eur-lex.europa.eu/legal-content/EN/TXT/?qid=1447773803386&uri=CELEX:52015DC0192 [accessed on 12th October 2017]

European Commission, *Towards quality inland waterway transport NAIADES II* – Communication from the Commission to the European Parliament, the Council, the European Economic and Social Committee and the Committee of the Regions – http://eur-lex.europa.eu/legal-content/EN/TXT/?qid=1490187231782&uri=CELEX:52013DC0623 [accessed on 12th October 2017]

European Federation of Inland Ports – http://www.inlandports.eu [accessed on 12th October 2017]

Evaluation of RIS Implementation for the period 2006–2011, Main Report. Zoetermeer, July 2014

External Portal WEB, Energy Efficiency for Ships, A95, www.eagle.org/eagle [accessed on 12th October 2017]

Fazi S., *Mode selection, routing and scheduling for inland container transport TU/Eindhoven*, http://www.nextlogic.nl/uk/ [accessed on 12th October 2017]

Fertsch M. (ed.), *Podstawy logistyki*, Wydawnictwo Biblioteka Logistyka, Poznań 2008

Ficoń K., *Logistyka morska. Statki, porty, spedycja*, Wydawnictwo Bel Studio, Warszawa 2010

Filina-Dawidowicz L., M. Kaup, A. Wiktorowska-Jasik, *Zintegrowany transport wodny i lądowy*, Wydawnictwo Uczelniane Zachodniopomorskiego Uniwersytetu Technologicznego w Szczecinie, Szczecin 2014

Gerigk M., *Kompleksowa metoda oceny bezpieczeństwa statku w stanie uszkodzonym z uwzględnieniem analizy ryzyka*, Monografie Politechniki Gdańskiej (101), Gdańsk 2010

Global Maritime Security Cooperation In The Age Of Terrorism And Transnational Threats At Sea, Maritime Security Primer, Copenhagen 2008

Global Terrorism Database, University of Maryland, Strategic Trends 2014 [in:] Center for Security Studies, ETH Zurich www.css.ethz.ch [accessed on 12th October 2017]

Grabski F., *Semimarkowskie modele niezawodności i eksploatacji,* Polska Akademia Nauk, Badania Systemowe, vol. 30, Warszawa 2002

Grabski F., *Stochastyczne modele awarii*, the proceedings of the 5th Symposium on Maritime Transport Safety, Iława 2009

Grant R.M., *Współczesna analiza strategii*, Oficyna Wolters Kluwer Business, Warszawa 2011

Grudzewski W.M., I. Hejduk, *Zmiany paradygmatów kształtujących systemy zarządzania* [in:] J. Rokita, W. Grudzewski (eds), *Zarządzanie strategiczne w warunkach nowej gospodarki*, Wydawnictwo GWSH, Katowice 2007

Grzelakowski A., *Globalizacja i jej wpływ na rozwój transportu morskiego i globalnych łańcuchów dostaw* [in:] *Wyzwania gospodarki globalnej*, Prace i Materiały Instytutu Handlu Zagranicznego Uniwersytetu Gdańskiego 2012, no. 31

Grzelakowski A.S., *Globalny rynek morskich przewozów kontenerowych i jego wpływ na światowy rynek frachtowy i logistyczny*, "Logistyka" no. 5/2013

Grzelakowski A.S., *Model zarządzania polskimi portami morskimi jako element systemu regulacji działalności sektora portowego* [in:] H. Klimek, J. Dąbrowski (eds), *Kierunki rozwoju polskiego transportu morskiego – programy i rzeczywistość*, Fundacja Rozwoju Uniwersytetu Gdańskiego, (the Foundation for the Development of Gdańsk University), Gdańsk 2010

Grzelakowski A.S., *Polityka transportowa* [in:] *Liberalizacja i integracja rynków przemysłów sieciowych. Trwały rozwój. Biała Księga 2003.* part III. Polskie Forum Strategii Lizbońskiej, (Lisbon Strategy Forum in Poland) Gdańsk-Warszawa 2003

Grzelakowski A.S., *Polityka transportowa UE na II dekadę XXI wieku*, "Przegląd Komunikacyjny" no. 1–2/2010

Grzelakowski A.S., *Rynek żeglugi kontenerowej na tle globalnego rynku frachtowego w 2012/2013 r. Jego dynamika i podstawowe tendencje zmian*, "Biuletyn Polskiej Izby Spedycji i Logistyki" no. 04–05/2013

Grzelakowski A.S., *Wyzwania gospodarki globalnej*, Prace i Materiały Instytutu Handlu Zagranicznego Uniwersytetu Gdańskiego no. 31, Gdańsk 2012

Grzybowski M, R.K. Miler, *Bezpieczeństwo na europejskim i globalnym rynku transportu morskiego*, ZN AMW no. 1, Gdynia 2008

Grzybowski M., R. Miler, *MSA as the Aspect of European Maritime Security*, The Romanian Economic Journal, Bucharest 25/2007

Gucma S., *Inżynieria ruchu morskiego*, "Okrętownictwo i Żegluga", Gdańsk 2001

Guidelines for Formal Safety Assessment (FSA) for Use in the IMO Rule – making Process, MSC/Circ. 1023, MEPC/Circ. 392, April 5, 2002

Haffer J., *Skuteczność zarządzania projektami w przedsiębiorstwach działających w Polsce*, TNOIK, Toruń 2009

Hobbes T., *Lewiatan*, PWN, Warszawa 1954

Hołowiński G., *Modelowanie procesów wymiany informacji w środowisku portowym i z jego otoczeniem z zastosowaniem sieci PETRIego* [in:] Wpływ portów morskich na funkcjonowanie i rozwój otoczenia, on-line publication [accessed on 12th October 2017]

Hołowiński G., *Zastosowanie elektronicznej wymiany danych EDI w transporcie morskim na przykładzie Portu Szczecin-Świnoujście – analiza uwarunkowań i korzyś*ci, Master of Busienss Administration, I Edition, the University of Szczecin, Szczecin 2000

http://morzaioceany.pl/zegluga-morska/pzm/1070-mniej-gazów-cieplarnianych-ze-statków-pżm.html [accessed on 8th February 2017]

http://blogs.worldbank.org/endpovertyinsouthasia/india-great-yet-unexplored-potential-inland-water-transportation [accessed on 27th July 2017]

http://co-opcoast.blogspot.com/2010_10_01_archive.html [accessed on 2nd June 2014]

http://direct.bargelink.com/Index [accessed on 20th July 2018]

http://ec.europa.eu/eurostat/statistics-explained/index.php?title=File:Transport_performance_of_goods_in_loaded_containers_by_type_of_transport_-_1000_TEUkm.png [accessed on 7th February 2018]

http://ec.europa.eu/eurostat/statistics-explained/index.php/Inland_waterway_transport_statistics [accessed on 4th January 2018]

http://ec.europa.eu/fisheries/cfp/control_enforcement/vms_en.htm [accessed on 3rd March 2015] http://emsa.europa.eu/operations/maritime-monitoring/86-maritime-monitoring/1520-integrated-maritime-data-environment-imdate.html [accessed on 5th February 2015]

http://emsa.europa.eu/operations/safeseanet.html [accessed on 17th July 2015]

http://fisheries.polestarglobal.com/vessel-monitoring [accessed on 22nd February 2015]

http://geopolityka.net/wp-content/uploads/2013/07/BG_no._2.pdf *Globalny układ sił ujęcie ilościowe* [in:] Biuletyn Polskiego Towarzystwa Geopolitycznego IV–VI 2013, pp. 6–9 [accessed on 17th July 2015]

http://gpsworld.com/rolls-royce-joins-partnership-to-develop-autonomous-ships/ [accessed on 24th March 2018]

http://hydrogeo.am.szczecin.pl/projekty/ris/4-ris [accessed on 27th July 2017]

http://kontenerypolska.blogspot.com/2012/01/sztauowanie-kontenerow-na-statku.html [accessed on 26th July 2014]

http://mminakowski.republika.pl/tanie_bandery_i_itf.htm, 28.12.2008 [accessed on 17th July 2015]

http://oceansbeyondpiracy.org/attack-timeline/somalia [accessed on 13th May 2016]

http://sevenseas-marine.com/services/ [accessed on 5th February 2015]

http://sophico.onet.pl/2006/10/28/na-strazy-morskiego-ladu-pakiet-dyrektyw-ke-erika-iii/ [accessed on 27th July 2016]

http://sunseaboats.com/how-does-the-falkirk-wheel-work/ [accessed on 4th June 2018]

http://tide.act.nato.int/mediawiki/index.php_30_Mar_07 [accessed on 19th July 2008]

http://tide.act.nato.int/mediawiki/index.php/NATO MSA Architecture [accessed on 11th September 2007]

http://www-old.solvo.ru/en/products/tos/ [accessed on 24th March 2018]

http://www.aislive.com [accessed on 3rd December 2008; 28th March 2015]

http://www.alphaliner.com/top100/index.php [accessed on 16th April 2018]

http://www.apiss.pl/produkty/kalkulator-zuycia-paliwa-dla-statkw/ [accessed on 27th July 2017]

http://www.architravel.com/architravel/building/falkirk-wheel/ [accessed on 4th June 2018]

http://www.avantetech.com/products/shipping/ [accessed on 12th November 2013; 23rd September 2014]

http://www.bbn.gov.pl/portal/pl/577/2630/updated on_10_January_2011_r.html [accessed on 25th June 2016]

http://www.chartworld.com [accessed on 27th July 2017]
http://www.cmr.gov.pl [accessed on 24th March 2015]
http://www.cmr.gov.pl/files/cmr/ogloszenia/pno.pdf [accessed on 22nd February 2015]
http://www.commercial-fishing.org/electronics/vms-reimbursement-program-news-a174.html [accessed on 13th September 2017]
http://www.containerhandbuch.de/chb_e/stra/index.html?/chb_e/stra/stra_01_03_01_01.html [accessed on 23rd May 2016]
http://www.containerhandbuch.de/chb_e/stra/index.html?/chb_e/stra/stra_01_03_01_02.html [accessed on 15th May 2018]
http://www.dsw.edu.pl/fileadmin/user_upload/wszechnica/04.pdf, p. 23 [accessed on 13th August 2016]
http://www.ekologia.pl/srodowisko/ochrona-srodowiska/najgrozniejsze-katastrofy-morskie-w-xx-i-xxi-wieku,12330.html [accessed on 8th August 2014]
http://www.eti.pg.gda.pl/mapy_cyfrowe_2009_2010/Instrukcja_laboratoryjna_ECDIS_KB_v4.pdf [accessed on 8th February 2016]
http://www.exactearth.com [accessed on 23rd May 2016]
http://www.fisa.cl/marketing/EXPONAVAL_2008/PDF%20Conferencistas/GMDA/dia_4/2nd%20WHMDAWS%20CDR%20Dave%20Wirth.pdf [accessed on 17th July 2015]
http://www.fulcrum-maritime.com/lritsystems.nsf [accessed on 24th April 2015]
http://www.furuno.com/en/business_product/merchant/product/ecdis_fmd/carriage/index.html [accessed on 5th May 2016]
http://www.gatehouse.dk/Maritime-Surveillance-479.aspx [accessed on 17th March 2014]
http://www.gmdss.com.au/navtex.htm [accessed on 16th June 2012]
http://www.gmdss.republika.pl/index.html [accessed on 16th March 2015]
http://www.gospodarkamorska.pl/Stocznie,Offshore/statki-autonomiczne-rozwiazaniem-przyszlosci.html [accessed on 11th May 2018]
http://www.iala.bp8.pl [accessed on 8th March 2015]
http://www.inlandnavigation.eu [accessed on 23rd March 2018]
http://www.inmarsat.com [accessed on 4th April 2013; 27th July 2016]
http://www.internationaltransportforum.org/IntOrg/ecmt/cm/pdf/IT-F200706e.pdf [accessed on 17th July 2015]
http://www.keytonetech.com/solutions/Cargo-Monitoring.html [accessed on 30th January 2015]

http://www.lhg.is/media/vaktstod_siglinga/SafeSeaNet_January_2009.pdf [accessed on 20th February 2010]

http://www.lrit.de/pol.php [accessed on 17th July 2015]

http://www.marinetraffic.com/ais/ [accessed on 16th February 2014]

http://www.modootel.co.kr/en/LRIT/ContentIndex?contentID=122&groupID=228 [accessed on 28th May 2018]

http://www.nmfs.noaa.gov [accessed on 11th February 2015]

http://www.opensped.pl/kontenerownia/o-kontenerze [accessed on 26th July 2014]

http://www.pcnavigo.com [accessed on 28th June 2017; 26th July 2017]

http://www.pcnavigo.com/charts/enc-and-ecdis/?lang=en [accessed on 9th February 2016]

http://www.polsteam.com/mniej-gazow-cieplarnianych-ze-statkow-pzm [accessed on 8th February 2017]

http://www.polsteam.com/mniej-gazow-cieplarnianych-ze-statkow-pzm [accessed on 23rd December 2014]

http://www.ris.eu/services/european_reference_data_management_system__erdms [accessed on 20th June 2018]

http://www.rncom.mod.uk/template.aspx?pageid=764&terms=namesis [accessed on 17th July 2015]

http://www.shippipedia.com/ship-automation-control-system/ [accessed on 24th March 2013]

http://www.solvo-europe.com/en/solutions/solvo-tos-container-en [accessed on 2nd August 2018]

http://www.spectec.net/catalogue.asp?t=maritime_businesssuite&lang=en [accessed on 28th August 2014]

http://www.spedcont.com.pl/kont_abc2.htm [accessed on 26th July 2014]

http://www.terroryzm.com/terroryzm-morski/ [accessed on 17th July 2015]

http://www.tideworks.com/products/spinnaker/ [accessed on 29th January 2013; 29th January 2016]

http://www.transas.com/products/vtms/coastal-surveillance-systems/CoastalMonitoring [accessed on 27th July 2017]

http://www.transport-research.info [accessed on 27th July 2017]

http://www.umgdy.gov.pl [accessed on 14th February 2016]

http://www.umgdy.gov.pl [accessed on 17th June 2014]

http://www.ums.gov.pl/modules.php?name=News&file=article&sid=366 [accessed on 27th July 2017] http://www.vesseltracker.com/en/Home.html [accessed on 16th February 2014; 23rd May 2016]

http://www.vilp.de/Plpdf/p120.pdf [accessed on 14th July 2014]

http://www.vtmis.odessa.ua/Demo/03_port1.aspx [accessed on 23rd May 2016]

http://www.wartsila.com/en/condition-monitoring-provides-key-to-ship-automation [accessed on 24th March 2013]

http://www.washingtonpost.com [accessed on 22nd January 2015]

http://www.worldshipping.org/about-the-industry/global-trade/ [accessed on 15th April 2018]

http://www.worldshipping.org/about-the-industry/global-trade/trade-routes [accessed on 14th April 2018]

http://zegluga.home.pl/zdjecia_ris/index2.html [accessed on 2nd April 2018]

https://bics.com [accessed on 24th June 2017]

https://cargofromchina.com/sea-air-rail/ [accessed on 19th May 2018]

https://mssis.volpe.dot.gov/Main/home/ [accessed on 20th February 2014]

https://portal.emsa.europa.eu/web/imdate [accessed on 24th January 2015]

https://safety4sea.com/fuel-management-system-upgraded-with-emissions-monitoring-tech/ [accessed on 24th March 2018]

https://shipoftheday.blogspot.com/2007/04/rhine-forest.html [accessed on 15th May 2018]

https://supplychaingamechanger.com/industry-4-0-ecosystem-infographic/ [accessed on 23th May 2018]

https://szczecin.uzs.gov.pl/zahsf_komunikaty_dla_kapitanow_statkow_nts_notices_to_skippers.htm [accessed on 2nd April 2018]

https://szczecin.uzs.gov.pl/zsic7_elektroniczne_mapy_nawigacyjne.html [accessed on 2nd April 2018]

https://thelexicans.wordpress.com/2014/03/10/infographic-maersk-triple-e-class/ [accessed on 17th May 2018]

https://tomvanwoensel.com/2015/02/26/externalities-of-freight-transportation/ [accessed on 27th July 2017]

https://unicen.smu.edu.sg/sites/unicen.smu.edu.sg/files/Intelligent-Maritime-Traffic-Prediction-Coordination.pdf [accessed on 26th July 2016]

https://unicen.smu.edu.sg/sites/unicen.smu.edu.sg/files/Intelligent-Maritime-Traffic-Prediction-Coordination.pdf [accessed on 14th May 2018]

https://www.ajot.com/news/shipping-accident-statistics-identify-loss-hotspot-in-asian-seas [accessed on 30th June 2018]

https://www.am.szczecin.pl/userfiles/File/instrukcje_ecdis.pdf [accessed on 8th February 2016]

https://www.americanexpress.com/us/content/foreign-exchange/articles/port-congestion-an-international-trade-challenge/ [accessed on 17th July 2015]

https://www.bureauvoorlichtingbinnenvaart.nl/assets/images/Over/BVB_01l.jpg [accessed on 15th April 2018]

https://www.bureauvoorlichtingbinnenvaart.nl/inland-navigation-promotion/basic-knowledge/waterways [accessed on 15th May 2018]

https://www.chip.pl/2017/05/autonomiczny-statek-towarowy-zasilany-energia-elektryczna/ [accessed on 13th May 2018]

https://www.google.pl/search?q=Intelligent+Awareness+System+(IAS)&tbm [accessed on 24th March 2018

https://www.jltspecialty.com/our-insights/publications/cyber-decoder/nis-directive-explained [accessed on 17th March 2018]

https://www.km.kongsberg.com [accessed on 15th March 2018]

https://www.marineinsight.com/marine-navigation/how-to-handle-a-ship-in-congested-high-traffic-waters/ [accessed on 26th March 2016]

https://www.researchgate.net/figure/Overview-of-European-in-land-waterways_fig1_273446772 [accessed on 15th April 2018]

https://www.rolls-royce.com/products-and-services/marine/ship-intelligence.aspx#section-intelligent-asset-management [accessed on 24th March 2018]

https://www.sevencs.com/navigation-software/orca-river/ [accessed on 15th May 2017]

https://www.ttclub.com [accessed on 5th May 2018]

https://www.zegluga.wroclaw.pl/news.php [accessed on 16th September 2016; 2nd August 2018]

IALA NAVGUIDE, 5th Edition, 2006; IALA Recommendation 0-134 on the IALA Risk Management. Tools for Ports and Restricted Waterways, May 2006

Igliński H., *Wybrane sposoby ograniczenia zużycia paliwa w transporcie morskim*, "Logistyka" 5/2012

IMO, Guidelines for Voluntary use of the Ship Energy Efficiency Operational Indicator (EEOI), MEPC, 1/Circ. 684, 17.08.2009

IMO, Reduction of GHG Emission from Ships, MEPC61/INF.18, 23.07.2010

Inland Waterway Transport Funding, Funding guide for inland waterway transport in Europe. European Commission, Energy and Transport DG, Brussels 2008

International Association of Classification Societies, 2002 Document No. W_WP6_4_01 Document Access: *Public Methods to Quantify Maritime Accidents for Risk-based decision-making* Deliverable No. W WP6 4 01 [accessed on 24th January 2012]

Isenberg D., *The rise of Private Maritime Security Companies*, "Policy Watch" 05(2012)

Jacyna M., *Modelowanie i ocena systemów transportowych*, Oficyna Wydawnicza Politechniki Warszawskiej, Warszawa 2009

Januszewski J., *Satellite Navigation Systems in the Transport, Today and in the Future*, Archives of Transport, vol. 22, No. 2, Katowice 2010

Jastrzębski T., Graczyk T., Iwańkowicz R., Rutkowski R., Sekulski Z., Taczała M., European Commission Project No GRD1-2000-26812 "Innovative Barge Trains for Effective Transport on Inland Shallow Waters" – INBAT, internal studies of KKMiTO, WTMiT, ZUT in Szczecin

Jorish A., *Today's Pirates Have Their Own Stock Exchange*, "The Wall Street Journal", 16 June 2011

Jóźwiak Z., *Wpływ rozszerzenia UE na transport regionu Morza Bałtyckiego*, Portal Morski no. 25/2004

Jurdziński M., *Innowacje technologiczne na statkach morskich w celu redukcji zużycia energii i emisji CO_2*, Zeszyty Naukowe Akademii Morskiej no. 77, Gdynia 2012

Jurdziński M., *Ocena metod i możliwości zmniejszenia zużycia paliwa oraz emisji gazów na statkach morskich*, Prace Wydziału Nawigacyjnego Akademii Morskiej w Gdyni no. 26, Gdynia 2013

Jurdziński M., *Podstawy nawigacji morskiej*, Wydawnictwo AM Gdynia, Gdynia 2003

Kachel J.M., *Particularly Sensitive Sea Areas. The IMO's Role in Protecting Vulnerable Marine Areas: Threats to the Marine Environment: Pollution and Physical Damage*, Springer Berlin Heidelberg 2008

Kacperczyk R., *Transport i spedycja*, Wydawnictwo Difin, Warszawa 2010

Kaczmarek A., *Strategie działania małych i średnich przedsiębiorstw*, Zeszyty Naukowe Małopolskiej Wyższej Szkoły Ekonomicznej, no. 2(13)/2009 vol. 2, Tarnów 2009

Kalitowski S., *Terroryzm morski*, http://www.terroryzm.com/terroryzm-morski [accessed on 27th July 2017]

Kamińska-Szmaj I., M. Jarosz [et al.] *Słownik Wyrazów Obcych*, Wydawnictwo Europa, Wrocław 2001

Kaup M. (2010), *Znaczenie systemów telematycznych w funkcjonowaniu transportu wodnego śródlądowego na przykładzie Systemu Informacji Rzecznej (RIS),* "Logistyka" no. 6/2014

Ketchen D.J., B.K. Boyd, D.D. Bergh, *Research methodology in strategic management: past accomplishments and future challenges*, "Organizational Research Methods", vol. 11, No. 4, 2008

Koliński A., *Przegląd metod i technik oceny efektywności procesu produkcyjnego*, "Logistyka" no. 5(2011)

Komunikat Komisji Do Parlamentu Europejskiego i Rady Europy, *Poprawa orientacji sytuacyjnej dzięki wzmocnionej współpracy między organami nadzoru morskiego: kolejne kroki w ramach wspólnego mechanizmu wymiany informacji dla obszaru morskiego UE*, (Communication of the European Commission to the European Parliament and European Council on the improvement in the situational awareness through the improved cooperation between maritime surveillance bodies: the next steps under the common mechanism of information exchange for the maritime areas of the EU), COM(2014) 451 final, Brussels 2014

Komunikat Komisji do Rady i Parlamentu Europejskiego w sprawie "Projektu planu działania na rzecz ustanowienia wspólnego mechanizmu wymiany informacji do celów nadzoru obszarów morskich UE", (Communication of the European Commission to the European Parliament and European Council on the project related to the plan of action for the establishment of a common mechanism of information exchange for the surveillance of the EU maritime areas; COM (2010) 584 ultimate version), Brussels 2010

Kontovas C., H. Psarafitis, *Marine Environment Risk Assessment: A Survey on the Disutility Cost of Oil Spills*, 2nd Int. Symposium on Ship Operations, Management and Economics, Society of Naval Architects and Marine Engineers 2(2008)

Kopacz Z., Morgaś W., Urbański J., *Wybrane zagadnienia międzynarodowego systemu bezpieczeństwa morskiego i bezpieczeństwa nawigacji*, ZN AMW, Gdynia 2005

Kopacz Z., W. Morgaś, J. Urbański, *Europejski system monitoringu ruchu statków i informacji*, Zeszyty naukowe AMW 2/2007, Gdynia 2007

Kopacz Z., W. Morgaś, J. Urbański, *Formalne i nieformalne oceny bezpieczeństwa morskiego*, Zeszyty Naukowe AMW no. 3 (162), Gdynia 2005

Kopacz Z., W. Morgaś, *Krajowy System Bezpieczeństwa Morskiego w zintegrowanej polityce UE*, Zeszyty Naukowe AMW, no. 2(185), Gdynia 2011
Kopaliński W., *Słownik wyrazów obcych*, Wiedza Powszechna, Warszawa 1983
Korcz K., *Radiokomunikacyjne aspekty planu implementacji strategii e-nawigacji*, "Elektronika" no. 11/2010
Korczak J., *Logistyka. Infrastruktura. Sieci. Strategie*, Wydawnictwo Politechniki Koszalińskiej, Koszalin 2013
Kościelski M., R. Miler, M. Zieliński, *European Maritime Safety Agency (EMSA) in the way to enhance safety at EU seas*, Zeszyty Naukowe AMW, Gdynia no. 2 / 2008
Kościelski M., R. Miler, M. Zieliński, *Maritime Situational Awareness (MSA)*, Zeszyty Naukowe AMW, Gdynia, no. 4 / 2007
Kościelski M., R. Miler, *Naval Co-operation and Guidance for Shipping (NCAGS)*, "Baltic Transport Journal", Gdynia no. 3 /2009
Kościelski M., R.K. Miler, M. Zieliński, *Automatic Identification System (AIS) as a Main Tool of NCAGS ADP Systems*, Zeszyty Naukowe AMW no. 3 (170), Gdynia 2007
Kostępski M., *Morskie zastosowania geograficznych systemów informacyjnych GIS*, Prace Wydziału Nawigacyjnego Akademii Morskiej no. 17, Gdynia 2005
Koziej S., *Bezpieczeństwo: istota, podstawowe kategorie i historyczna ewolucja*, "Bezpieczeństwo Narodowe" no. II/2011/18
Koźlak A., *Ekonomika transportu – teoria i praktyka gospodarcza*, Wydawnictwo Uniwersytetu Gdańskiego Gdańsk 2008
Krasucki Z. (ed.), *Transport i spedycja w handlu zagranicznym*, Wydawnictwo Uniwersytetu Gdańskiego, Gdańsk 2000,
Krawiec F., *Strategiczne myślenie w firmie*, Difin, Warszawa 2003
Królikowski A., *Shipping data collection and utilization*, Maritime Office Gdynia (ppt.), Gdynia 2014
Krupski R., J. Niemczyk, E. Stańczyk-Hugiet, *Koncepcje strategii organizacji*, PWE, Warszawa 2009
Krużelewska E., *Stosunki międzynarodowe i ich uczestnicy* [in:] J. Stachura (ed.), *Współczesne międzynarodowe stosunki polityczne*, Wydawnictwo WSE, Białystok 2010
Krystek R. (ed.), *Zintegrowany System Bezpieczeństwa Transportu. Vol. I. Diagnoza bezpieczeństwa transportu w Polsce*, WKŁ – PG, Warszawa – Gdańsk 2009

KSBM Etap I, Studium wykonalności, dokument wewnętrzny UM Gdynia, Gdynia 2007

Kubiak K., *Bałtyckie wyzwania dla polityki Unii Europejskiej i Sojuszu Północnoatlantyckiego* [in:] K. Kubiak, P. Mickiewicz, J.A. Schön (eds), The Baltic Congress of European Safety, Sopot 2007

Kubiak K., *Piractwo i terroryzm morski. Nowe wyzwania dla bezpieczeństwa międzynarodowego*, on-line publication http://www.dsw.edu.pl/fileadmin/user_upload/wszechnica/04.pdf [accessed on 27th July 2017]

Kubik T., *GIS. Rozwiązania sieciowe*, PWN, Warszawa 2009

Kuriata A., *Systemowość pojęcia międzynarodowego transportu multimodalnego (MTM) oraz jego bezpieczeństwa* [in:] R. Miler, A. Mytlewski, B. Pac (eds), *Wymiary logistyki – aspekt transportowy*, Prace Naukowe WSB vol. 35, CeDeWu, Warszawa 2014

Kwiatkowski T., *Wykłady i szkice z logiki ogólnej*, Wyd. UMCS, Lublin 2004

Lee B., N. Park, *Performance Improvement Based Authentication Protocol for Intervessel Traffic Service Data Exchange Format Protocol Based on U-Navigation System in WoT Environment*, Journal of Applied Mathematics, vol. 2014 (2014), Article ID 734768

Lenon J., J. Frąckiewicz, *Systemy sprawnego działania. Teoria i praktyka*, Wydawnictwo Antyk Marcin Dybowski, Warszawa 2000

Liberadzki B., L. Mindur, (ed.), *Uwarunkowania rozwoju systemu transportowego Polski,* SGH i Wydawnictwo Instytutu Technologii Eksploatacji – PIB, Warszawa – Radom 2007

Lloyd's List publication, Fairplay and Containerisation International, Maritime Security: Elements Of An Analytical Framework For Compliance Measurement And Risk Assessment, United Nations New York – Geneva 2006

LNG nowe paliwo w żegludze bliskiego zasięgu, Det Norske Veritas Poland, www.dnv.pl [accessed on 27th July 2017]

Łysik A., *Terroryzm morski. Współczesne zagrożenie w transporcie międzynarodowym*, Studia i Materiały ITi HM UG, wydawnictwo UG, Gdańsk 2012

Madej M., *Zagrożenia asymetryczne bezpieczeństwa państw obszaru atlantyckiego*, Polski Instytut Spraw Międzynarodowych, Warszawa 2007

Majzner P., W. Piszczek, *Investigation of vessel traffic processes at waterway intersections*, Scientific Journals Maritime University of Szczecin, vol. 21(93), Szczecin 2010

Makowski A., *Siły morskie współczesnego państwa*, Impuls Plus Consulting, Gdynia 2000

Marciniak-Neider D., J. Neider (eds), *Podręcznik spedytora*, Wydawnictwo Polskiej Izby Spedycji i Logistyki, Gdynia 2011

Maritime Cloud Conceptual Model (2016), http://maritimecloud.net/docs/IALA%20Input%20-%20Maritime%20Cloud%20conceptual%20model.pdf [accessed on 27th July 2017]

Maritime Security: ISPS Code Implementation, Costs And Related Financing, *Report by the UNCTAD secretariat*, UNCTAD/SDTE/TLB/2007/1

Maritime Surveillance In Practice – Using Integrated Maritime Services, EMSA Publication, Lisbon 2014

Markusik S. (ed.), *Infrastruktura logistyczna w transporcie (vol. 3 Infrastruktura liniowa)*, Wyd. Politechniki Śląskiej, Gliwice 2013

Matwiejczuk R., *Efektywność – próba interpretacji*, "Przegląd Organizacji" no. 11(2000)

Mazurkiewicz A., *Sprawność działania – interpretacja teoretyczna pojęcia*, Zakład Ekonomiki Inwestycji i Zarządzania Strategicznego, Wyd. Uniwersytetu Rzeszowskiego, Rzeszów 2009

Mesarović M., *Ludzkość w punkcie zwrotnym: Drugi Raport dla Klubu Rzymskiego*, PWE, Warszawa 1977

Michałowska M. (ed.), *Efektywność transportu w teorii i w praktyce*, Wydawnictwo Akademii Ekonomicznej w Katowicach, Katowice 2010

Michałowska M. (ed.), *Transport w gospodarce opartej na wiedzy*, Wydawnictwo Akademii Ekonomicznej, Katowice 2009

Michlowicz E., *Systemy logistyczne a teoria systemów*, (the proceedings of the 13th Conference on Applied Logistics) "Total Logistic Management", TLM 2009, Zakopane 2009

Miciuła I., *Model systemu wspomagania decyzji w oparciu o dane geoprzestrzenne na potrzeby rzecznego systemu informacyjnego z wykorzystaniem dynamicznej domeny 3D*, Zachodniopomorski Uniwersytet Technologiczny, Szczecin 2010

Miler R., *Bezpieczeństwo transportu morskiego*, WN PWN, Warszawa 2015

Miler R., *Monitoring bezpieczeństwa transportu morskiego – Modelowanie systemów – Strategie Ekonomizacji*, Wydawnictwo UG, Gdańsk 2016

Miler R., T. Szczepaniak, *EU's GHGs International Shipping Policy Impact on the BSR Seaborne Trade Competition* [in:] R. Miler, A. Mytlewski, B. Pac (eds), *Wymiary logistyki – aspekt transportowy*, Prace Naukowe WSB Gdańsk, CeDeWu Wydawnictwa Fachowe, Warszawa 2014

Miler R., *Use of Vessel Monitoring Systems in the Logistics Chain of Supply within the EU Seas* [in:] J. Jaworski, A. Mytlewski (eds), *Funkcjonowanie Systemów Logistycznych*, Prace Naukowe WSB w Gdańsku, vol. 2 / 2009

Miler R.K., A. Bujak, *exactEaarth-AIS as a One of the Most Advanced Shipping Monitoring Systems* [in:] J. Mikulski (ed.), *Activities of Transport Telematics*, Springer, Berlin Hildeberg 2013

Miler R.K., A. Bujak, *Implementacja systemów monitoringu zużycia paliw i efektywności energetycznej statku jako pochodna dyrektyw proekologicznych UE w transporcie morskim*, "Logistyka" no. 2/2014

Miler R.K., A. Bujak, *The concept of the Integrated Maritime Data Environment (IMDatE) as a framework for European integrated and comprehensive shipping monitoring data exchange system* [in:] J. Mikulski (ed.), *Telematics – Support for Transport*, Springer, Berlin 2015

Mindur L. (ed.), *Współczesne technologie transportowe,* Wydawnictwo Politechniki Radomskiej, Warszawa 2002,

Miszczyk H., *Telematyka jako nowoczesne narzędzie wsparcia działalności firm branży TSL*, Zeszyty Naukowe "SCENO 2", Kielce 2006

Mocker M., A. Teubner, *Towards a Comprehensive Model of Information Strategy*, ECIS 2005 Proceedings

Montewka J., *Predicting risk of collision for oil tankers in the Gulf of Finland*, Journal of KONBiN 3–4 (11,12) Espoo 2009

Muczyński M., *Afryka w stosunkach międzynarodowych* [in:] J. Stachura (ed.), *Współczesne międzynarodowe stosunki polityczne*, Wydawnictwo WSE, Białystok 2010

Mylly M., *EMSA's Integrated Maritime Environment – A tool for Maritime Awareness*, EMSA Publication, Lisbon 2014

Naus K., *Możliwości wykorzystania systemu wymiany informacji bezpieczeństwa żeglugi w zarządzaniu kryzysowym,* Zeszyty Naukowe AMW 4/2008, Gdynia 2008

Neider J., *Rozwój polskich portów morskich*, Wydawnictwo UG, Gdańsk 2013

Neider J., *Transport międzynarodowy* (the 2nd revised edition), Polskie Towarzystwo Ekonomiczne, Warszawa 2012

Nerć-Pełka A., P. Pełka, *Informatyzacja procesów zaopatrzenia na statkach morskich* [in:] M. Chaberek, A. Jezierski (eds), *Informatyczne narzędzia procesów logistycznych,* CeDeWu Wydawnictwa Fachowe, Warszawa 2010

Niedzielski E., *Determinanty sprawnego działania w zarządzaniu*, Zeszyt Naukowy "Ekonomia i Zarządzanie" no. 1(2011)

Niemczyk J., *Strategia. Od planu do sieci*, Wydawnictwo UE we Wrocławiu, Wrocław 2013

Nowakowski T., *Niezawodność systemów logistycznych*, Oficyna Wydawnicza Politechniki Wrocławskiej, Wrocław 2011

Nowicka A., *Współczesna nawigacja morska oraz nawigacja przyszłości*, Wydawnictwo Akademii Morskiej w Szczecinie (Biblioteka Cyfrowa), Szczecin 2010

Obłój K., *Strategia organizacji*, PWE, Warszawa 2009

Odlanicka-Poczobutt M., *Prakseologia a klasyczne kryteria oceny sprawności systemów logistycznych*, Zeszyty Naukowe Politechniki Śląskiej "Organizacja i Zarządzanie", no. 70, Gliwice 2014

Oziemski S., *Entropia układów logistycznych*, Materiały X Konferencji Logistyki Stosowanej "Total Logistic Management", Zakopane 2006

Paine R.T. (*et al.*), *Trouble on oiled waters: Lessons from the Exxon Valdez oil spill*. Ann. Rev. Ecol. Syst. 27 (197) 1996

Pazdan M. (ed.), *Międzynarodowe prawo handlowe*, Wydawnictwo Difin, Warszawa 2006

Peterson Ch.H., S.D. Rice, J.W. Short, D. Esler, J.L. Bodkin, B.E. Ballachey, D.B. Irons, *Long-term ecosystem response to the Exxon Valdez oil spill*, "Science" 302 (2082) 2003

Pietraszek R., *Problemy zanieczyszczenia mórz olejami wynikłe z katastrofy zbiornikowca Torrey Canyon*, "Technika i Gospodarka Morska" no. 8–9 (373) 1967

Pilotażowe wdrożenie RIS Dolnej Odry, Studium wykonalności, vol. 1. Część główna. Praca wykonana przez zespół specjalistów konsorcjum w składzie: Akademia Morska w Szczecinie oraz Lemtech Konsulting sp. z o.o., Szczecin 2011. (The Main Section. The Study developed by the panel of consortium experts of Maritime Academy of Szczecin and Lemtech Konsulting Ltd.) Szczecin 2011

Planowanie strategiczne jako narzędzie realizacji celów organizacji, (materials published by the Chancellery of the Polish Prime Minister) on-line publication http://dsc.kprm.gov.pl/sites/default/files/pliki/79.pdf [accessed on 15th May 2017]

Pluciński M., *Polskie porty morskie w zmieniającym się otoczeniu*, CeDeWu, Warszawa 2013

Polityka morska Rzeczypospolitej Polskiej do roku 2020, dokument Międzyresortowego Zespołu do Spraw Polityki Morskiej Rzeczypospolitej Polskiej, Warszawa 2013

Powierża L., *Elementy inżynierii systemów*, Oficyna Wydawnicza Politechniki Warszawskiej, Warszawa 1997

Przyszła unijna polityka morska: europejska wizja oceanów i mórz, Zielona księga Komisji Europejskiej 2006 r.

Raport KE: *Obliczanie kosztów zewnętrznych transportu*, Publikacje PE, Bruksela 2009 (the Report of the European Commission on the calculation of the external costs of transport; Publications of the European Parliament; Brussels 2009)

Report of The Swedish Club http://europedirect-szczecin.eu/index.php/29-newsletter/kwiecien-2014-2-9-2014/347-innowacje-w-gospodarce-morskiej-szczecinski-navdec [accessed on 15th May 2017]

Report on *Safety and Shipping 1912–2012. From Titanic to Costa Concordia. An insurer's perspective from Allianz Global Corporate & Specialty*, Seafarers International Research Centre (SIRC), Cardiff University, Cardiff 2012

Report on Shipping Accidents in the Baltic Sea Area During 2010, HELCOM, Baltic Marine Environment Protection Commission

Review of Maritime Transport 2013, UNCTAD/RMT/2013, (data supplied by Clarkson Research Services), New York – Geneva 2013

River Information Services, Modernizing inland shipping through advanced information technologies, Transport Research Knowledge Centre (TRKC), European Commission's Directorate-General for mobility and transport, Belgium 2010

Rodrigue J.-P., Dept. of Global Studies & Geography, Hofstra University, non-published materials https://www.hofstra.edu, New York 2016, pp. 37 [accessed on 15th May 2016]

Rogowska J., J. Namieśnik, *Wraki jako źródło zanieczyszczenia środowiska morskiego*, "Inżynieria Morska i Geotechnika", no. 1 (3) 2009

Rogowska J., *Wpływ wraków na środowisko na przykładzie s/s Stuttgart*, Politechnika Gdańska, Gdańsk 2011

Rokiciński K., *Możliwości sił morskich w zapobieganiu zagrożeniom asymetrycznym w rejonach przybrzeżnych* [in:] M. Żuber (ed.), *Katastrofy naturalne i cywilizacyjne. Zagrożenia i reagowanie kryzysowe, part II. Wojna, terroryzm i inne zagrożenia asymetryczne*, Wrocław 2006

Rokiciński K., *Wybrane aspekty zagrożeń asymetrycznych na morzu w funkcji wykorzystania sił morskich*, Zeszyty Naukowe AMW no. 1, Gdynia 2005

Romanowska M., *Planowanie strategiczne w przedsiębiorstwie*, PWE, Warszawa 2009

Rosenberg D., *The political economy of piracy in the South China Sea*, "Naval War College Review", Summer 2009, Vol. 62. No. 3
Rosicki R., *O pojęciu i istocie bezpieczeństwa*, "PP" no. 3/2010
Rosquist T., R. Tuominen, *Qualification of Formal Safety Assessment. An Explamatory Study*, "Safety Sciences" 2003
Rozporządzenie Komisji (WE) nr 414/2007 z dnia 13 marca 2007 r. w sprawie wytycznych usług informacji rzecznej (RIS), o których mowa w art. 5 dyrektywy 2005/44/WE Parlamentu Europejskiego i Rady w sprawie zharmonizowanych usług informacji rzecznej (RIS) na śródlądowych drogach wodnych we Wspólnocie. Bruksela 2007. The Ordinance of the European Commission (EC) no. 414/2007 of 13th March 2007 on the guidelines for RIS services referred to in Art. 5 of the Directive 2005/44/EC of the European Parliament and European Council on the harmonised river information services (RIS) along the EU inland waterways; Brussels 2007
Rutkowski G., *Problemy bezpieczeństwa w przemyśle off-shore*; part 1. *Rozlewy olejowe oraz inne rodzaje zagrożeń rejestrowane w sektorze off--shore*, Prace Wydziału Nawigacyjnego Akademii Morskiej w Gdyni no. 26, Gdynia 2011
Rydzkowski W., K. Wojewódzka-Król (ed.), *Transport*, PWN, Warszawa 2007
Salomon A., *Spedycja w handlu morskim. Procedury i dokumenty*, Wydawnictwo Uniwersytetu Gdańskiego, Gdańsk 2003
Salomonowicz H., *Morskie systemy transportowe początku XXI wieku*, "Logistyka", 2010, no. 4
Santala M., *Condition monitoring provides key to ship automation*, Wärtsilä Corporation e-publication 2012, http://www.wartsila.com/en/condition-monitoring-provides-key-to-ship-automation [accessed on 15th May 2017]
Sawicki J.K., *Polskie Ratownictwo Okrętowe 1951–2001. Zarys działalności*, Pracownia Historii Wyższej Szkoły Morskiej w Gdyni (publication series "Księgi Floty Okrętowej", vol. 13), Gdynia 2001
Second IMO GHG Study, IMO Publications, London 2009
Skorupski J., *Sieci Petriego jako narzędzie do modelowania procesów ruchowych w transporcie*, Prace Naukowe Politechniki Warszawskiej, Z 78, Warszawa 2011
Sousa J. de, *EMSA's Integrated Maritime Environment – a tool for improved Maritime Domain Awareness*, EMSA Publication, Lisbon 2012

Sprawozdanie z inicjatyw i opinii Komisji Europejskiej we wrześniu 2013 r., (2013), (transport, ochrona środowiska, turystyka, sport, przemysł, badania naukowe, innowacje, agenda cyfrowa, ochrona zdrowia, polityka zatrudnienia), Bruksela, wrzesień 2013. (The report on the initiatives and opinions of the European Commission in September 2013 (transport, protection of natural environment, tourism, sport, industry, scientific research, innovation, digital agenda, health care, employment policy; Brussels, September 2013)

Stapford M., *Maritime Economics*, Routledge, London 2009

Stareńczak P.B., *Statystyki wypadków na morzu w 2012 roku, a utrzymujący się trend spadkowy* http://www.portalmorski.pl/statki/wypadki-ratownictwo-sar/34012-poprawa-bezpieczenstwa-statkow-i-zeglugi, publikacja on-line [accessed on 12th August 2014]

Stateczny A., W. Kazimierski, N. Wawrzyniak, *Analiza funkcjonalności geoinformatycznego systemu ochrony portu*, "Archiwum Fotogrametrii, Kartografii i Teledetekcji", Vol. 23, 2012

Strategia Unii Europejskiej dla Regionu Morza Bałtyckiego EUSBSR http://www.balticsea-region-strategy.eu; Commission (2009) "European Union Strategy for the Baltic Sea Region"

Strategia Unii Europejskiej w zakresie bezpieczeństwa morskiego, dokument RADY UNII EUROPEJSKIEJ, (European Union Strategy for Maritime Safety; a document by the Council of the European Union) 11205/14, Brussels 2014

Sustainability and Efficiency of Visions for CISE, Report for European Commission – Directorate General Maritime Affairs, 330012909 – Version 1.0

Szcześniak J., A. Weintrit, *Europejskie systemy kontroli i śledzenia ruchu statków – geneza, zasady funkcjonowania oraz perspektywy rozwoju*, Zeszyty Naukowe Akademii Morskiej w Gdyni, no. 77, Gdynia 2012

Szczyrbowski K., *Zmiany w strukturze ochrony wybranych podmiotów gospodarki morskiej na przykładzie wprowadzenia kodu ISPS*, (non-published materials of the Polish Naval Academy) Gdynia 2008

Szpyrka M., *Sieci Petriego w modelownaiu i analizie systemów współbieżnych*, WNT, Warszawa 2008

Szubrycht T., *Transport morski – ekonomiczne wymiary współczesnych zagrożeń asymetrycznych,* Logitrans VII Konferencja Naukowo-Techniczna, (the proceeding of the 7th Scientific and Technological Conference Logitrans) Szczyrk 2010

Szyszko M., *Rozwój generacji portów morskich*, Wydawnictwo Akademii Morskiej, Szczecin 2010

The Baltic Transport Outlook 2030 (BTO2030 Executive Report – final version)

The development of the CISE for the surveillance of the EU maritime domain and the related Impact Assessment PART 1: INDIVIDUAL ANALYSIS, EUROPEAN COMMISSION, DG MARE, Lyngby 2013

The development of the CISE for the surveillance of the EU maritime domain and the related Impact, Assessment, PART 2: COMBINED ANALYSIS, EUROPEAN COMMISSION and DG MARE, Lyngby 2014

Tołkacz L. *Infrastruktura transportu wodnego,* Wydawnictwo Zachodniopomorskiego Uniwersytetu Technologicznego w Szczecinie, Szczecin 2010

Transport wodny śródlądowy w Polsce w 2014 r., opracowanie sygnalne GUS, (developed by the Central Statistics Office in Poland) Warszawa 2015

Trojanowski J., *Rzeczne służby informacyjne czynnikiem poprawy bezpieczeństwa żeglugi na śródlądowych drogach wodnych w Polsce*, Biblioteka Cyfrowa AM w Szczecinie, Szczecin 2010

Urbański J., Cz. Dyrcz, W. Morgaś, *Bezpieczeństwo morskie oraz bezpieczeństwo morskie państwa – podobieństwa i różnice*, ZN AMW; 172K/1/2008

Urbański J., W. Morgaś, C. Specht, *Bezpieczeństwo morskie – ocena i kontrola ryzyka*, Zeszyty Naukowe AMW no. 2 (173), Gdynia 2008

Urbański J., W. Morgaś, Z. Kopacz, *Bezpieczeństwo morskie i ochrona żeglugi oraz zarządzanie nimi; próba identyfikacji i specyfikacji przedmiotu*, "Przegląd Hydrograficzny", Gdynia 4/2008

Vanroye K., B. van Mol, *Zmieniająca się rola portów morskich UE w globalnej logistyce morskiej – możliwości, wyzwania i strategie*, ekspertyza *Buck Consultants International* dla PE i KE, Brussels 2009

Vinnem J., *Offshore Risk Assessment*, Springer, London 2007

Wardin K., *Piractwo XXI wieku zagrożeniem bezpieczeństwa transportu morskiego*, Logitrans VII Konferencja Naukowo-Techniczna, Szczyrk 2010

Wawruch R., *Możliwości wykorzystania statkowych systemów automatycznej identyfikacji (AIS) do monitorowania ruchu morskiego*, Wyd. AM, Gdynia 2007

Wawruch R., *Usługa Informacji Rzecznej (RIS) na Śródlądowych Drogach Wodnych Unii Europejskiej i Zasadność jej Ustanowienia w Polsce*, Prace Wydziału Nawigacyjnego Akademii Morskiej w Gdyni, no. 24, 2010

Weintrit A., *Elektroniczna mapa nawigacyjna. Wprowadzenie do nawigacyjnych systemów informacyjnych ECDIS*, Wydawnictwo Fundacji Rozwoju Wyższej Szkoły Morskiej, Gdynia 1997

Weintrit A., H. Nitner, *Stan prac legislacyjnych Międzynarodowej Organizacji Morskiej dotyczących wykorzystania na morzu Elektronicznych Map Nawigacyjnych ENC oraz systemów ECDIS*, "Przegląd Hydrograficzny" no. 3, BHMW, Gdynia 2007

Weintrit A., H. Nitner, *Czy jesteśmy już w epoce map elektronicznych?*, "Przegląd Hydrograficzny" no. 5, BHMW, Gdynia 2009

Wilczyński P., *Monitorowanie transportu ładunków niebezpiecznych a bezpieczeństwo portów morskich*, Prace Wydziału Nawigacyjnego AM w Gdyni no. 17, Gdynia 2005

Winiarska M., *Wybrane problemy dotyczące transportu morskiego w kontekście ochrony jego bezpieczeństwa*, Colloquium Wydziału Nauk Humanistycznych i Społecznych, IV/2012

Wójcik A., P. Banaś, Z. Pietrzykowski, *Schema of Inference Processes in a Preliminary Identification of Navigational Situation in Maritime Transport* [in:] J. Mikulski (ed.), *Telematics - Support for Transport*, Springer, Berlin Heidelberg 2014

Wojewódzka-Król K., R. Rolbiecki, *Infrastruktura transportu*, Wydawnictwo UG, Gdańsk 2008

Wojtowicz J., *Żegluga morska bliskiego zasięgu – uwarunkowania i problemy rozwoju*, materiały pokonferencyjne VI Kongresu Spedytorów, (the proceedings of the 6th Congress of Forwarders) PISIL, Warszawa 2005

Woś K., *Analiza nawigacyjna dolnego Odcinka Odry objętego obowiązkiem wdrożenia systemu RIS,* "Logistyka", no. 6/2011

Wspólny Komunikat Do Parlamentu Europejskiego i Rady w sprawie otwartego i bezpiecznego światowego obszaru morskiego: elementy strategii Unii Europejskiej w zakresie bezpieczeństwa morskiego, publikacja KE i Wysokiego Przedstawiciela Unii Europejskiej do Spraw Zagranicznych i Polityki Bezpieczeństwa, Join (2014) 9 final, Bruksela 2014 (Joined Communication to the European Parliament and European Council on the open and safe global maritime area: the elements of the European Union strategy on maritime safety; publication of the European Commission and the High Representative of the EU for Foreign Affairs and Security Policy; Join (2014) 9 final, Brussels 2014

www.con-nect.nl [accessed on 13th July 2016]

www.dnv.pl [accessed on 26th July 2016]

www.sheltermar.com.br/en/vts/vtmis/ [accessed on 19th April 2018]

Wydro K., *Usługi i systemy telematyczne w transporcie*, "Telekomunikacja i Techniki Informacyjne" 2008, no. 3–4

Wymagania dotyczące technicznego i pozatechnicznego bezpieczeństwa żeglugi jednostek pływających Straży Granicznej, Baza Aktów Prawnych, Dz. U. z 2002 r., Nr 108, poz. 956 (Border Guard requirements of technical and non-technical safety of vessels; the Database of Legal Acts; Journal of Laws 2002, no. 108; item 956)

Yamada Y., *The Cost of Oil Spills from Tankers in Relation to Weight of Spilled Oil*, Marine Technology, 46(2009)

Załącznik 1 do Rozporządzenia Rady Ministrów z dnia 7 maja 2002 r. w sprawie klasyfikacji śródlądowych dróg wodnych. (2002) Dziennik Ustaw no. 77, item. 695

Załoga E. (ed.), *Europejska przestrzeń transportu. Infrastruktura i środki transportu.* Zeszyty Naukowe Uniwersytetu Szczecińskiego no. 743, publication series "Problemy Transportu i Logistyki" no. 20, Szczecin 2012

Załoga E., T. Kwarciński, D. Milewski, A. Wysocka, *Sustainable development as the economical challenge* [in:] H. Witte (ed.), Nachlatigkeit, ökonomische Gleichgewichte und Wertschöpfung, Serie Extensión, Colección Estudios Sociales y Humanidades, Bahía Blanca, Argentina 2014

Załoga E., *W kierunku nowego paradygmatu rozwoju transportu w Unii Europejskiej*, "Logistyka" no. 2/2014

Zmieniająca się rola portów morskich UE w globalnej logistyce morskiej – możliwości, wyzwania i strategie, Ekspertyza, PE 2009 (The Changing Role of the EU Sea Ports in Global Maritime Logistics – Opportunities, Challenges and Strategies; an expert opinion, European Parliament, 2009)

LIST OF FIGURES

LIST OF TABLES

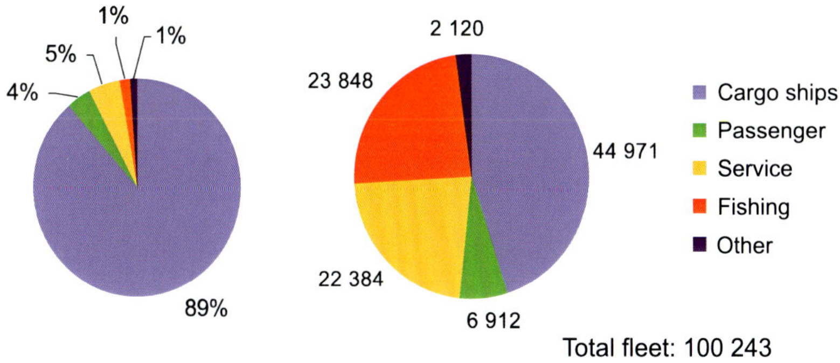

Figure 1-18. The structure and volume of the global fleet (2007)

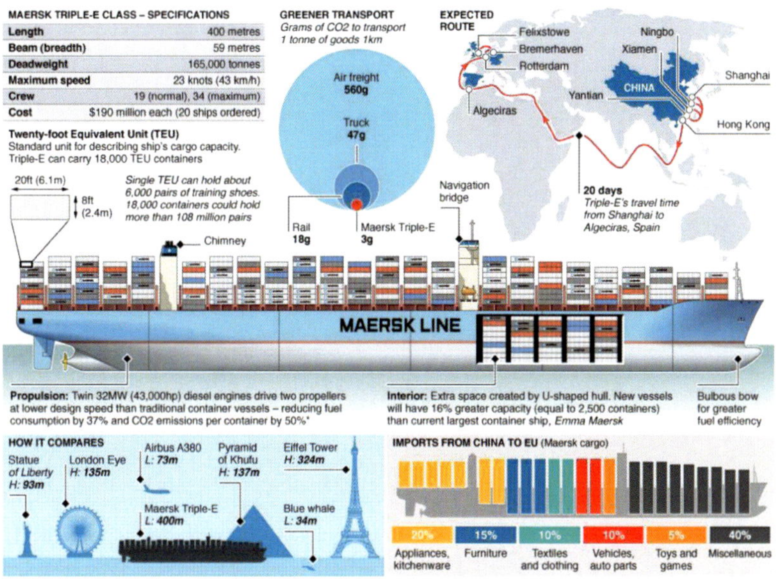

Figure 1-19. Basic technical and comparative parameters of an oceangoing container vessel (MAERSK)

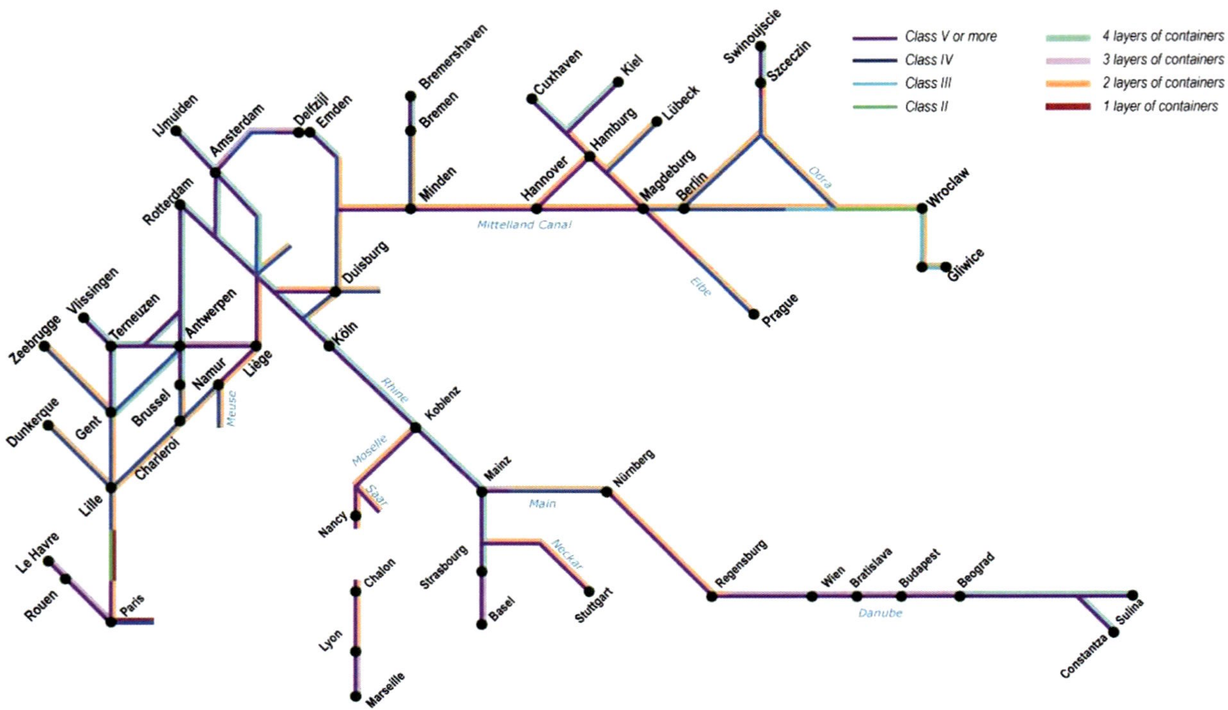

Figure 1-27. The network on inland waterways and the number of container layers which can be transported by water

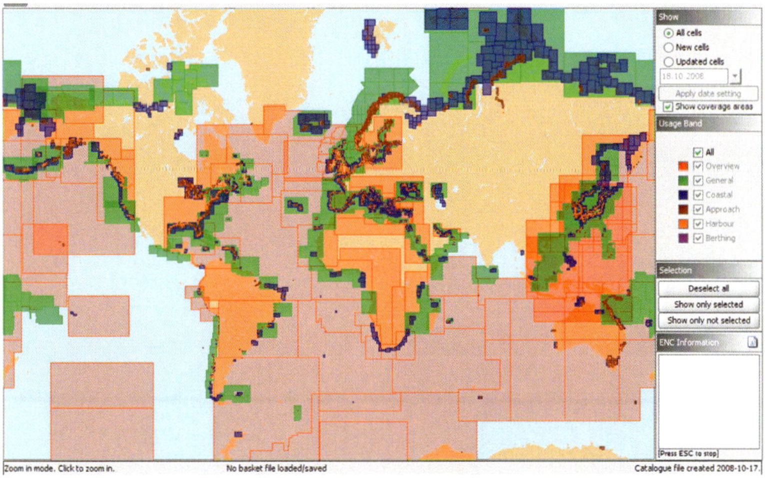

Figure 2-9. The areas covered by the ENC, according to the PRIMAR catalogue – all frequencies [2008]

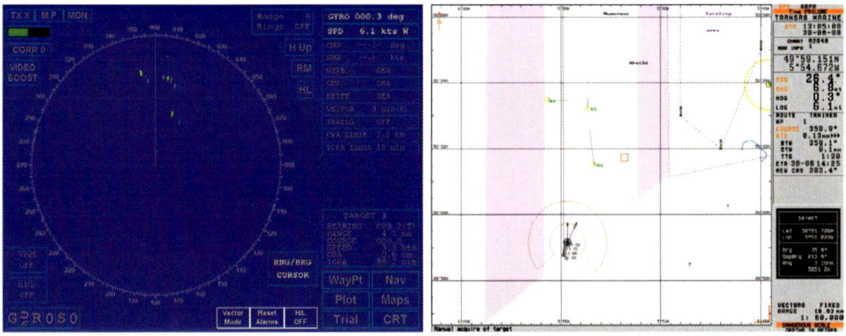

Figure 2-11. An example of a radar image overlaid on an ECDIS display

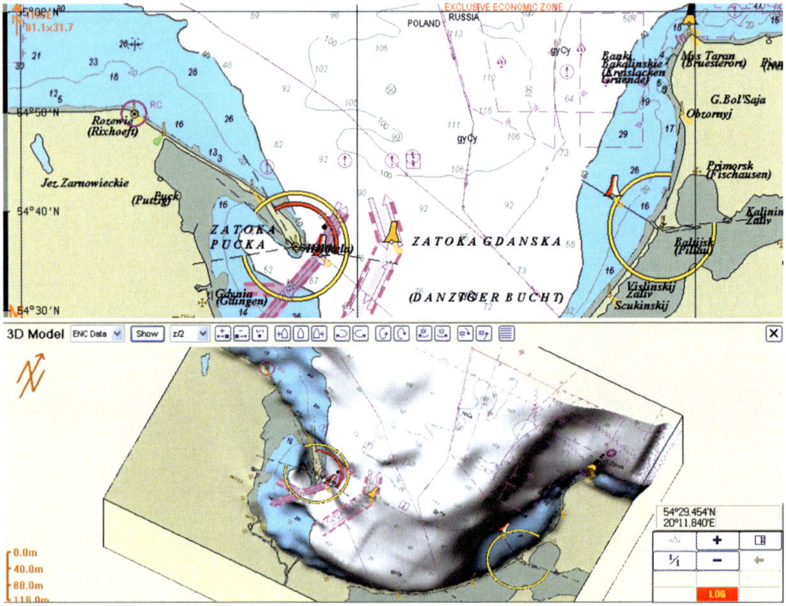

Figure 2-12. 2D/3D display taken from an ECDIS chart in the dKart Navigator system

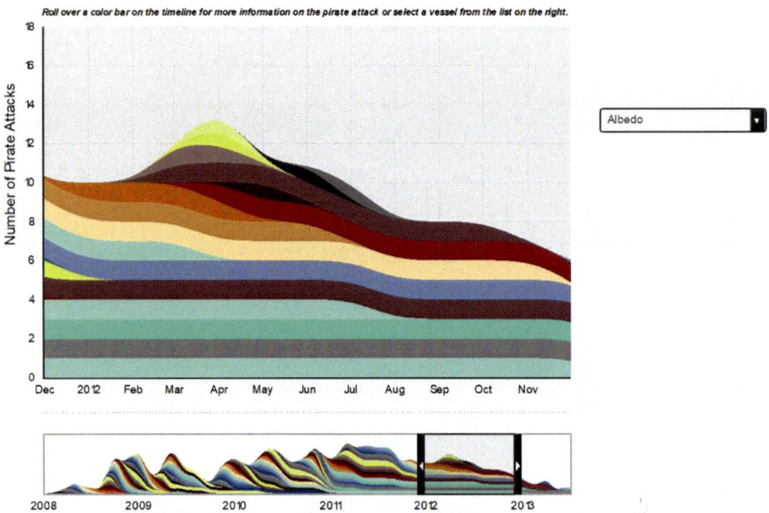

Figure 2-16. Piracy incidents at the Somalian coast in 2008–2013

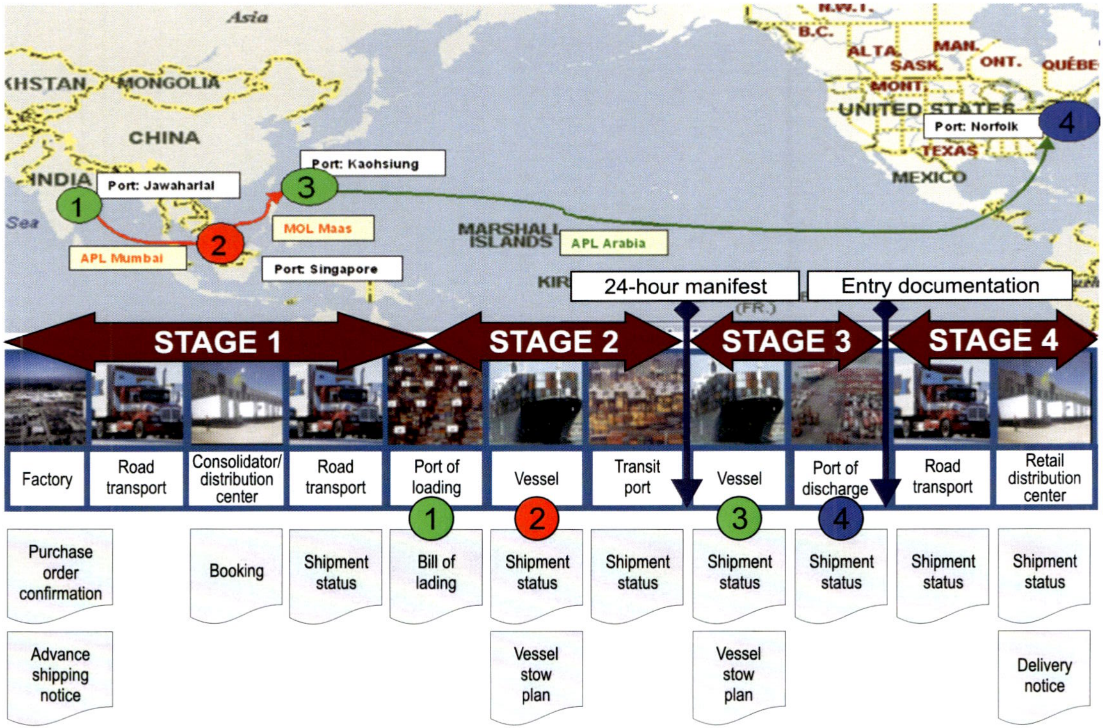

Figure 2-33. A concept of a land-sea logistical chain

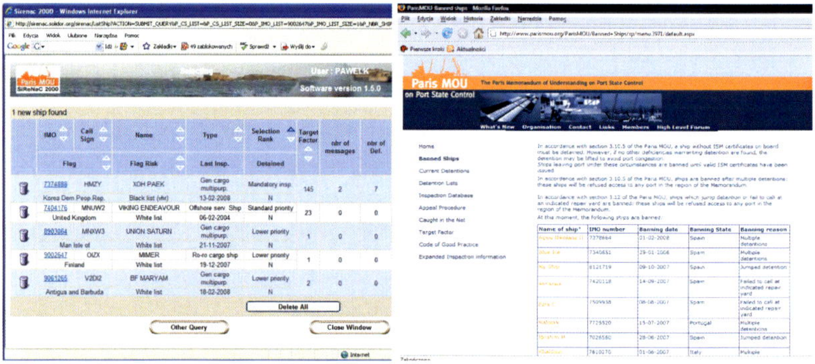

Figure 3-5. The Paris MoU – SIReNac application used for the implementation of the PSC and FSC functions

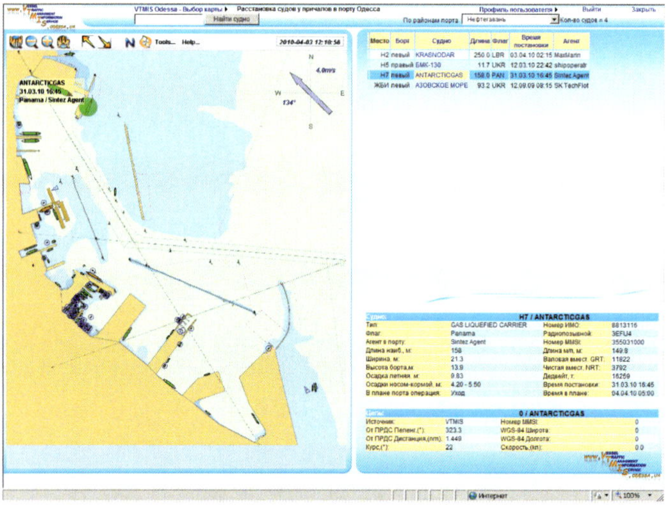

Figure 3-10. The VTS system of the port of Odessa displayed on the operator's console

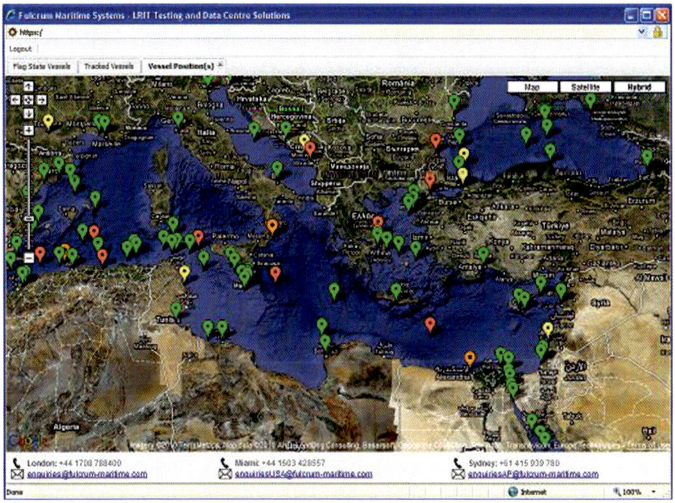

Figure 3-12. The interface of the LRIT system

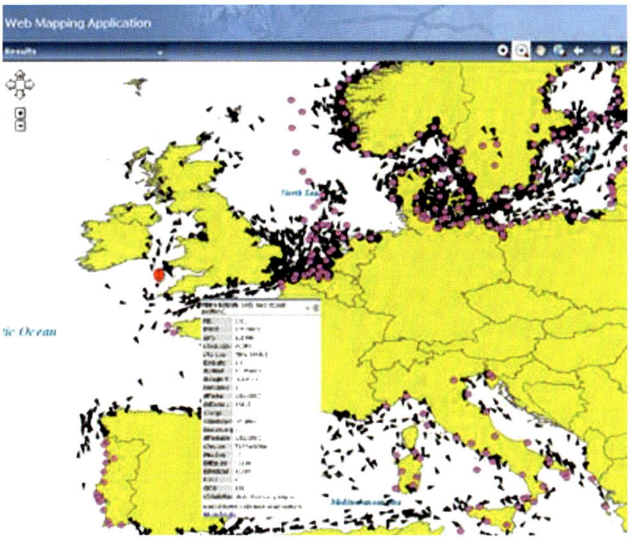

Figure 3-13. Navigation traffic displayed in the STIRES application

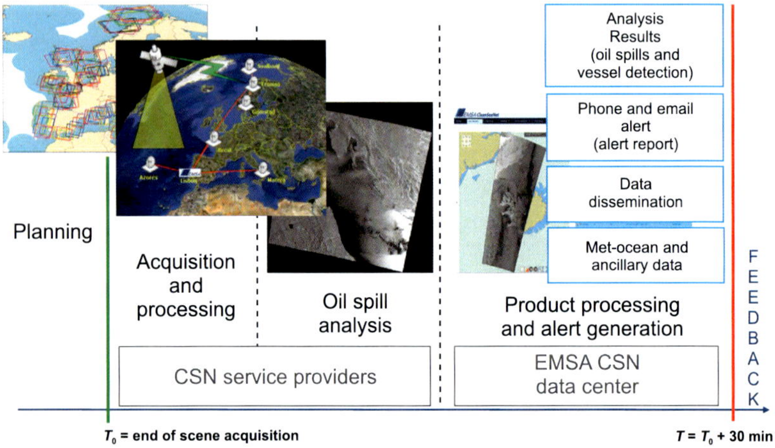

Figure 3-14. The concept presenting operation of the CleanSeaNet system

Figure 3-15. An example of the SMDE-class system at the national level – the interface of the SWIBŻ-PL application

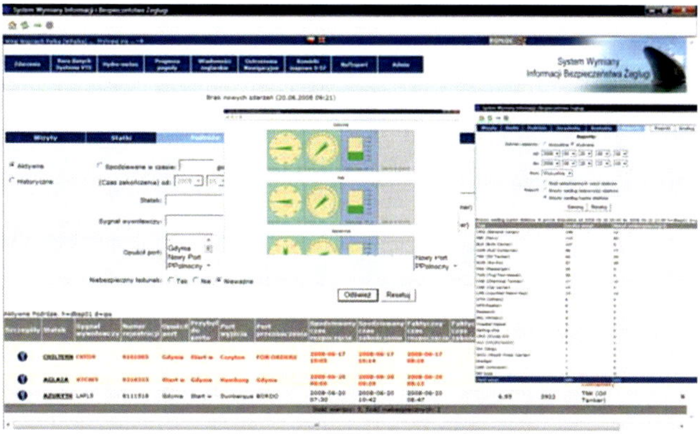

Figure 3-16. Specific functions performed by the SMDE-class system – the interface of the SWIBŻ-PL application in the tab of nautical information exchange

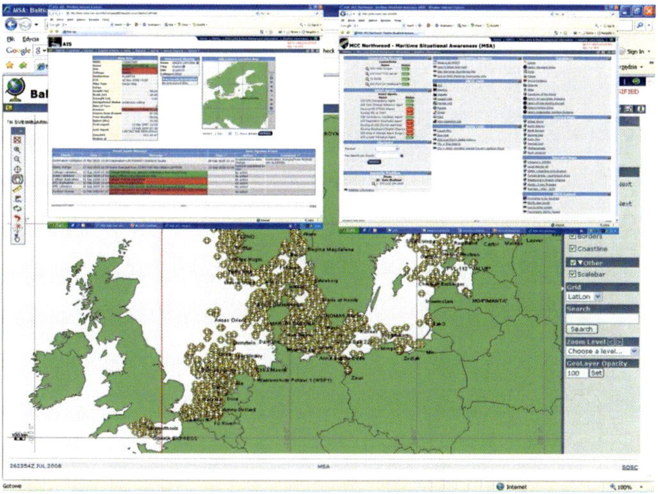

Figure 3-18. The interface of the BRITE application

Contact Listing filtered with "Anomalies past 24 Hours"							
MMSI	Timelate	Name	IMO	Callsign	Position	Destination	ETA
13374	000:02:38	SANTA MARIA		EBFL	432940N 0083703W		
2576099	000:01:15	NCIT		LM9170	592523N 0102909E	HORTEN	
25862000	016:16:22	ANICIA	8115552	LAJY6	535350N 0141601E	IGGESUND	12 Feb 2007 10:28
100000000	000:00:59	SJOKRIGSKOLEN		SKSK	602337N 0051602E	HOMEAA	
123123123	000:01:52	CPT CIA	147258369	0	363735N 0061938W		
205007400	000:00:44	PIETER PAUL RUBENS		OT2513	512513N 0033233E	ZEEBRUGGE	
205037000	000:05:18	NORMA	833007200	ORTK	512025N 0031323E	ZEEBRUGGE	20 Sep 2006 04:00
205107000	000:02:53	PILOT 1		ORBA	512200N 0024923E	OSTEND	21 Dec 2006 12:00
205108000	000:01:05	ZEELEEUW		ORBB	511058N 0024442E	OOSTENDE	08 Nov 2006 08:00
205109000	000:01:13	PILOT 6		ORBF	511337N 0025603E	WESTPOST_VLISSINGEN	23 Feb 2007 08:00
205202000	000:02:58	WESTDIEP		ORJS	512019N 0031219E	ZEEBRUGGE	04 Dec 2006 00:00
205206000	000:02:26	ASTER		ORGA	512007N 0031158E	ZEEBRUGGE	23 Feb 2007 11:45
205216000	000:33:53	VALCKE		ORIV	512012N 0031202E	ZEEBRUGGE	14 Nov 2006 16:00
205218000	000:03:03	BNS BELGICA	8222563	ORGQ	512016N 0031225E	ZEEBRUGGE	06 Dec 2006 10:00
205394000	000:00:21	WATERWAYS 1		ONCJ	511927N 0031247E	ROTTERDAM	03 Jun 2007 03:00
207133000	017:21:11	AMETHYST	6817857	LZFH	542659N 0135158E	VARNA	03 Mar 2007 20:07
209001000	000:01:02	AMRUM TRADER	9150080	C4QW2	282942N 0135121W	FERROL	04 Mar 2007 14:30
209125000	005:43:13	ZEPHYR	8521488	C4PD2	514802N 0030328E	IMMINGHAM	05 Mar 2007 22:00
209351000	000:05:51	MAERSK NAIROBI	9231494	P3CF9	360849N 0052608W	ALGECIRAS	03 Mar 2007 14:00
209357000	000:52:56	ANNERDIEP	925617100	P3QA9	404015N 0085233W	AVEIRO	04 Mar 2007 07:00

Figure 3-19. The interface of the BRITE system for indication of anomalies (positions in red)

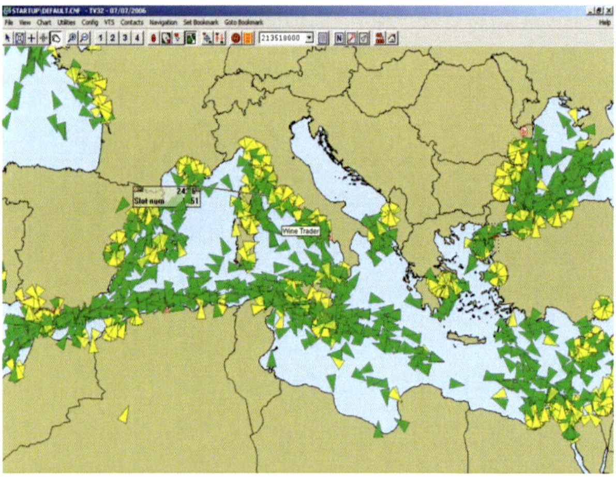

Figure 3-23. The interface of the MSSIS application

Figure 3-25. An image of vessel traffic in a selected area, as displayed in the TV 32 application for the MSSIS system

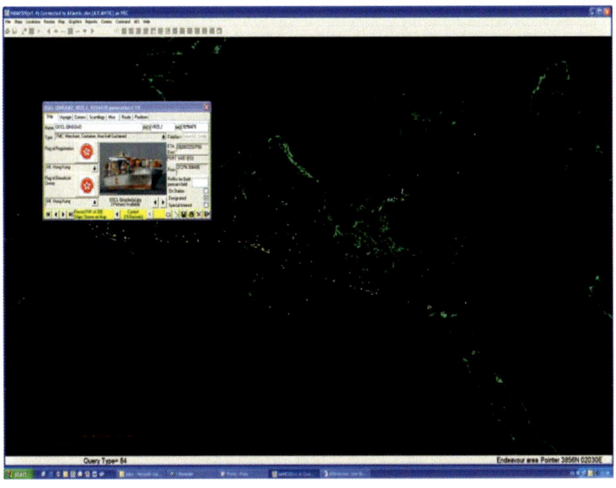

Figure 3-26. The interface of the NAMESIS application

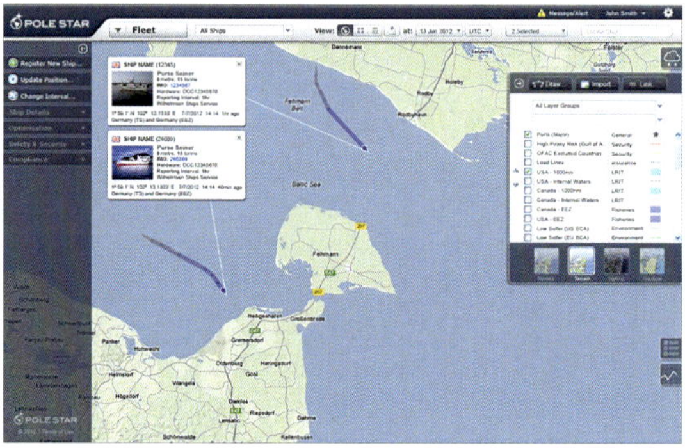

Figure 3-29. The VDS operator's screen

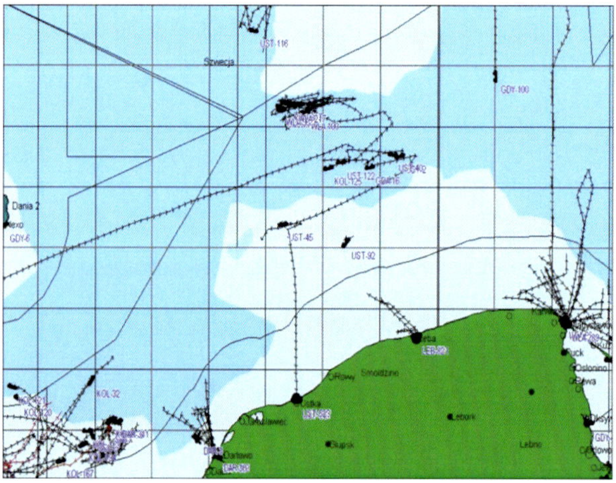

Figure 3-31. The interface of the VMS application

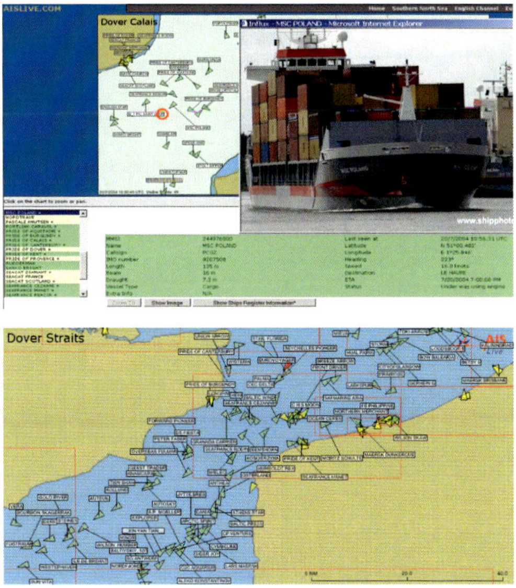

Figure 3-33. The interface of the AIS Live application, Dover Straits traffic from AIS Live

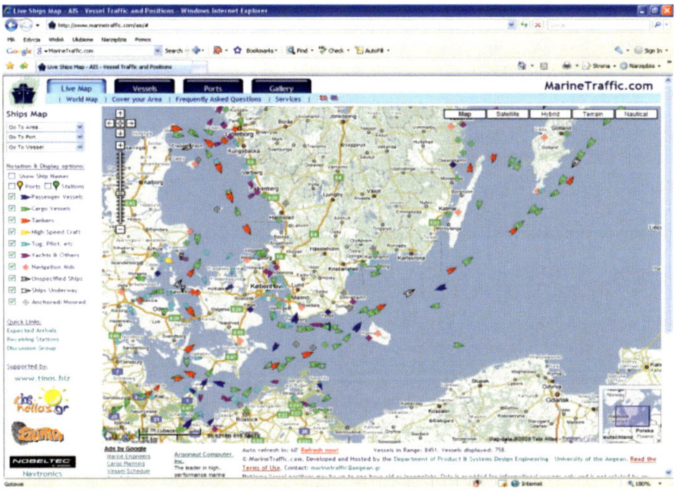

Figure 3-34. The interface of the Marine Traffic application

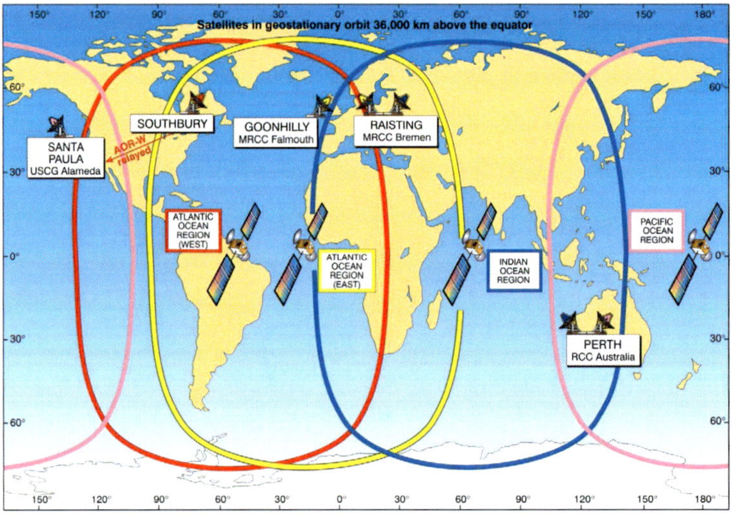

Figure 3-40. The range of satellite connection provided by the INMARSAT system

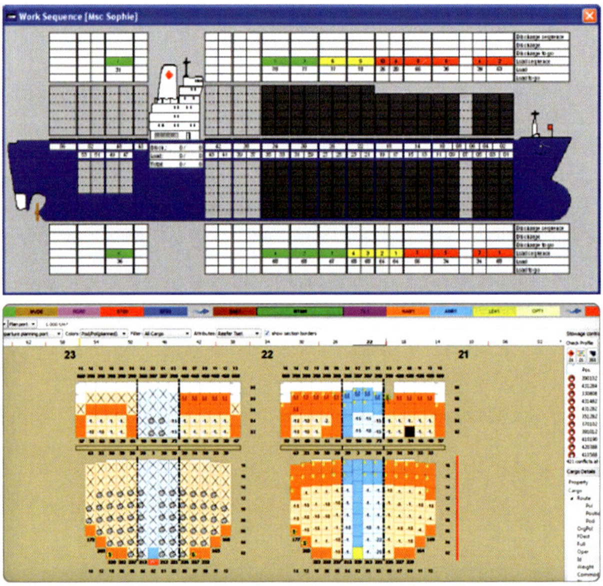

Figure 3-44. The SPINNAKER application window in the vessel planning function; the stowage plan

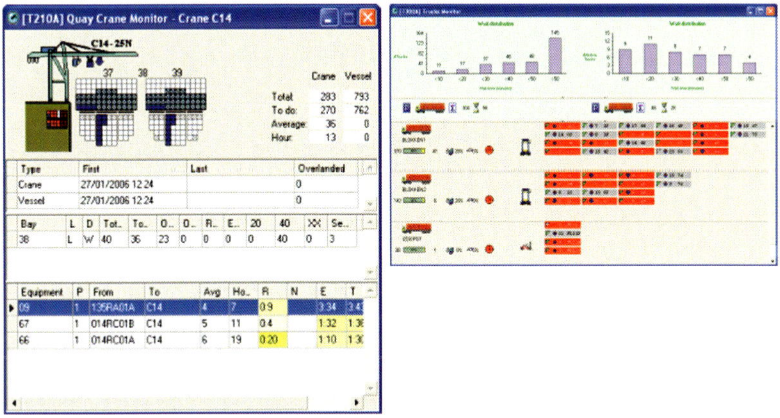

Figure 3-45. The windows of the Spinnaker Planning Management System® application

Figure 3-54. The AVANTE system of container location in maritime transport

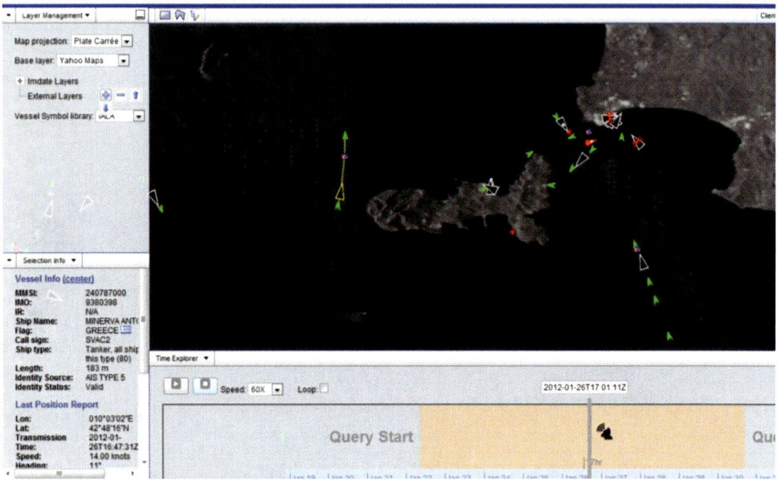

Figure 3-58. The interface of the time slider graphic function

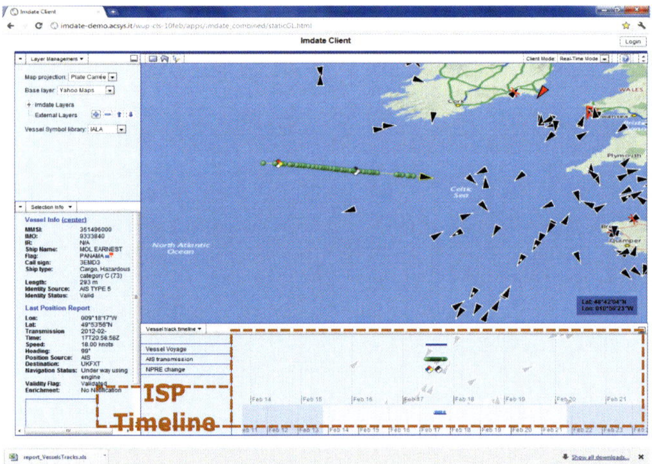

Figure 3-59. The interface of the timeline function

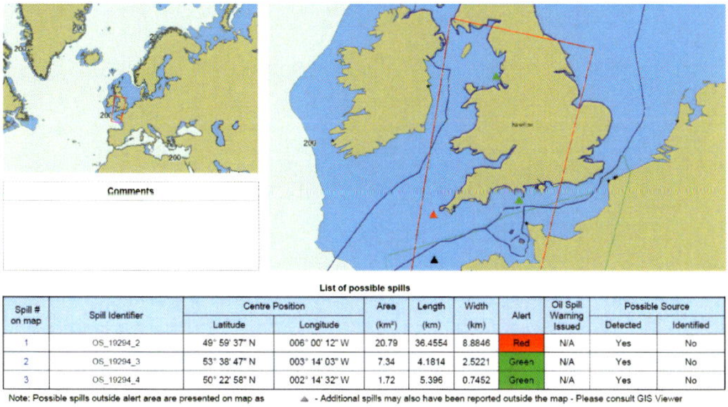

Spill # on map	Spill Identifier	Centre Position		Area	Length	Width	Alert	Oil Spill Warning Issued	Possible Source	
		Latitude	Longitude	(km²)	(km)	(km)			Detected	Identified
1	OS_19294_2	49° 59' 37" N	006° 00' 12" W	20.79	36.4554	8.8846	Red	N/A	Yes	No
2	OS_19294_3	53° 38' 47" N	003° 14' 03" W	7.34	4.1814	2.5221	Green	N/A	Yes	No
3	OS_19294_4	50° 22' 58" N	002° 14' 32" W	1.72	5.396	0.7452	Green	N/A	Yes	No

Figure 3-60. Detection of an oil spill in the SSN tab of the IMDatE system

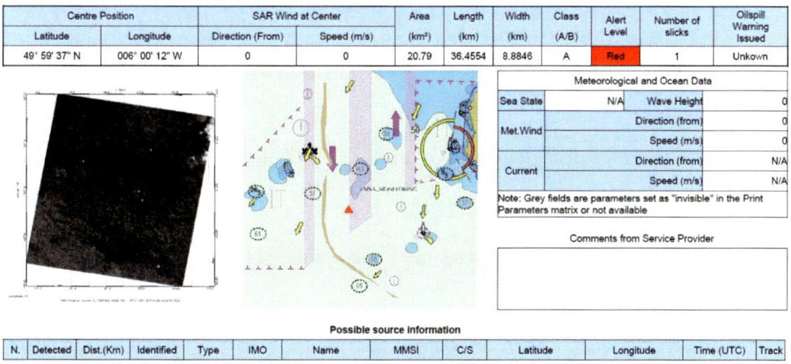

Centre Position		SAR Wind at Center		Area	Length	Width	Class	Alert Level	Number of slicks	Oilspill Warning Issued
Latitude	Longitude	Direction (From)	Speed (m/s)	(km²)	(km)	(km)	(A/B)			
49° 59' 37" N	006° 00' 12" W	0	0	20.79	36.4554	8.8846	A	Red	1	Unkown

Meteorological and Ocean Data			
Sea State	N/A	Wave Height	0
Met.Wind		Direction (from)	0
		Speed (m/s)	0
Current		Direction (from)	N/A
		Speed (m/s)	N/A

N.	Detected	Dist.(Km)	Identified	Type	IMO	Name	MMSI	C/S	Latitude	Longitude	Time (UTC)	Track

Figure 3-61. Integration of the information layer referring to an oil spill with the layer of movement and other data in the SSN tab of the IMDatE system

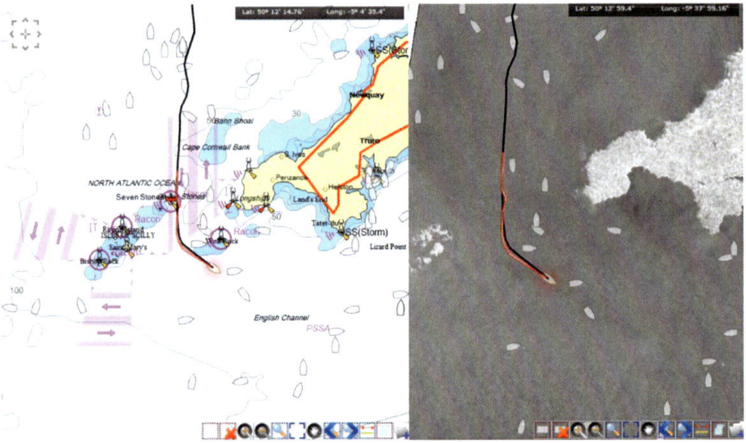

Figure 3-62. The analysis of the information layer referring to an oil spill with the layer of movement and indication of the vessel responsible for the spill in the SSN tab of the IMDatE system

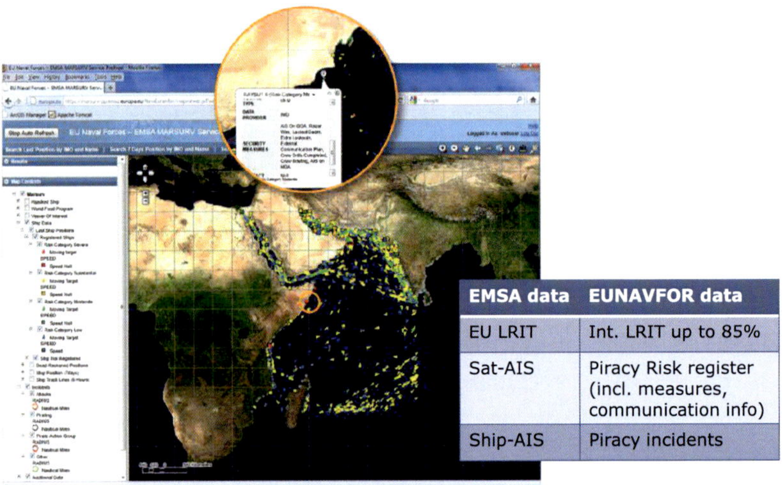

Figure 3-63. The anti-piracy function interface in the IMDatE system

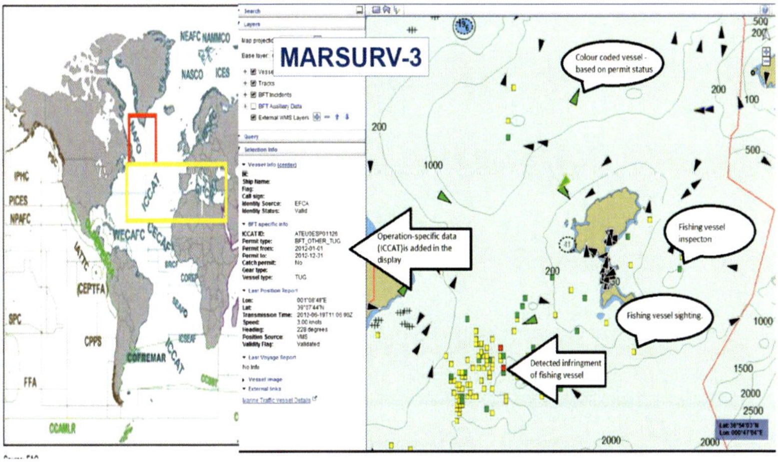

Figure 3-64. The interface of Illegal Fishing function in the IMDatE system

Figure 4-3. A concept of a River Hopper pallet carrier

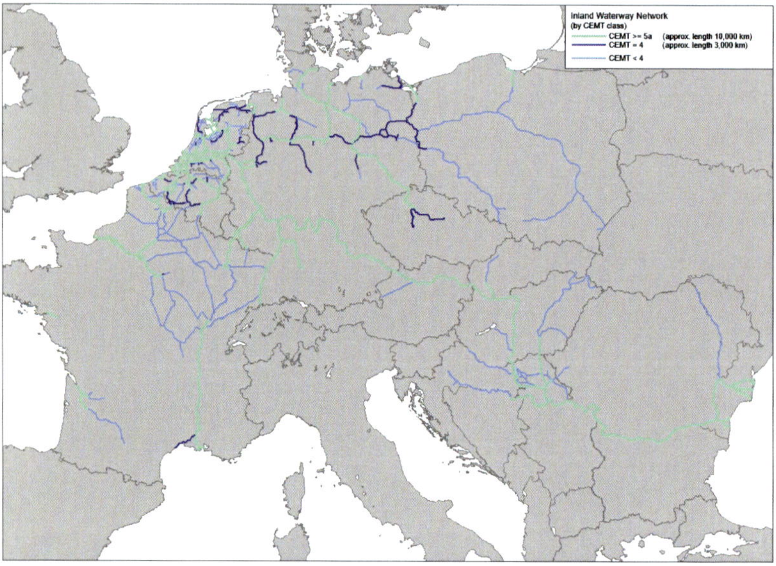

Figure 4-13. European waterways covered by the RIS Directive [2014]

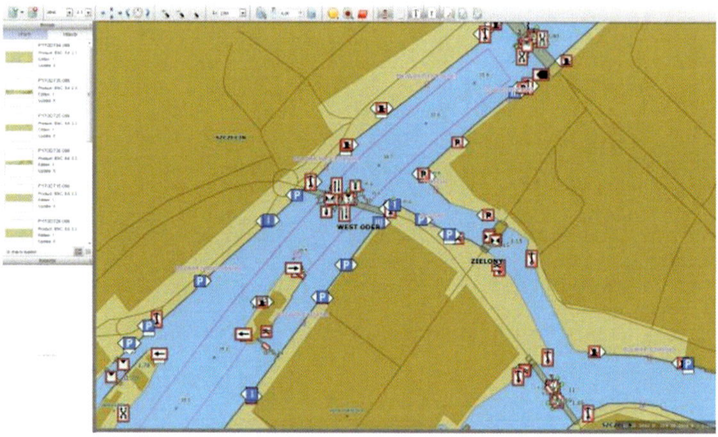

Figure 4-16. An image presenting the use of the Inland ECDIS system in the RIS system: a display that can be observed in a screen installed on a vessel (for the Oder River, in the vicinity of Szczecin)

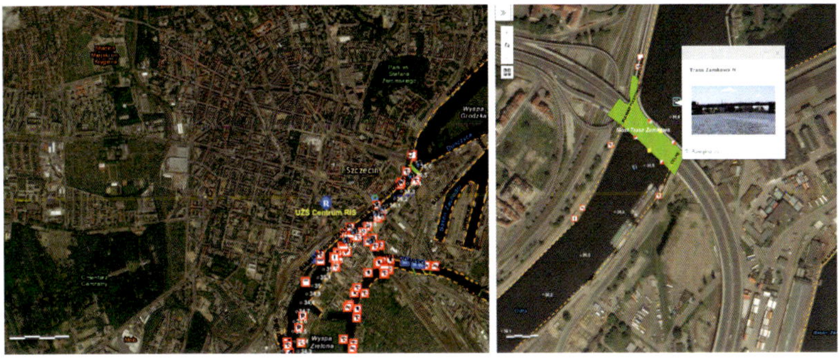

Figure 4-17. An interactive chart of the Oder River area

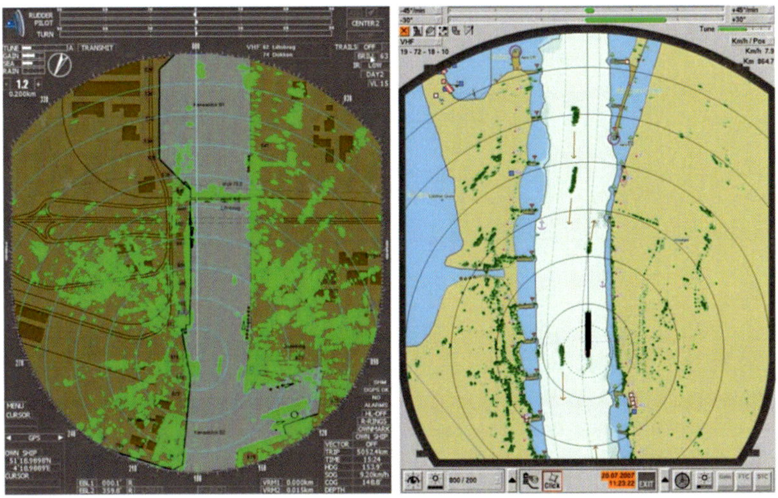

Figure 4-18. A display of the situation in a water area presented with the use of the ARPA-type radar (left) and a radar display overlaid on the ENC chart (right)

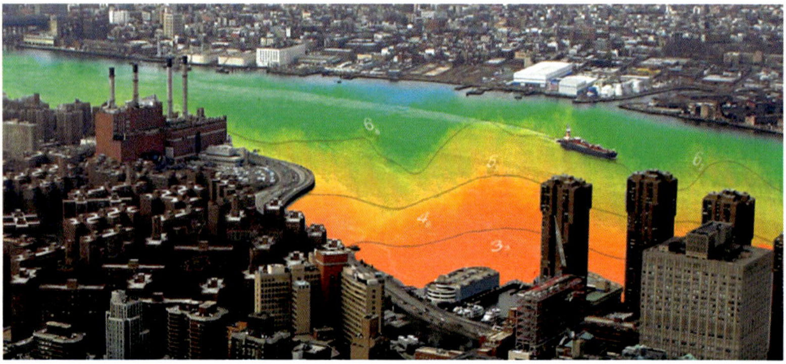

Figure 4-19. Safety isobaths for a vessel navigating along an inland waterway

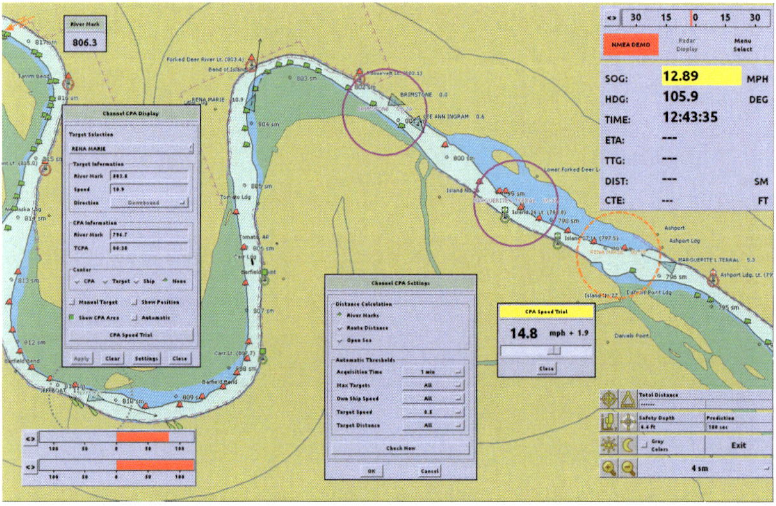

Figure 4-24. The interface of the ORCA™ River user in the CPA/TCPA function

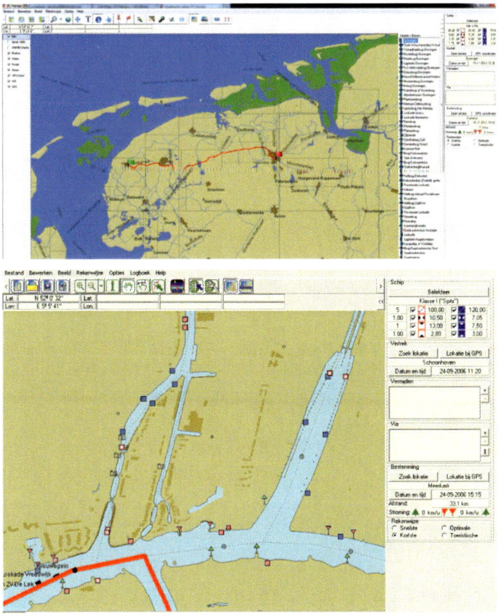

Figure 4-25. A display of a voyage in the PC-Navigo for the Planning mode (top) and the Navigation mode (bottom)

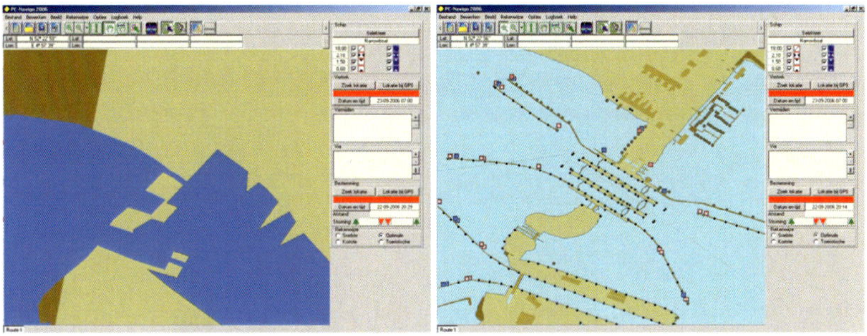

Figure 4-26. A display of a navigational chart in the PC-Navigo without the information from the Inland-ECDIS database system (left) and with such information provided (right)

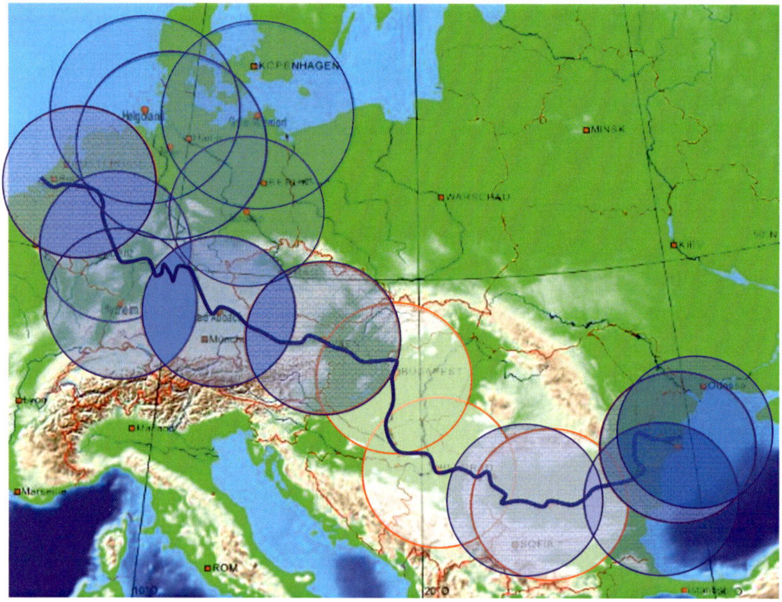

Figure 4-27. IALA DGNSS beacon coverage (planned or already in service) along major inland waterways in Europe [2014]